FROM PLANE TO SPHEROID

Determining the Figure of the Earth from 3000 B.C. to the 18th Century
Lapland and Peruvian Survey Expeditions

*In commemoration of the 250th anniversary of the expeditions to
Lapland and Peru*

FROM PLANE TO SPHEROID
Determining the Figure of the Earth from 3000 B.C. to the 18th Century
Lapland and Peruvian Survey Expeditions

*In commemoration of the 250th anniversary of the expeditions to
Lapland and Peru*

BY
JAMES R. SMITH

LANDMARK ENTERPRISES
10324 Newton Way
Rancho Cordova CA 95670
United States of America

ISBN 0-910845-29-8

*Dedicated to all my
relatives and friends*

TABLE OF CONTENTS

		Page
PREFACE		xii
INTRODUCTION		xiv
CHAPTER		
1	THE SHAPE AND POSITION OF THE EARTH	1
2	ASTRONOMERS AND MATHEMATICIANS: Pre - Eratosthenes	4
3	THE GNOMON AND SCAPHE	7
4	ERATOSTHENES - POSEIDONIUS - PTOLEMY	10
	Eratosthenes	10
	Poseidonius and Ptolemy	12
5	ASTRONOMERS AND MATHEMATICIANS: Eratosthenes to I-Hsing	13
6	THE ARCS OF I-HSING AND AL MAMUN IN ARABIA	16
	I-Hsing and his arc measurement	16
	The Arabian arc of Al-Mamun	19
7	ASTRONOMERS AND MATHEMATICIANS: I-Hsing to Ulugh Beg	21
8	CIRCLE SUBDIVISION AND TRIGONOMETRIC FUNCTIONS, TRANSVERSALS, VERNIER, MICROMETER AND LOGARITHMS	24
	Circle subdivision	24
	Trigonometric functions	25
	Transversals	29
	The vernier	34
	The micrometer	34
	Logarithms	35
9	ASTRONOMERS AND MATHEMATICIANS: Peurbach to Pascal	38
10	EARLY INSTRUMENTATION	42
	Angle measurning devices	
	The groma	42
	The dioptra	43
	Triquetrum	43
	Cross staff to quadrant	43
	The barometer	53

TABLE OF CONTENTS CONTINUED

CHAPTER 10 (Continued)

Mechanical computing devices		
The abacus	52
The slide rule	52
Calculating machines	52
Ancillary equipment		
The thermometer	52
The barometer	53
The Plain table (Plane table)	55
11	THE FIRST TRIANGULATION	57
12	THE ARCS OF FERNEL, NORWOOD, RICCIOLI, AND GRIMALDI	67
	Fernel	67
	Norwood	68
	Riccioli and Grifmaldi	68
13	INTERNATIONAL SCIENTIFIC BODIES	69
	The Royal Society	69
	L'Academie Royal des Sciences	69
14	PICARD'S ARC MEASURE	71
	Instruments and method	71
15	THE CASSINIS AND SIR ISAAC NEWTON	78
	The Cassinis	78
	Sir Issac Newton	79
16	THE ARC OF THOMAS IN CHINA	84
17	ASTRONOMERS AND MATHEMATICIANS - Cassini I to Cassini III	87
18	THE UNIVERSAL UNIT OF MEASURE	90
19	THE CONTROVERSY	92

TABLE OF CONTENTS CONTINUED

CHAPTER

20   THE EXPEDITION TO PERU   95

          Louis Godin   95
          Pierre Bouguer   95
          Charles Marie de La Condamine   97
          Jorge Juan y Santacilia   97
          Antonio de Ulloa y de la Torre Giral   100
          Joseph de Jussieu   101
     The mountains and Paramos, or Deserts of the Area   110
     Survey stations   111
     The baseline at Yarouqui   118
     Reduction of observations   120
     The baseline at Tarqui   121
     Abstract of observations   122
     Connection of heights to sea level   132
     Heights of stations above sea level   141
     Astronomical observations   142
     The extra work of Godin and Juan   145
     Recomputation   146
     Equipment   148
          Quadrant   148
          The sector   151
          Standard toise   152
          The barometer   152
          Speed of sound   153
          Horizontal angles   157
          Sides   157
     Commemoration   157
     The death of Seniergues   159
     Return to Europe   161
     Brief chronology   165

TABLE OF CONTENTS CONTINUED

CHAPTER

21   THE EXPEDITION TO LAPLAND   168

                    Pierre Louis Moreau de Maupertuis   168
                    Alexis-Claude Clairaut   170
                    Anders Celsius   170
                    Reginald Outhier   170
                    Pierre-Charles Le Monnier   170
                    Charles-Etienne-Louis Camus   170
                    Anders Hellant   171
            The survey stations   176
            Angular observations   177
            Base measurement   180
            Meridian direction   181
            Linear length of the arc   181
            Astronomy at Pello   183
            Astronomy at Torneå   183
            The arc length   183
            Heights   185
            Verifying the sector   185
            Refraction   185
            Comparison of refraction at Torneå andParis   186
            Pendulum experiments   186
            Svanberg and the remeasurement of the arc.   187
            The repeating circle   190
            Observations   190
            Computed sides   191
            Conclusions   191
            Brief Chronology   192

BIOGRAPHIES   194

REFERENCES   207

INDEX   211

LIST OF FIGURES

FIGURE   Page

1	The gnomon.	7
2	The scaphe.	8
3	An obelisk as a gnomon.	9
4	Geometry of a shadow observation.	17
5	Use of chords.	25
6	Geometry of chords.	25
7	Multiple angles using chords.	27
8	Diagonal scale.	29
9	Circular transversals.	30
10	Construction of circular transversals. (a)	32
11	Construction of curcular transversals. (b)	32
12	Napier's bones.	35
13	Basis of logarithms.	36
14	Solution of a triangle.	37
15	The cross staff.	44
16	The backstaff.	44
17	Bouguer's quadrant.	44
18	The zenith sector.	51
19	Standard bar.	56
20	Diagram of the scheme of Gemma Frisius.	58
21	Diagram of the scheme of Tycho Brahe.	59
22	Diagram of the scheme of Snellius.	60
23	Diagram of the baselines used by Snellius.	61
24	Diagram of the computational scheme of Snellius.	62
25	Riccioli's method of arc measure.	68
26	Diagram of the scheme of Picard.	75
27	Oblate or prolate - The oblate situation.	80
28	Oblate or prolate - The prolate situation.	81
29	The basis of Newton's reasoning.	82
30	A Chinese gnomon.	84
31	The Chinese arc.	85
32	Diagram of the scheme of Bouguer and La Condamine.	116
33	Baseline reduction. (a)	119
34	Baseline reduction. (b)	119
35	Bouguer's method of baseline reduction.	120
36	The connection to Mama Tarqui observatory.	121
37	Eight Tables of the observed angles of the Peru schemes.	122 through 135
38	Diagram of the connection of heights to sea level.	140
39	Diagram of the scheme of Godin and Juan.	144
40	Scheme of points for the experiments at Chimborazo.	156
41	Diagram of the scheme of Maupertuis.	178
42	Diagram of the computational scheme.	182
43	Diagram of the scheme of Svanberg.	188
44	Delambre's method of reduction.	190
45	The repeating circle.	190

LIST OF PLATES

PLATE   Page

1   The use transversals on a mural quadrant made by Bleau for Snellius.   31
2   The Digges theodolitus.   46
3   Digges instrument topographicall.   47
4   An early English theodolite of the 17th century.   48
5   A zenith sector used by Picard.   49
6   Zenith sector as used in an observatory.   50
7   The quadrant of Picard.   72
8   The field use of the quadrant.   73
9   Title page of "Mesure des Trois Premiers Degre's".   96
10 Charles Marie de La Condamine   97
11 Jorge Juan.   98
12 Antonio de Ulloa.   99
13 Engraving illustrating some 17th century instruments.   100
14 Map of the routes between France and Peru.   102
15 Inscription left at the Equator,   104
16 Plan of Quito as drawn by members of the expedition.   106
17 Panorama of the mountains as seen from Quito.   107
18 Panorama of the site of the Yarouqui baseline.   107
19 Representation of the volcanic and astronomical phenomena
         at a mountain station.   109
20 Map of the territory of Peru from the coast to the mountains.   114
21 Map of the area of triangulation scheme in Peru.   115
22 Base line measurement.   117
23 Details of the quadrant.   149
24 Method of use of the quadrant.   150
25 The zenith sector as used in Peru.   154
26 The zenith sector mounted in an observatory.   155
27 Diagram of the pyramids.   158
28 Inscription on the wall of the Jesuit church in Quito.   160
29 Pierre Louis Moreau de Maupertuis   169
30 Map of the area of the triangulation scheme in Lapland   173
31 View of the town of Torneå   193
32 The neighborhood of the observatory at Kittis   193

PREFACE

Although this volume has been written basically to commemorate the 250th anniversary of the expeditions to Peru and Lapland, an attempt has been made to do rather more than just describe their work. So that all the technical aspects of mounting a survey expedition in the early 18th century can be appreciated the history of them has been briefly traced. However, it must not be taken as a complete history of surveying since that would require several volumes.

The terms *'Peru'* and *'Lapland'* are used throughout since these are the descriptions normally used in English language publications. In fact the area of the Peru expedition is now wholly in Ecuador and the area of the Northern expedition was entirely South of the true Lapland.

In general terms the contents have been kept in chronological order rather than in any other of the many possible arrangements. Where dates are concerned, particularly the early ones. these cannot be guaranteed since different references often vary by several years.

Units of measurement are a problem in such a work as this. Since everything happened before the metric era, it would ruin the flavour to convert the myriad of old units into some form of metric equivalent. Thus, generally the unit of the original work has been kept throughout with equivalents quoted where relevant. A list of approximate conversion factors is given at the end. This should only be used as an aid and should not be considered as exact. Even the spellings can be controversial. Here the term stade has been used throughout although some would use either stadia or stadium. In addition the French ligne (1/12 inch) is spelt line.

The spelling of place names presents a particular headache. Often the original writings used a phonetic form and can differ even within the same publication. Whilst any one name is spelt the same way throughout the text no great significance should be read into that. It is more a case of which spelling appeared to be the most common. In relation to the Lapland expedition modern spellings have been adopted wherever possible. In both cases various alternative spellings are listed where the survey stations are described or where a name appears for the first time.

In preparing this work a vast number of sources have been used. How best to list these was given considerable thought. In studying some works it became obvious that certain forms of tabulating references could be very annoying and time consuming. Hence the fact that they are in one composite, alphabetical section, but with a list of appropriate numbers at the end of each chapter. No doubt this also will appear illogical or unhelpful to some readers and, for that, apologies are offered but many alternatives were considered and rejected.

Why the reason for commemorating the expeditions? They were such a landmark in geodesy that it was felt that the opportunity should not be missed to record in detail an appreciation of the efforts of early practitioners of the survey profession. It is just unfortunate that it was not thought of some years earlier so that a more thorough investigation could have been done. All the research and writing involved here has been done in just one year when some five times that could have produced a much more polished result. Nevertheless it is hoped that it will come somewhere near to what it set out to achieve.

Considerable help has been provided in compiling this work by many friends. In particular, mention must be made of Mrs. P. Lane-Gilbert of the Royal Institution of Chartered Surveyors Library, Mr. M. Malkin and Mr. R. McDougall of the Geodetic Library, Directorate of Military Survey, Mr. K. Mildren of the Portsmouth Polytechnic Library, Dr.R. V. Turley of the Southampton University Library, Mr. D. Wileman of the Royal Geographical Society Library and their respective colleagues.

The whole exercise would not have come about for the encouragement and help of Mr. Neil Franklin of Texas, Chairman of FIG Commission I, and M. Jan de Graeve of Brussels, Chairman of the Commission I Working Group that covers the history of surveying. The recomputation was the work of Miss C. Hudson, a student of the Department of Photogrammetry and Surveying, University College London, under the guidance and encouragement of Dr. A. L. Allan and Mr. D. P. Mason.

Particularly in relation to the Lapland expedition there has been extensive correspondence with Erik Tobé of Uppsala without whose help that section would have been considerably the poorer.

My sincere thanks to these and the many others I have had correspondence or discussion with, or help in various forms.

It would also be appropriate here to record thanks to those who are assisting in organising related symposia and exhibitions. Particularly Dr. A. L. Allan, Mr. P. Clark, Mr. N. Franklin, Brig. G. Gathercole, Mr. P. Gilbert, M. J. De Graeve, Mr. J. C. Hulbert, Miss J. Insley, Capt. R. C. Martinez, Dr. A. McConnell and Mr. D. Wallis and the numerous others at different venues.

Reference should be made to the help of Roy and Jean Minnick of Landmark Enterprises and Mr. T. Hooper of Surveyors Publications.

Photography was by the Photographic Unit and the diagrams by the Cartographic Unit, both of Portsmouth Polytechnic.

The Head and members of staff of the Department of Civil Engineering, Portsmouth Polytechnic have been encouraging and helpful throughout.

Sincere thanks are expressed to the following for permission to use their copyright material:

Archives de L'Académie des Sciences de Paris - for use of plates and other material from original publications relating to the expeditions to Peru and Lapland.

Deutsche Geodätische Kommission - for diagrams from *Beitrag zur Entwicklungsgeschichte des Theodolits*' by Max Engelsberger, 1969.

Museo Naval-Madrid for the photographs of Jorge Juan, Antonio de Ulloa and the quadrant of Ulloa.

Netherlands Geodetic Commission - for use of material from *Gemma Frisius, Tycho Brahe and Snellius and their triangulations*' by N. D. Haasbroek, 1968.

The Royal Society, London for various extracts from their Transactions of the 18th century, the references to which are noted in the text.

The Publishers and Editor of *Vistas in Astronomy*' for much of the material in the chapter relating to the work of I-Hsing in China as originally published by Arthur Beer et al in 1961. See reference [14].

Nevertheless any errors that have crept into the text from mis-interpretation of the original material or otherwise are solely of my making and it is hoped that they are not so numerous as to mar the use of the publication.

J. R. Smith
Portsmouth, U.K.
1986

INTRODUCTION

The aim of this volume is to commemorate the 250th anniversary of two famous survey expeditions sent to Peru and Lapland in the 18th century by the French Academy of Sciences. However, to adequately put the expeditions of 1735 and 1736 into perspective it seems prudent to trace the development up to that time of the main factors that were of relevance. In this context consideration needs to be given to the development of thoughts on the shape of the earth; the observational instruments available; the mathematical aids of the day; units of linear measure; previous determinations of the size and shape of the earth and general scientific development from the earliest times. Looking back to the early eighteenth century one might be inclined to assume all too readily that today's facilities were available then and to condemn particular aspects of the work when in fact, in the light of what was available at the time, it is probable that they all did a magnificent job.

For any triangulation scheme there is a requirement for portable angle measuring instruments that are sub-divided to very fine limits. Such instruments need to have the facility to sight accurately on distant targets. What sort of instruments were there and how were they divided?

Triangulation implies full knowledge of the geometry of triangles and of the trigonometric functions. How did these develop and were the functions of sufficient accuracy? Did all the necessary arithmetic have to be carried out by long hand methods or were there other facilities?

Electronic distance measurement (EDM) equipment obviously had not been heard of. How then were baselines measured and in what sort of linear unit since it was the pre-metric era?

What was the shape of the earth? Was it still considered flat at the end of the seventeenth century or had scientific thought crystallized on to some sort of spherical form? How near were estimates at that time for the size of the earth?

How well known was the practice of triangulation? Was it observed, computed and adjusted as it is today or was it very much in its infancy?

In tracing the development of these and other aspects leading to the expeditions, brief details are given of the personalities involved so that a flavour of the times and the spread of scientific thought can be obtained by the reader. Many of the names may be new in the context of surveying and others, while well known, may normally be connected with some other branch of science. Indeed it is very surprising how many individuals have contributed in some way, large or small, to the developments required to successfully measure arcs of the earth's surface.

In fact, more than 200 are mentioned in this volume, and that is by no means exhaustive.

In such an investigation as this a researcher can be easily led up many interesting byways and it becomes difficult to know just where to draw the line. Hopefully all the important milestones have been detailed without too much digression or too many irrelevances. As many dates as possible are included to keep the chronology correct. However different references often differ by a few years on a particular happening especially for early events. Only where considerable differences occur are specific mentions made. Where references to scientific aspects differ then that which appears to be the most authoritative has been accepted and where possible that nearest in time to the original material has been given priority although some of the alternatives are mentioned as well.

Chapter One

THE SHAPE AND POSITION OF THE EARTH

The story might be said to begin some 5000 years ago, probably in the present day Iraq, where the ancient civilizations of Sumer, Akkad and Elam occupied the plains around the rivers Tigris and Euphrates in Central and Southern Mesopotamia. However with the rise of Babylon, about 1800 B.C., astronomy had its beginnings. Records on clay tablets have been deciphered to indicate the interest at that time in the changing arrangement of the heavens.

It was a time when the earth was considered as a circular island where the centre rose considerably above the rim. This was surrounded by a circular sea which was itself bounded by a range of mountains that was able to support the sky. Even then it would seem that the circle was considered to be the ideal shape.

By the 8th and 9th centuries B.C., Homer described the earth as a convex disc surrounded by a constantly moving ocean. Resting on the rim of the disc was a high vault forming the heavens and shaped as an inverted hemisphere and supported by many tall pillars. Maybe this was one way of describing far distant mountains since their peaks do appear to be holding up the sky.

Anaximandros of Miletus (ca.611-545 B.C.) had a somewhat similar idea with a flat disc, of thickness about a third of the diameter, suspended in space and surrounded by ocean. His ideas of the habitable earth were larger than those of Homer; extending from Southwest England in the West to the Caspian Sea in the East. The vicinity of the Aegean Sea and Greece was the centre of the earth with the whole in a state of equilibrium and at the centre of the world. Anaximandros is credited by some with the discovery of the obliquity of the ecliptic although others are inclined towards the later Oenipedes (ca.439 B.C.).

A contemporary of Anaximandros was Thales (ca.624-547 B.C.) who is described as the first geometer. However as far as he was concerned the earth was similarly a short cylinder but floating on a vast ocean. Yet others of the same era, particularly from the Indian subcontinent, described it as a flat tray supported on the backs of 3 elephants who in turn were on the back of a huge tortoise.

Anaximenes of Miletus (ca.545 B.C.), a pupil of Anaximandros, was the first of the Greeks to think in terms of the stars as situated on a sphere that rotated around the earth. However for the shape of the earth he preferred a rectangle supported on air. Additionally he had the novel idea that night came as a result of the sun going below an unidentified range of mountains. His contemporary, Hecataeus, also from Miletus, stuck with the disc shape.

Pythagoras (ca.580-500 B.C.) was the first to consider the earth as a sphere in space but this could well have been solely because a sphere was, at that time, thought to be the most perfect of regular solids. He envisaged a central spherical earth with the sun, moon and planets revolving in concentric circles. A further adherent to the spherical form was Parmenides of Elea (ca.515-450 B.C.) but there is no record as to whether he simply followed Pythagoras or whether he reached his own conclusions. He considered the earth to be stationary at the centre of a series of concentric spheres. He went further by dividing the earth into five żones, akin to the present day ideas — two polar areas, two temperate areas and one equatorial area. He was the only philosopher of his time, other than the followers of Pythagoras, to think of the earth as a sphere.

A few years later Anaxagoras of Clazomenae (ca.499-428 B.C.), the last philosopher of the Ionian School, reverted thought to a flat earth supported on air in a hemisphere, one of many backward steps taken in this story. Some suggest that he thought eclipses to be shadows of the earth but he must have interpreted them to be from a circular plate rather than from a sphere.

Neither idea however would tally with those of Empedocles of Acragas (ca.490-435 B.C.) who had many strange conceptions of the universe. In particular he favoured the heavens as an egg-shaped surface made of crystal with the stars held on it in fixed positions by upwards air pressure but such that the planets were still able to move.

Plato (ca.427-347 B.C.) was a distinct follower of Pythagoras and felt that the heavens moved around a spherical earth once every 24 hours. This was probably because the sphere was said to be the most perfect shape. On the other hand one reference says that Plato thought of the earth as a cube. Whichever it was he thought of it as immobile and kept at its central position by virtue of its symmetrical location within the heavens. Hicetas, around the same time, taught that the earth rotated on its axis every 24 hours which was quite the opposite to Plato's ideas.

The Pythagorean philosophy was continued by Philolaos, also in the 5th century B.C., who wrote of a spherical universe of finite size at the centre of which was a fire, but not the sun. He was the first to explain the daily rotation of the stars and sun by suggesting that the earth circled the central fire every 24 hours. In total he had 10 bodies orbiting — the earth, moon, sun, six planets and a counter earth that was always out of sight. The stars were fixed on a surface against which the others moved.

A very important character in the development of ideas on the shape of the earth was Eudoxus of Cnidos (ca.408-355 B.C.). His consideration of spherical motion was earth-centred and he proposed the existence of 27 concentric spheres each of which rotated on its own axis at a definite speed. He was, in effect, combining a geocentric system with a homocentric one and with rotating spheres. After gathering a group of like minded individuals together he started a school at Cyzicos and later built an observatory at Cnidos.

During the 4th century B.C. Heracleides of Pontos, a pupil of Plato, was suggesting a compromise between the earth centred and sun centred systems. In effect he founded the geoheliocentric system and proposed epicycles to describe the motions of Mercury and Venus around the sun. The earth rotated from West to East on its axis with the sun and planets rotating round the earth.

Around the same time as Heracleides came Aristotle (384-322 B.C.). He taught of a spherical universe of finite size with a stationary spherical earth at its centre with the other planets circling round the earth. For this he had to double, to 55, the number of homocentric spheres, compared with Eudoxus, to account for all possible heavenly movements. As with his predecessors, it is likely that his spherical earth was again based more on the enduring thoughts of symmetry than on any reasoned scientific explanation. It is, however, reported that he noticed when travelling north to south, the pattern of the heavens changed. Because of the rate of this change he felt that the earth must be small in relation to the heavens. In the same way that Anaxagoras noticed the circular form of eclipses so did Aristotle but he related this to the sphere rather than a disc.

One of the first critics of the homocentric spheres was Callipos of Cyzicos (ca. 370 B.C.). The ideas of Eudoxus did not seem quite right to Callipos and he added further spheres to give a total of 33, each rotating on its own axis and at its own speed. This was a sort of compromise between the Eudoxus and Aristotle results. Additionally he made reforms to the calendar and estimated the lengths of the seasons.

A pupil of Aristotle, Dicaiarchos of Messina taught of a spherical earth and made efforts to determine the heights of mountains, possibly by use of an early version of the dioptra, but his results were usually excessive. His investigations extended to appreciating the effects of the moon and sun in generating tides.

Ecphantos of Syracuse, who flourished in the early 4th century B.C. continued to promote rotation of the earth but positioned it at the centre of the universe.

By the late 3rd century B.C. Autolycus of Pitane was assuming all the stars to be on a celestial sphere and he succeeded in determining geometric relations between them without any assistance from either plane or spherical trigonometry.

About this time globes were introduced. Many of the well known philosophers and astronomers are recorded as having one soon after the idea of a spherical earth became generally accepted. There is even mention of one in a Naples museum during the 4th century B.C.

Aristarchos of Samos (ca.300-230 B.C.) was the first to advance a heliocentric system where the earth was of negligible size in relation to the remainder of the heavens. The stars and planets were fixed but revolved around the earth which, in its turn, rotated on its own axis once every 24 hours. This continuation of revolutionary ideas was quickly debunked by the contemporary geocentric enthusiasts and the heliocentric theories were sunk for some 1800 years until the appearance of Copernicus.

The best known mathematician and astronomer of this time was Eratosthenes (276-195 B.C.) who figures prominently later in this story. He accepted the spherical earth of the followers of Pythagoras and with his subsequent observations settled the argument between flat and spherical supporters.

2

Another major character in the story came about a century later — Poseidonius of Apamea, a Stoic philosopher, geographer and one of the best know astronomers, who died in Rome about 50 B.C. He repeated, with refinements, the measurements of Eratosthenes.

Seleucos, a Babylonian astronomer was one of the few, and probably the last, ancient observer to accept the heliocentric ideas of Aristarchos.

At the turn from B.C. to A.D. the Greek geographer Strabo wrote ". . . I must take for granted that the universe is spheroidal . . . if the earth is spheroidal, just as the universe is, it must be inhabited all the way round." He accepted the spheroidal theory from the idea that bodies tend towards the centre and that each body inclines towards its own centre. To this he added the phenomena of curvature of the sea that gradually made lights appear or disappear as one sailed along. Besides horizontal movement he also appreciated that by raising the position of either the observer or the target, a previously non-intervisible line could become intervisible. He considered the habitable part of the earth to be the equivalent of 8000 by 3600 miles; longest in the East-West direction. Claude Ptolemy (100-178 A.D) continued the ideas of a geocentric planetary system and was also in favour of its shape being spherical. When compiling his famous atlas he made some of the first steps into map projections. When writing his 'Geography', he said "Drawing a map on a sphere gives the likeness of the shape of the earth. . . . Making a map on a plane surface. . . requires a certain adjustment to correspond to the spherical form, in order to make distances . . . real distances."

Even so, not everyone in astronomy and mathematics agreed that the earth was spherical. In the 4th century A.D. one of the worst, or strongest, of the adversaries of a sphere was Lactantius. In his "Divine Institutions" he devoted a whole chapter to ridiculing the possibility of a spherical earth. How would it be possible for men to hang head down or the oceans stay in place? Surely it was impossible for the starry heavens to be underneath the earth? In the same century the Bishop of Gabala referred to the Bible for his proof that the earth must be flat.

"It is he that sitteth upon the circle of the earth and the inhabitants thereof are as grasshoppers; that stretcheth out the heavens as a curtain, and spreadeth them out as a tent to dwell in." *(Isaiah 40:22.)*

"Let there be a firmament in the midst of the waters, and let it divide the waters from the waters." *(Genesis 1:6.)*

"And God made the firmament, and divided the waters which were under the firmament from the waters that were above the firmament and it was so." *(Genesis 1:7.)*

During the 6th century Cosmas Indicopleustes scorned the idea of a round earth by trying to prove that the tabernacle of Moses was a model of the universe. It had a flat bottom, vertical sides and an arched roof. The earth serving as the footstool of the Lord was a rectangular plane twice as long as it was wide and resting on the flat bottom of the universe. The Sun managed to be hidden at night as it could not possibly go below the earth. It is somewhat surprising that he took such an attitude since he had travelled widely before becoming a monk. As Lactantius, he queried how a spherical earth would be able to retain lakes and oceans.

Not all were of like mind. The Hindu astronomer, Aryabhata the Elder taught that the daily rotation of the earth on its axis provided the rising and setting of the heavens.

During the 8th century, The Venerable Bede, a monk from the North of England, was writing of a spherical earth although his ideas of its water and air were reminiscent to twelve centuries earlier. He tended to an egg shape with the earth as the yolk and water as the white; with air and fire as the shell.

By the middle of the 9th century another disbeliever, Hrabanus Maurus was relating to the scriptures and referring to a square earth because of its 'four corners'. At the same time he considered the horizon as a circle—which was somewhat difficult to correlate.

In the early 15th century Nicolaus de Cusa was having strange ideas about the earth but his fallacies were rejected by Regiomontanus, although in his turn he could not accept the idea of a rotating earth. The time was fast approaching however when the revival of scientific activity began to take off from where it had been in the time of Aristotle and Ptolemy.

One of the main characters to end the 2000 year old argument was Copernicus (1473-1543). He gave credit to Aristarchos and agreed that the sun was fixed at the centre of the universe and that the earth revolved around it. Such ideas did not find favour in the church and it was only after Galilei (1564-1642) had suffered the brunt of the recriminations that the ideas were accepted. In 1633 Galilei was tried in Venice for his heretical beliefs and made to sign a retraction and then kept under virtual house arrest until his death.

While the general form of the heavens was then generally settled, even after Galilei there was Giovanni Riccioli (1598-1671) propounding arguments both for and against the motion of the earth.

The era of Newton and the Cassinis was fast approaching. Sir Isaac Newton (1642-1727), in his theoretical investigations of the earth considered it to be not only some spherical shape but also non-symmetrical. He deduced that it must be oblate—or flattened at the poles. This was in complete contradiction to the views of his French counterparts, the Cassinis, who from their arc measurements were sure that the earth was prolate—flattened at the equator.

It was to take two major scientific expeditions to settle this argument once and for all.

_____*References:* *See in particular:* 1, 2, 3, 4, 10, 13, 29, 33, 38, 40, 76, 80, 98, 107, 127, 129, 145, 186, 216, 223, 226, 227, 245

Chapter Two

ASTRONOMERS AND MATHEMATICIANS
Pre - Eratosthenes

It is difficult to make sensible selection of landmarks for this period of history since all aspects of both astronomy and mathematics have some influence on the later, more pertinent topics. Here, an effort has been made to concentrate on only those topics that are particularly relevant. This is not intended as a history of astronomy or mathematics. It is concerned only with highlighting those specific aspects relevant to measuring an arc of the meridian.

At least 5000 years ago the Babylonian priests and philosophers of the day were observing the heavens and forecasting the future. They would have been aware of daytime and night, changing lengths of days and the effects of changing atmospheric conditions. They would have noticed that the sun is not always at the same elevation, that it moved during the day, and that the stars moved at night. They were not however, in a position to record anything other than general qualitative impressions. Quantitative recording was not possible.

Simple counting systems were developing where there was a need. This would be for the exchange of goods, buying and selling, quantities for constructions, and for setting and re-setting the field boundaries in the flood plains of the mighty rivers Nile, Tigris, Euphrates, Ganges and Indus.

By the time of the Sumerians around 2000 B.C. counting was based on a unit of 60 and this applied to both integers and fractions. The extension to a decimal system had to wait until 1585. It is thought that the unit of 60 could have been derived from early estimates of a year as 360 days. The day also at that time was divided into 360 but in the form of 12 equal lengths each of 30 gesh. By chance in modern terminology we have 1 gesh = 4 minutes of time = 1°.

Square roots of numbers could be extracted, the areas of simple figures calculated and the first approximations of π, such as 3⅛, were known. Quadratic and cubic equations could be solved and right angled triangles were of benefit to the surveyors (or agrimensores). It is said that Rameses II (ca. 1300 B.C.) had sufficient knowledge and expertise to subdivide areas of land into equal sized plots. This is quite feasible since the ''rope stretchers'' or first land surveyors are know to have been at work a 1000 years before Rameses - in the times of Amenemhat I.

Geometry first became a science with Thales (640-547 B.C.) and he applied himself to the circle, isosceles triangle, congruent triangles and the inscribed angle of a semi-circle. He founded the Ionic school and made early measurements of the heights of the pyramids from the lengths cast by their shadows. This was possibly achieved by also measuring the shadow of a stick of known length and then using similar triangles. He is said to have calculated the distance to a ship at sea using the principle of similar triangles.

Among the pupils of Thales was Anaximandros who succeeded to the leadership of the Ionian school when Thales died. Both were particularly interested in astronomy and gave impetus to use of the gnomon. Anaximandros. from solar observations, could have gained some idea of the ecliptic and maybe even of its obliquity but this is uncertain. He did however give some consideration to the distances of the sun and moon from the earth and suggested 27 and 19 times the earth radius respectively. As far as is known however, he had no idea of the value of the earth's radius.

Another user of the gnomon was Anaximenes (ca. 585-528 B.C.) who may have been a pupil of Anaximandros since he is known to have been at the Ionian school.

In the same century Theodoros invented some of the earliest known surveying instruments, used in building construction - the diabetes, libella, and square.

The renowned Greek philosopher and mathematician Pythagoras gave his name to a theorem that has ever since been of inestimable value to the surveyor. Nevertheless there are those who feel that the idea contained in the theorem was known and used in practice long before Pythagoras became involved. That was, however, by no means his only contribution to mathematics although it is almost impossible to separate his individual work from that of his followers. After his early travels, he settled in Crotona, S. Italy and gathered round him a number of his followers. Unlike most of the other schools this was much more of a closed community or brotherhood. All teaching was by word of mouth and any discoveries, hypotheses, and the like were shared in such a way that they are only known as being of the Pythagorean school. The school was short-lived and disbanded before his death.

Oenipedes (ca. 500-430 B.C.) furthered the work of Anaximadros on the eliptic although there are no records of it being quantified. He gave the length of a year as 365 + 22/59 days — remarkably close to the currently accepted value but his method of arriving at that figure is unknown. The fraction in particular is curious with its denominator of 59 rather than 60.

Herodotus (ca.484-425 B.C.) was principally an historian but is credited with the movement of Egyptian geometric philosophies to Greece. He introduced the idea of a meridian as a north-south reference line.

About 389 B.C., Plato founded his philosophic school which lasted almost 900 years. A philosopher and mathematician who had studied under Socrates, his ideas were generally after the Pythagorean school. He did little in the way of producing original developments except improve upon some of the existing geometric methods.

When Plato opened his school the only other geometer of note in Greece was Archytas who had been referred to by Horace as a measurer of earth and sea. It was at that time that Democritus was constructing plane figures in such a way that he felt the rope-stretchers (harpendonaptae) of Egypt would have been proud of him.

At about the same time Philolaos was explaining the daily rotation of the stars and putting forward suggestions as to the motion of the sun. In general he was a devoted follower of the Pythagorean teachings.

Of considerable importance was Eudoxus (ca.408-355 B.C.) who had studied under Plato and Archytas to be described as the founder of scientific astronomy. Considered the greatest mathematician and astronomer of his time he built an observatory at Cnidos. He proposed a series of concentric spheres to account for planetary motion in sup-

port of his heliocentric theory. He introduced the detailed study of mathematical astronomy into Greece where he founded a school in Athens about 368 B.C.

His method of exhaustion for the computation of areas, volumes and lengths of arcs was the first step towards calculus and he also produced a treatise on the theory of proportion. On the practical side he is credited with the development of a new form of sundial and possibly also of an astrolabe.

A development on the heliocentric ideas of Eudoxus was the geoheliocentric system and theory of epicycles as suggested by Heracleides. In addition he considered the earth to be rotating from west to east on its axis every 24 hours.

Aristotle (ca.384-322 B.C.), yet another student of Plato, taught over the whole field of knowledge but in particular was the founder of the science of logic. In 335-4 B.C. he founded a new philosophic school, the Lyceum, in Athens. In astronomy, however, his teachings were not always based on observation or even theory but tended to be ideas of how the universe ought to be.

He is credited, possibly incorrectly, with the oldest attempt at estimating the size of the earth with a circumference value of 400,000 stade. For his planetary theories he required 55 concentric spheres.

By this period two subjects were separating — with geodesy detaching itself from geometry. At that time, however, geodesy was not confined to land or earth measurement but would have better been called mensuration. According to Geminus it was a function of geodesy to measure, not a cylinder or a cone, but heaps as cones and tanks or pits as cylinders. He was the author of works on both astronomy and mathematics.

Autolycus, of the 3rd century B.C., was one of the most learned of astronomers and mathematicians of that time. He wrote two works on astronomy and with his death came the demise of the Hellenic domination.

The penultimate name for this period is the renowned Archimedes of Syracuse (ca.287-212 B.C.) who was active in arithmetic, geometry, mechanics, astronomy and physics and was particularly interested in the geometry of measurement. He anticipated by some 1900 years, some of the ideas later to be resurrected by Newton and others. From an inscribed, 96 sided polygon he derived a value lying between 3 1/7 and 3 10/71 (some quote 3 11/71) as the relationship between the circumference of a circle and its diameter that is between 3.142 and 3.141. At that time the symbol π was not used in Greece to denote this value.

After Aristotle, Archimedes is credited with the second estimate for the size of the earth as 300,000 stade although various authorities doubt that the calculation was his. He measured the diameter of the sun and used a form of Jacob staff or dioptra. He constructed an orrery to illustrate the movements of the sun, moon, and planets with sufficient

5

accuracy to forecast eclipses of the sun and moon.

When working with large numbers Aristotle mentioned what turned out to be the basis of logarithms — the addition of ''orders'' of numbers.

Lastly, mention must be made of Aristarchos who made numerous astronomical observations particularly in relation to the summer solstice and the sizes and distances of the sun and moon. This he did in terms of fractions of a quadrant since the 360° circle had yet to be introduced. Although his results were all quite wide of the present day values, nevertheless, he had made a useful step forward. For the angular diameter of the sun he gave 1/180 of a quadrant, or 1/2°. This compares with the quoted value of Archimedes of between 27' and 32' 56" — values which approximate to 1/192 and 1/168 of a quadrant — i.e. 1/8 and 1/7 respectively times 1/24 quadrant. It is noticeable that the 180 denominator of Aristarchos is midway between the 168 and 192.

Aristarchos appreciated the steady change in the ratio of an arc to its chord as the subtended angle changed but it was another century before the changes were tabulated.

References: See in particular: 13, 28, 40, 76, 80, 107, 123, 124, 127, 214, 223, 226, 227, 232,

Chapter Three

THE GNOMON AND SCAPHE

Shadow recording devices come in numerous shapes and designs. They range from the vertical stick in the ground to complex spherical, cylindrical or conical shapes. In fact, the variation of shape is limitless. In the context of this treatise two basic forms are of interest, the vertical column and the hemispherical bowl.

Over 1000 years B.C. there are references to regular observations of the position and length of the shadow cast by a vertical stick in the ground. Possibly invented by the early Babylonians and Egyptians, they would have appreciated that the shadow length varied during a day, and from day to day, and that there was a minimum at noon. Hence they accumulated information on the diurnal and annual variations and realized that the minimum length varied from noon at midsummer to noon at midwinter.

Figure 1

The gnomon

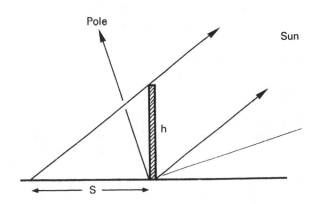

At the same time the philosophers were no doubt full of questions; why does it change length during a day, from day to day and throughout the year? At what sort of interval, in days, does the length repeat?

Probably shepherds were among the first to utilize the information for the seasonal movements of their flocks and also the tillers of the land for the planting and harvesting of crops.

The gnomon is said to have been introduced by Anaximandros of Miletus in the 6th century B.C. although crude forms of it could well have existed centuries earlier. The term "gnomon" was originally an astronomical instrument. Later it became a term for a perpendicular to a straight line and later still an instrument for right angles or a quarter circle. Yet another reference gives its derivation as from the Greek "one who knows". From the simple stick shadow and knowledge of the height of the stick and length of shadow at any given time it would not have been too long before gnomons were used for special purposes.

In order to use gnomons, knowledge of the obliquity of the ecliptic was required. Developing this knowledge is often credited to Anaximandros, although some give credit to Oenipedes, a contemporary of Anaximandros. The obliquity is the angular distance between the equator and either of the tropics. It is slightly variable and at that time would have been about 23°45′05″ whereas today it is about 23°26′45″.

In the time of Oenipedes it was estimated at 1/15 circle (24°) and by the time of Eratosthenes in the 3rd century B.C. he is credited with the value of 11/83 circle between the two solstices or twice the obliquity, so the obliquity would have been about 23°51'20", only some 15' greater than what it should have been. Considering that 1/83 is over 4° it was probably more luck than judgment to get so close. There is mention of the fraction of 11/83 in at least two books that discuss the 'Almagest' of Ptolemy. Whilst it could have been some strange division of the circle it is thought by Toomer [254] to be far more likely to have been derived from gnomon measurements. There is another possibility for the fraction of 11/83 of Eratosthenes as instanced by Delambre [66].

Eratosthenes could have arrived at this figure from either of graduations equivalent to 47° 40' or 47° 50'. This would assume the use of, say, an armilles solsticiale graduated, or at least readable by interpolation to 10'.

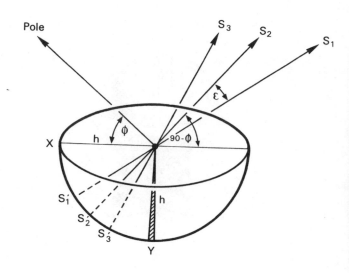

The scaphe

Figure 2

$$\text{Then } 47°40' = \frac{47\frac{2}{3}}{360} = \frac{143}{1080} = \frac{11 \times 13}{1080} = \frac{11}{83\frac{1}{13}}$$

$$\text{or} \quad 47°50' = \frac{47\frac{5}{6}}{360} = \frac{287}{2160} = \frac{13 \times 11\frac{1}{26}}{1080} = \frac{11\frac{1}{338}}{83\frac{1}{13}}$$

In the figure:
$$s_1 - s_1' = \text{winter solstice.}$$
$$s_2 - s_2' = \text{equinox}$$
$$s_3 - s_3' = \text{summer solstice}$$
$$\phi = \text{latitude}$$
$$\epsilon = \text{obliquity of the ecliptic}$$
$$h = \text{height of gnomon.}$$

In either case the result could be approximated to 11/83.

If only the values of "h" and "s" are determined, then a calculation is required to obtain the elevation of the sun. This difficulty was overcome by Aristarchos who devised a means of recording the angle directly. This he did by using a hemispherical bowl carved out of a block of stone, and containing a central vertical gnomon of height equal to the radius of the bowl. This arrangement came to be known as an hemisphaerium or scaphe and could well have been the first true angle measuring instrument. It was also known as skaphion and gnomon or skiotheron.

The inside of the bowl had a series of circles concentric around the gnomon to indicate the altitude of the sun. At that time it was usual to divide the circle into either 48ths or 96ths (i.e. divisions of either 7 ½° or 3 ¾°). It was almost certainly the instrument used by Eratosthenes at Alexandria to determine the sun's altitude and he made the result 1/50th of a circle. This could however have been a simple rounding off of 1/48. Note however the caution that appears later as to whether or not Eratosthenes should have any credit in this respect.

At the equinox the 'shadow' angle $y - s_2' = $ latitude of the point. The altitude at any time is the value of the arc from X to the end of the shadow.

His value of 11/83 for twice the obliquity would be

$$\text{arc } s_1' - s_3'$$

Among the sources of error in such a method, in addition to the crude subdivision of a circle, would be omission of any allowance for the diameter of the sun. Since this is of the order of 30' the error introduced would be 15'. This would compound the difficulty of determining the end of the shadow since there is a penumbra area of doubt. This is apparent whether it is a scaphe or any other form of gnomon that is being used.

The result of such uncertainty was the development of various shapes of gnomon and devices such as small holes near their tops in an effort to achieve a more distinct point for use in measuring.

Some gnomons were pointed obelisks where the bevel on the point was the same as the summer solstice angle.

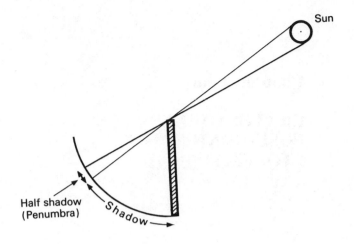

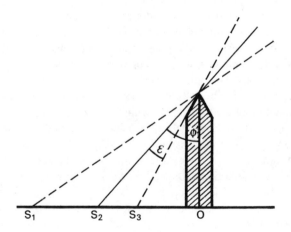

Figure 3. An obelisk as a gnomon.

Of course, there would be no need to wait for the solstices to use such devices to determine the difference of latitude. The difficulty would be to obtain the noon shadow angle on the same day at both places.

References
See in particular:
33, 66, 75, 92, 98, 107, 129, 223, 226 and 259.

Chapter Four

ERATOSTHENES
POSEIDONIUS
PTOLEMY

Eratosthenes

The dates for Eratosthenes life are usually given as 276-195 B.C. but his birthdate can be found varying from 296 B.C. to 276 B.C. His date of death can vary from 214 B.C. to 192 B.C.

Born in Cyrene (now Shahhat in Libya), he was the first notable figure in the story of measures of the size of the earth. He has been described as the greatest mathematical geographer of antiquity. Note should be taken, however, of reference [261] where doubt is cast upon his work.

In analyzing in depth the dimensions of the pyramids, Werner, in that paper concludes that the builders of them (ca 2600 B.C.) must have known the radius of the earth. This is deduced from the perimeter of the pyramid of Giza as 1760 cubits which, it was said, was to be equal to ½ minute of latitude. This it is to, within a centimetre. But such a statement must surely also depend on very good knowledge of the relation between the cubit and the metre. Readers are left to decide for themselves which is the more probable explanation.

Eratosthenes, son of Aglaos, received his early education at Cyrene under Lysanias and in Alexandria under Callimachus. His early career was spent in Athens. In 235 B.C. when he was about 40, Ptolemy III (Ptolemaios III Evergetes, ruled 247-222 B.C.) of Egypt invited him to Alexandria both as tutor to his son Philopator, Ptolemy IV, and as Royal librarian at the famous museum there as successor to Callimachus. This library was the most famous in antiquity but not the earliest. Founded in 284 B.C., Eratosthenes was its fifth librarian from 235-195 B.C. By the 5th century A.D. it was almost nonexistent but probably not completely destroyed until the Muslim invasion of 645 A.D.

His interests ranged over astronomy, geography, history, philosophy and poetry. He is known to have been nicknamed both "Beta" and "Pentathlus". The former suggesting perhaps that he was considered second best only to Plato and the latter relating to his prowess at athletics. With such a wide range of interests, and the position he held, it is thought that he was able to select from the existing doctrines, those parts that pleased him. Some also thought he was somewhat of an amateur and didn't take a deep interest in any particular subject.

Among his notable activities, mention should be made of the updating of the Ionian map which at that time was round in shape and centred on Delphi. Since he considered the inhabited world (oikoumene) to be rectangular and some 7800 stade in the east-west direction and 3800 stade in the north-south direction. He used meridians and parallels as the basis for the new map.

There is considerable uncertainty about the modern equivalent of the stade used by Eratosthenes. One reference [195] says there were several stade in use at the time but does not elaborate on this. References [143] and [74] quote 157.5m for the short stade, [260] and [3] give 185m as the Attic stade, and [80] gives 559 ft. (170.4m). Reference [176] gives 185.3m from the relation 1 stade = 6 plethora each of 100 Greek feet. Reference [2] relates 252,000 stade to 29,000 English miles or 1 stade = 185.2m which is said to be from the writings of Hipparchus. By the time of Ptolemy the Egyptian stade was given as 210m.

For the baseline he chose the parallel stretching from the Sacred Promontary in W. Spain (Gibraltar) to the Himalayas and passing to the south of Sicily, just touching Greece, through Rhodes, the south edge of Toros Daglari (the Taurus mountains) of Turkey, the Elburz mountains south of the Caspian and north of Tehran, and thence to the Hindu Kush and the Himalayas.

For the fundamental meridian he took that through Meroé (Merowe) 31°46′ E, Syene (Assuan) 32°57′ E, Alexandria (El Iskandariya) 30°00′ E, Rhodes (Rodos) 28°10′ E, Byzantium 29°00′ E, and the mouth of the river Borysthenes (Dnieper) possibly at Olbia 32°21′ E which had been bounded, and later rebuilt, by the Greeks from Miletus.

There seem to be notable discrepancies between the past and present geographical positions of various cities. Those quoted above are from a modern atlas while [177] gives the following as those used originally.

Meroé	33°	44′	E
Syene	32	50	
Alexandria	29	51	
Rhodes	28	14	
Byzantium	28	59	
Olbia	32	00	

Not a particularly straight meridian by today's standards but it was probably sufficient for all practical purposes. A further meridian passed through Carthage 10°11′ E., Sicily 12°25′ E., and Rome 12°30′ E..

It will be obvious that Rhodes was the origin of the system, probably because it was then at the peak of its importance and power.

About 230 B.C. (some references say 250 B.C.) Eratosthenes made his celebrated determination of the size of the earth from observations at Syene in Upper Egypt (probably Assuan) and Alexandria. Another [177] prefers Philae to Assuan as this was regarded as the farthest limit of the Roman empire up to the days of Trajan about 100 A.D.

Although not at all accurate for a number of reasons, it was the first value to approach the present day value and it used principles that have been retained to the present. Basically two quantities are required—the ground distance between two points that are situated on the same meridian, and the difference in latitude between those points. Inaccuracies were due to:

a. the, necessarily crude, measure of the vertical angle.

b. the method of estimating the distance between the cities.

c. assuming that both Alexandria and Assuan were on the same meridian. Assuan is some 3° E of Alexandria.

d. the apparent addition of an arbitrary quantity to make the result divisible by 60.

e. assuming that Syene was on the Tropic of Cancer whereas the location is thought to be some 37 miles north of the Tropic at 24° 05′ 30″ N.

Although his approach to the problem was essentially the same as that used up to modern times the equipment and computational techniques have since been refined considerably.

Some persons, including Hipparchus, thought that Eratosthenes did not make any observations himself, and that he accepted the statements of travellers (what reliability can one place on such data?) that, at the summer solstice, the sun filled a well from top to bottom i.e. it was directly overhead. But did it reflect in the whole of the water surface or simply cover say three quarters of it? In reference [3] it is pointed out that 2200 years ago at noon in midsummer the sun would not fall vertically into a well at Syene.

If the well were 1m diameter and 10m deep and the illumination covered only three quarters of the water surface then the sun could be up to 1½° out of the vertical. Thus there could be an unquantifiable error from this source.

At Alexandria, Eratosthenes is thought to have used a sciotheron which had a pointer or style in the centre of a hemispherical bowl (scaphe) to determine the angle of the sun from the shadow cast. This he gave as 1/50 of a circle (since division of the circle to 360° was not then known) — equivalent to 7°12′. In other words the shadow came just under 1/12th of the way up the side of the bowl. Division of a circle into 96 parts or 48 parts could be marked off by successive divisions of angles. Might it then have been 1/12th that he recorded and then, after multiplying by 4 to get 1/48th of a circle, rounded it to 1/50th? This would give a difference equivalent to 0.3°.

To complete the calculation the distance between the terminals was required. This was generally quoted as 5000 stade arrived at as a camel journey of 50 days at 100 stade per day. Alternatively another reference refers to it as a figure supplied by the Royal road surveyors. It is not inconceivable that the surveyors then were camel-mounted and that the two possibilities refer to the same process. There was however a type of surveyor called a Bematistes, who was trained to walk with equal paces and to count them as he went. But for some 1000 km? Ref. [218] analyzed the speed of camels on several long treks and obtained surprisingly consistent results of 2½ British miles per hour or about 18 miles per day not allowing for halts. Considering the varying values for the stade, 100 stade is probably equivalent to about 18.5 km or about 12 miles. So, allowing for halts the figure quoted of 100 stade per day would seem reasonable.

However, besides the wanderings of any route between the terminals there was also the matter of 3° difference in longitude. This in itself would reduce the measured meridian distance from 5000 stade to around 4675 stade, or a 6½% error. It is possible that allowance had been originally made for the wanderings of the route and some accidental account made for all or part of the 3°.

Eratosthenes, using the figures of 1/50th of a circle and 5000 stade found the circumference as 250,000 stade, and this is the figure reported by Cleomedes in 175 A.D. However, Hipparchus who was chronologically much nearer to Eratosthenes, accepted a figure of 252,000 stade. It has been suggested that Eratosthenes added the extra 2000 stade to give a result divisable by 60 so that $1° = 700$ stade.

A variation on this figure can be found in a statement in ref. [179], "If man travelled north 43 leagues the pole would be 1° higher i.e. the circumference would be 15,750 leagues or 252,000 stade."

With regard to the southern terminal of the arc there is considerable doubt as to the actual position of the observation. Syene (or Assuan) is just below the first cataract at latitude 24°05′ N and where the obliquity of the ecliptic was then 23°43′. However the "well" is thought not to have been in Syene proper but in Elephantine, an island in the river Nile opposite Syene which was an important religious and military centre in Pharaonic times. Alternative names for this island are Yebu in Egyptian and Jazirat Aswan in Arabic.

To make comparisons with modern times the sources of error already mentioned are compounded by the uncertainty as to the length of the stade used. If the value of 1 stade = 185m is used the results become:

- Arc 7.2° (1/50), length 5,000 stade, circumference 250,000 s = 46,250 km

- Arc 7.2° length 4,675 stade, circumference 233,750 s = 43,240 km

- Arc 7.5° (1/48), length 4,675 stade, circumference 224,400 s = 41,510 km

These can be compared with the 18th century approximation of 40,000 km to get a feel of the probable order of accuracy (or of compensating errors) of Eratosthenes' observations depending on how one interprets the original measurements. In fact one might express wonder at the result being as close as it is considering all the possible sources of error.

Among his other achievements he derived the obliquity of the ecliptic from the difference between the zenith distance of the sun at both winter and summer solstices as 0.5 × 11/83 of a circle. This interesting fraction is mentioned in detail in the previous chapter.

He also recorded a value for the distance from Rhodes to Alexandria of 5½° or 3,850 stade (if a degree is taken as 700 stade). Was it really 5½° or maybe 1/64th of a circle = 1/16th of a quadrant, a quantity much more readily found from crude astronomical observations? 1/64 of a circle = 5.6° or about 3,920 stade.

At the end, when he was about 80 years old he is said to have been blind and to have committed suicide by voluntary starvation.

References for Eratosthenes.
See in particular:
1, 2, 3, 6, 28, 33, 38, 40, 74, 80, 81, 98, 125, 143, 176, 177, 179, 183, 186, 195, 197, 218, 227, 232, 245, 249, 260, and 261.

Poseidonius and Ptolemy

After Eratosthenes the next measure, or estimate of the size of the earth was by Poseidonius (135 - 50 B.C.) He was head of the Stoic school and well versed in geography and astronomy. Based in Rhodes, he was the first to make allowance for atmospheric refraction and to explain the effects of the sun and moon in producing tides.

The fact that he measured the distance to the sun and the circumference of the earth has only come down to the present day through the writings of Cleomedes. It is said that in his extensive travels he noticed that the star Canopus just touched the horizon in Rhodes whereas at Alexandria he estimated its meridian altitude as 7½° or in terms of units of the day, 1/48th part of the zodiac, or ¼ of a sign. He considered the distance from Rhodes to Alexandria to be 5,000 stade, giving a circumference of 240,000 stade. Equivalent to 667 stade per degree.

In fact the latitude of Alexandria was 31°12′ N and Rhodes was 36°27′ N. The difference between the two was 5°15′ instead of 7½°. In addition there was a difference of longitude of nearly 2°.

The length of the stade that was used varies from 157m to 210m. Thus while one reference equates 5,000 stade to 925 km and says it should be 580 km this is invalid without knowledge of the length of the stade. The Attic stade of 185 m. gives 925 km while the short stade of 157.5m would give 787.5 km.

From the writings of Strabo the value found by Poseidonius was 180,000 stade as an improved estimate taken from the work of Eratosthenes. It was this value that was later accepted by Ptolemy and said to be in terms of the official Egyptian stade of 210m. This equates exactly to the 240,000 stade of Poseidonius in terms of the short stade of 157.5m, 180,000 × 210m = 37,800 km and 240,000 × 157.5m = 37,800 km.

Others estimated the distance as only 4,000 stade, while Eratosthenes had found it as 3,850 stade. The result was obviously a poor one and of less worth than that of Eratosthenes even though it was used for many centuries.

References for Poseidonius and Ptolemy.
See in particular:
3, 6, 33, 76, 98, 107, 232, and 260.

Chapter Five

ASTRONOMERS AND MATHEMATICIANS
Eratosthenes to I-Hsing

Of the same vintage as Eratosthenes was Hypsicles, who flourished in Alexandria. He is thought to have been the author of "De ascensionibus" which suggested the division of a circle into 360 parts, at the same time continuing the use of sexagesimal fractions.

He was soon followed by Hipparchus (180-125 B.C.) who was one of the greatest among ancient astronomers and described by some as the father of trigonometry. Those who doubt the claim of Hypsicles to introducing the 360° circle credit it to Hipparchus.

Hipparchus was the first to appreciate precession of the equinoxes by comparing his star observations with those of Timocharis, some 150 years previously. Some also credit him with the invention of the dioptra which was described in much detail by Heron.

He is credited with the invention of a system of latitudes and longitudes and constructed the first celestial globe. In so doing he probably introduced the theory of the stereographic projection.

Hipparchus confirmed the estimation of Eratosthenes that the distance between the solstices was 11/83 of a circle but this did not stop him from being critical of much of the work of Eratosthenes and correcting it.

Hipparchus equated several groups of places as being on the same parallels but in so doing tended to be considerably in error. He put Alexandria on the same parallel as Cyrene; Alexandria (Troas) an ancient port south of Tenedos, on the same parallel as Apollonia on the Gulf of Arta and as Amphipolis; Nicaea on the same as Byzantium and Massilia (Marseilles). The following Table of modern latitudes indicates the inaccuracies.

Alexandria	31°	11′ N.
Cyrene	32	50
Alexandria (Troas)	39	40
Apollonia (Epirus)	39	03
Amphipolis (Trace)	40	48
Nicaea	40	25
Byzantium (Istanbul)	41	00
Massilia (Marseilles)	43	18
Syene (Philae)	24	01
Assuan	24	05

Among his other findings was the length of the tropical year as 1/300 day less than the previously accepted 365¼ days. Above all however he was the first person to make regular use of trigonometry and is considered to have founded both plane and spherical trigonometry. He earned the title of father of trigonometry particularly as a result of his computation of a table of chords that was equivalent to modern day sine functions or in fact, double sines of half angles. These were at 15′ intervals from 0° to 90°. Details are given in a separate chapter.

Poseidonius (135-50 B.C.) was one of the first astronomers to take atmospheric refraction into account although some give this credit to Ptolemy and yet others specifically say that it was unknown to Poseidonius. He was the first to explain tides as due to the joint action of the sun and moon and to notice spring and neap tides.

One of the earliest mentions of the use of sighting tubes in instruments was by Huai-Nan-Tzu, a Chinese prince of the second century B.C. Of the same vintage, the stoic philosopher, Crates constructed a terrestrial globe while he was in Rome in 168 B.C.

Around 70 B.C. Geminus produced works on astronomy and mathematics which gave a good insight into early Greek mathematics.

Where Cleomedes is put chronologically is open to debate since he is variously said to date in the first century B.C. up to the second century A.D. The best that can be done is to put him after Poseidonius (135-50 B.C.) and before Ptolemy (100-178 A.D.). He is, however, of particular importance since he is the source of all references to the works of both Eratosthenes and Ptolemy. This he did in *Cleomedis de motu circulari corporum caelestium libri duo* which was first translated into Latin in 1488 and into Greek in 1539.

He remarked on refraction (as cataclasis) and included comments on atmospheric refraction. This he described in the sense that it makes the sun visible even though it is actually below the horizon. He referred also to the bending when light goes through water say, in a jug. If a ring is placed on the bottom of a jug when it is empty and viewed at an angle it will not be visible. When the jug is filled with water the ring comes into view.

Another person difficult to date was Heron with different authorities varying from 150 B.C. to 250 A.D. He did much to establish a scientific foundation for engineering and land surveying. In particular he wrote a treatise called *Dioptra* where he described an ancient form of theodolite and a large number of geometric problems with their solutions. He was essentially a practical surveyor as well as physicist and mathematician. His formula for the area of a triangle is well known

$$A = [s(s - a)(s - b)(s - c)]^{1/2}$$

He worked on chord tables and although sexagesimal fractions were common he continued using unit fractions.

Among his other developments were a level where two transparent dishes were connected by a long tube, a leveling staff to go with it, and a forerunner to the thermometer.

The first and second centuries in China saw Pan Ku constructing a primitive form of abacus and Chang Heng building a celestial hemisphere designed to rotate about an equatorial plane. The latter also wrote on both astronomy and geometry.

Ptolemy in the second century A.D. was the last great Greek astronomer. His greatest contribution was his writings that came to be known as the *Almagest*. This contained all established Greek knowledge on astronomy and was in effect an encyclopedia which remained in vogue until 1543. His atlas similarly lasted into the 16th century with most of its errors resulting from the use of the value for the earth radius given by Poseidonius.

He developed the use of sexagesimal fractions and the table of chords of Hipparchus, deduced longitudes from the differences of apparent times of eclipses and used a rectangular grid on maps. He was aware of refraction but did not know how to quantify or allow for it.

Among the instruments he described were the astrolabe, the earliest reference to a quadrant, a triquetrum, and other instruments for determining latitude by transits of circumpolar stars. However, there was difficulty with longitude because of the lack of portable, accurate timepieces. Thirteen centuries later Copernicus is found building instruments much as designed by Ptolemy. These were specifically for meridian altitudes of the sun and of other bodies — the triquetrum and quadrant respectively.

Ptolemy also explained plane and spherical trigonometry and this was taken further by his contemporary Menelaus who defined a spherical triangle as the area included by arcs of great circles where each arc was less than a semi-circle. His work on spherical triangles was similar to that of Euclid on plane triangles. He wrote several books on the calculation of chords and derived his *Theorem of Menelaus* in relation to a spherical triangle.

Marinus who flourished around the same time as Ptolemy and Menelaus was the founder of mathematical geography. He was more successful than Hipparchus in defining positions by two coordinates. He used a series of meridians and parallels as straight lines perpendicular to each other. His prime meridian was through Fortunatae Insulae, which are thought to have included the Canary Islands, Madeira, and the Azores. He later took the meridian through Ferro, one of the Canary Islands. He took 36° N as the base parallel and marked off degrees of longitude along this according to their proportion to degrees of latitude at that latitude. He did not try to preserve similar correct proportions along other parallels.

The surveyor Hyginus Gromaticus (Gromatici were those who used the groma) described use of what were called Indian circles for determining a meridian line for boundary positioning.

Liu Hung a Chinese astronomer, who flourished just before 200 A.D. showed that the equator and ecliptic were not coincident. A compatriot Lu Chi constructed a celestial map and a little later Wang Fan replaced the celestial hemisphere of Chang Heng by a complete celestial sphere with the earth at the centre and with the equator and ecliptic as separate circles. Hsu Yueh gave some of the earliest writings on abacus arithmetic.

During the apparent hiatus from about 200 A.D. to 700 A.D. there were some notable developments. Theon of the 4th century A.D. commented on the likeness between the chorobates of Carpos and the diabetes of Theodoros. He described a method of finding the square root of a number expressed in sexagesimal fractions.

Moving to India, the Hindu mathematician and astronomer Aryabhata of the 5th century A.D. was the first to give

14

a special name to the sine function and to compile a table of values at 1° intervals, although a century earlier in the *Paulisa Siddhanta* they were similarly tabulating angular functions.

For this Aryabhata used a radius value of 3438 whereas Brahmagupta 150 years later used 3270. He also produced a table of versed sines and gave π as $3\frac{177}{1250}$ or 3.1416.

His contemporary Varahamihira derived formulae which, if transformed to modern terminology, contain both sines and cosines, for example —

$$\sin \frac{a}{2} = \left[\frac{1 - \cos a}{2}\right]^{1/2} \text{ and } \sin^2 a + \text{versin}^2 a = 4 \sin^2 \frac{a}{2}$$

When Indian scholars travelled to China from the 7th century onwards, among their scientific introductions was an early form of trigonometry. The publication *Khai-Yuan Chan Ching* included a table of sines, the first appearance of a zero symbol and a sexagesimally divided circle to minutes and seconds. Previously the circle in Chinese science had been divided into 365½ parts. The first appearance of the decimal symbol was, however, probably around 330 B.C. This obviously went hand in hand with the division of lengths into powers of ten although even earlier, in the Chou dynasty of the 6th century B.C., a bronze foot rule was subdivided into tenths and one of these into further tenths. At that time the length of the foot was equivalent to 0.231 m.

Mention has already been made of the various values for the length of the stade. The Chinese foot was somewhat different. It showed a gradual and continuous increase over the period of 3000 years from the Chou dynasty to that of Ching. Quoted values increase from the equivalent of 0.195 m to 0.308 m.

This was very much a period for measurement of the changing lengths of the shadows cast by gnomons.

References
See in particular:
1, 13, 28, 40, 76, 80, 107, 123, 129, 145, 177, 186, 197, 214, 223, 226, 227, 232 and 260

Chapter Six

THE ARCS OF I-HSING AND AL MAMUN IN ARABIA

I-Hsing and his arc measurement

A Tantric Buddhist monk, I-Hsing (682-727 A.D.) was one of the greatest mathematicians and astronomers of Chinese history and his recorded contributions embrace both sciences.

For many centuries the Chinese accepted that for every 1000 li one travelled north or south the length of a gnomon shadow would change by 1 Chinese inch (1 li was approximately equal to 1/2 km). They further considered that the centre of the earth was where the shadow of an 8 foot gnomon was 1.5 feet long at the summer solstice.

By 445 A.D. actual measurements were proving this to be wildly incorrect since it appeared that 3.6 inch per 1000 li was nearer the mark.

In 604 A.D. one Liu Chhuo felt that the time was ripe for proper measurements to be made over several hundred li but unfortunately the Emperor of the time was not impressed and so the situation rested.

Although it is indicated above that one li was about 1/2 km., there were two possible values at that time from the following relationships -

1 Chinese foot	=	0.2957m. It was divided into 10 inches.
6 feet	=	1 pu or double pace.
300 pu	=	1 li (Long Thang version)
	=	532 m.

1 Chinese foot	=	0.2456 m.
6 feet	=	1 pu
300 pu	=	1 li (Short Thang version)
	=	442 m.

These were by no means the only variations because equivalents as low as 193 m have been recorded.

The gnomons in references [186] and [14] would appear to have been constructed on the basis of the Short Thang since from the diagram:

8 Thang ft. = 1.98m or 1 Thang ft. = 0.2475 m. and
1.5 Thang ft. = 0.37m or 1 Thang ft. = 0.2467 m.

which agrees closely with the above value of 0.2456 m.

During the more enlightened Thang dynasty (618-906 A.D.) an expedition was organized in 724 A.D. by the Astronomer Royal Nankung Yüeh and the monk, I-Hsing, for the making of observations at a series of gnomon spread over some 11,400 li in the vicinity of the 114° E line of longitude. The number of gnomons can vary from 9 to 11 depending on how much credibility is placed on the more remote points. It is thought, even proved, that the figures relating to some stations would seem to have been extrapolated rather than observed.

Each gnomon was 8 feet high and of identical shape and established vertical by use of a water level and plumb line. Their latitude ranged from 17° N, not far from Hue, to 40°N near the Great Wall and then a further one at 52° N near Lake Baikal which is one of those considered to have been extrapolated.

The stations and positions were:

	Name		Latitude N³		Longitude E°
1	Thieh-Lo	≅	52.00	≅	114.
2	Wei-chou		39.80		114.5
2a	Thai-yuan		37.67		112.5
3	Hua-chou		35.55		114.7
4	Pien-chou		34.80		114.55
5	Yang-chhêng⁴		34.43		113.03
6	Hsu-chou		34.15		114.38
7	Yu-chou		33.60		114.35
7a	Hsiang-chou		31.99		112.1
8	Lang-chou		29.00		111.60
9	Chiao-chou		(20.73)²		106
10	Lin-i		(17.25)²		106 ± 1

Notes:

[1] Numbers 1 to 8 are the same as those used in
    [14] p 21.

[2] Values derived from polar altitudes whereas
    all others are scaled from a 1934 Chinese
    atlas.

[3] Sexagesimal degrees NOT the Chinese divi-
    sion of 365¼ tu to a circle. From a 1934
    Chinese atlas.

[4] This station is also recorded as 34° 26′ N
    113° 02′ E and 34.63 N 113.00 E.

The station descriptions were:

1. Thieh-Lo was near Lake Baikal
2. Hêng-yeh Chun Army post in Wei-chou was NE of
    the present Wei-Chou district in Chahar. It was near
    the modern Ling-chhiu and the Great Wall in North
    Shansi.
3. Hua-chou (Pai-ma Hsien) was originally on the
    south bank of the Yellow River but since the river
    changed its course it is now well north of it.
4. Chün-i or Khai-fêng in Pien-chou.
5. Yang-chheng - ancient central astronomical obser-
    vatory. Several slightly different positions are
    quoted. The gnomon was set up in 723 A.D. by
    Astronomer Royal Nankung Yüeh on Imperial edict.
    Although it was not set up by Hsing it is the only one
    of the gnomons that still exists.
6. Fu-khou Hsien in Hsü-chou
7. Wu-chin near Shang-tshai in Yü-chou
8. Lang-chou (Wu-ling)
9. Chiao-chou (Hu-fu, capital of An-nan) in northern
    Vietnam where its longitude cannot be greater than
    106°. it is now Hanoi in Tongking.
10. Lin-i in southern Vietnam similarly about 106° E.
    Now Indrapura in Champa near the modern city of
    Hue.

From station Wei-chou to station Lin-i was 7973 li
(approximately 2500 km). Simultaneous measurements
were taken of the shadow lengths at the summer (and
winter?) solstices. The two chief observers were Ta-
Hsiang and Yuan-Thai who were probably also Buddhist
monks. On the completion of the observations they were
commanded to inter-compare the results in the presence of
I-Hsing.

In recent years close scrutiny of the observations by
Arthur Beer et al (1961) [14] cast doubt on some of the
results.

Suggestions are:

a. that distances between the more remote stations were
    not directly measured.
b. that winter solstice shadow lengths were calculated
    from the summer values.
c. that polar altitude values were similarly derived from
    the summer solstice values.
d. that all computations were accurate to about 1 part in
    1000 - which would only have been possible if trigon-
    ometric tables of an accuracy of 1 part in 500 and
    intervals of 5 minutes of arc were available. Whilst
    just a possibility this is not considered a realistic
    conclusion.
e. it would be virtually impossible to measure the
    shadow lengths accurately to 1 part in 1000 which
    would be equivalent to 0.1 inch.
f. to achieve angles resolvable to 2 minutes of arc would
    require a circle of some 17 metres radius with 1 cm. =
    2 minutes.

With regard to d, if such tables had just been developed
it would not have been unreasonable to demonstrate their
power by computing required elements from the minimum
of observations.

The method of reduction from shadow lengths to calculated latitude is as follows from [14] p 17.

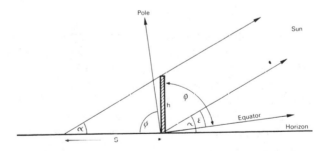

Figure 4

Geometry of a shadow observation.

s = shadow length
h = height of gnomon = 8 Chinese feet
α = angle of sun at summer solstice.
ε = obliquity of the ecliptic.

$$\tan \alpha = \frac{h}{s}$$

At the summer solstice

$$\phi = 90° - \alpha_s + \epsilon = 90 - \alpha_E$$

[14] concludes that the value used for ϵ was

$$23° 40' = 24 \text{ tu}$$

Thus $\phi = 113° 40' - \alpha_s$

Where 1 tu = 1/365.25 of a circle.

Line					
3 to 4	198 li	179 pu	= 198.60 li	Approximately	100 km
4 to 6	167	281	= 167.94	"	85
6 to 7	160	110	= 160.37	"	80
3 to 7		Total	526.91 li	"	265 km

Difference in $\phi°$, li per degree

3 to 4	0° 16'	744.75
4 to 6	0° 38'	265.17
6 to 7	0° 31'	310.39
3 to 7	1° 25'	371.94

Station.	Polar Altitude°	Length of summer shadow (ft)	Calculated $\alpha°$	Calculated $\phi°_s$	Calculated ϕ tu
1.)	51° 15'[3]	4.13[3]	62° 42'	50° 58'	51.71
2.)	39° 25'[3]	2.29[2]	74° 02'	39° 38'	40.21
2a.)	—	6.0[1]	53° 08'[1]	36° 52'	37.40[1]
3.)	34° 47'[3]	1.57[2]	78° 54'	34° 46'	35.27
4.)	34° 18'[3]	1.5312[2]	79° 10'	34° 30'	35.00
5.)	34° 12'[3]	1.4788[2]	79° 32'	34° 08'	34.63
6.)	33° 48'	1.44[2]	79° 48'	33° 52'	34.36
7.)	33° 19'[3]	1.365[2]	80° 19'	33° 21'	33.84
7a.)	—	4.8[1]	59° 02'[1]	30° 58'[1]	31.42[1]
8.)	29° 04'[3]	0.77[2]	84° 30'	29° 10'	29.59
9.)	20° 06'[3] 21° 17' }	− 0.33[2]	92° 21'	21° 19'	21.63
10.)	17° 09'[3]	− 0.57 − 0.91[2] }	94° 05' 96° 29'	19° 35 17° 11'	19.87 17.43

Notes:

[1] Derived from equinox values in absence of summer solstice values.

[2] Measured

[3] Estimated

There would appear to be fundamental errors in some of the observations since, for example, between 3 and 4 the difference of polar altitude is 0° 29' with a difference computed from shadow lengths of 0° 16' while between 4 and 6 the comparable figures are 0° 30' and 0° 38'. The distances apart are: 3 to 4, 199 li and 4 to 6, 168 li.

With the points reasonably close to one another it could be expected either:

a. that if the differences in polar altitudes were the same the distances would be the same.

b. that if the shadow lengths were correct then the distance 3 to 4 would be shorter than 4 to 6 which is not so.

There are thought to have been only 3 measured distances:

Whilst [14] went to great lengths to derive from the observations a value for the length of a degree the values are so susceptible to large errors that no worthwhile conclusion can be drawn.

a. The polar altitudes are recorded to 0.1 tu (approx. 6' of arc), so that even assuming perfect observations the altitudes could be in error by 0.05 tu (3' of arc.)

b. If the instrument construction was such that 0.5 Chinese inch represented 0.1 tu, its radius would need to be of the order of 28 Chinese feet.

c. Shadow lengths are recorded to 0.01 Chinese feet, equivalent to about 4' of arc in α or ϕ. This would be very difficult to achieve because of the indefiniteness of the end of the shadow. As is now commonly appreciated with the shadow of almost any object the intensity of shadow decreases gradually rather than suddenly.

In 1279 Kuo Shou-Ching developed a special pinhole device for focussing the image of the cross bar on the top of his 40 ft. gnomon.

d. Assuming that distances between some stations were measured (since they are recorded to the nearest 1 pu or about 1.6 m) there would appear to be an error in either the distance to station 7 or in the shadow length at that point.

Using only 'certain' measures (although there is no indication of how the distances were determined) —

Line	Shadow length (in.)	Distance (li)	li/inch
3 to 6	1.3	366.54	281.95
3 to 7	2.05	526.91	257.03
3 to 4	(0.388)	198.60	(511.86)
4 to 6	(0.912)	167.94	(184.14)

This gives a mean value of 269.5 li/inch for 3 to 6 and 3 to 7.

18

Working backwards from the mean value over lines 3 to 6 and 3 to 7 then the shadow differences for 3 to 4 and 4 to 6 ought to be about 0.74 and 0.62 respectively. In which case they would suggest that the shadow length at 4 should be 1 ft. 5.0 inch which is 5/16 inch less than quoted in [14]. Whence the tabulation could be:

Station	Shadow length (ft)	Latitude $\phi°$
3	1.57	34° 46'
4	1.50	34° 17'
6	1.44	33° 52'
7	1.365	33° 21'

Line	$\Delta\phi°$	Distance (li)	li/degree
3 to 6	0° 54'	366.54	407.27
3 to 7	1° 25'	526.91	371.94
3 to 4	0° 29'	198.60	410.90
4 to 6	0° 25'	167.94	403.06
6 to 7	0° 31'	160.37	310.39
		Mean	380.71

Accepting the overall value of 3 to 7 despite the doubt about 7 considering the consistency between 3 to 4 to 6, then 1° would be about 372 li or 1 tu about 366.6 li.

It appears however, in the face of all this indefiniteness, that I-Hsing settled for a relationship of:

$$1 \text{ tu} = 351 \text{ li } 80 \text{ pu} = 351.27 \text{ li} = 155.26 \text{ km}$$
$$\text{or } 1° = 356 \text{ li } 117 \text{ pu} = 356.39 \text{ li} = 157.52 \text{ km}$$

For the short Thang.

Values which [14] shows can be obtained if the polar altitude differences 3 to 4; 4 to 6 and 6 to 7 are each 0.5 tu and the average is taken of the results. Whence the difference in shadow length for 1000 li would be nearly 4 inches rather than the value of 1 inch that had been accepted by the early Chinese.

References
See in particular:
14, 186

The Arabian arc of Al-Mamun

Abu-al-'Abbas 'Abdullah al-Mamun (786-833 A.D.), the seventh Abbasid Caliph, known as *The Trustworthy* reigned from 809 to 833 A.D. After the death of Harun al-Rashid (Aaron The Just) who reigned from 786 to 809 (some references give 813 instead of 809) the Empire was torn with conflicts among his two sons, one of whom was Al-Mamun and the other, his half brother Al-Amin. As a result the caliphs lost their political might and had to be satisfied with the spiritual leadership of the faithful.

In 762 A.D. Baghdad was established by the Arabs as the capital and during his reign Al-Mamun, who numbered astronomy among his interests, founded an observatory there in 829. It was furnished with instruments made to Ptolemaic design and included a 20 foot radius quadrant and a 56 foot radius stone sextant. Although earlier Omayyad Caliphs had had an observatory near Damascus that was entirely eclipsed by the new one.

About 820 A.D. Al-Mamun requested his astronomers Al-Farghânî and Al-Khowârizmi to arrange the measurement of an arc on the Plain of Sinjar (Zinjar) near Baghdad, but there are said to be records of other arcs as well. There is, however, some confusion, depending on the references used. The locations of three seem quite definite. However, one of these was divided into two parts that could well account for mention of a fourth.

The first measure appears to be along the caravan route from Tadmor near Palmyra (Palmyre) at latitude 34° 33' N, and longitude 38° 17' E to Raqqah (Rakka, Ragga) at 35° 57' N, and 39° 02' E where Sanad bin 'Ali and Khalid bin 'Abdul Malik Al-Marwarrudhi made the length 57 Arab miles for 1 degree. The recorded distance in one reference is 187 km. (about 95.5 Arab miles depending on the conversion factor used). This would appear to be the skew distance between the points. If it is reduced to the equivalent length of a meridian degree it approaches the quoted figure of 57 Arab miles.

A rough check might be made by using Clarke 1858 values, at 34° 30' N, 1° longitude = 301,327 feet. At latitude 34° 30' to 36° 00', the difference of latitude in feet is 545,983. thus the skew angle is of the order of 29° so that the meridian equivalent of 95.5 Arab miles would be 83.6 miles or 55.7 miles per 1°.

Yet another reference, [39] quotes the measure as 76 Arabian miles = 67 Black miles or 1° = 57 Arab miles.

This measure was checked by Al-Asturlabi and Al-Buhtari at some unknown location with the same result.

The third measure was on an arc from near Baghdad, at 33° 22' N and 44° 25'E, to a point near Al Kufah (Kufa) at 32° 05' N, and 44° 19' E. The length is recorded as about 75 Arab miles and 1° as 56 2/3 Arab miles. Thus the terminals here must have been taken as about 1° 20' apart and accepted as being on the same meridian.

The fourth measure was required because of the variations in the other results. It was on the Plain of Sinjar west and southwest of Mosul from latitude 34° 30' to 36° 30' N, at 42° E longitude, by Al Falaki and his brothers. The terminals were probably near Anah and Sinjar. It is thought that the two measuring parties started from a central point. One worked northwards and the other southwards, until their astronomical observations indicated they and gone 1 degree. Going north 1° equalled 56 Arab miles, and going south 1° equalled 56 2/3 Arab miles. This is recorded in [260] as 109 km (presumably the average), although this gives 55.57 Arab miles using the conversion figure given in [173] of 1 Arab mile = 1.961 371 km. On the other hand [195] gives 1 Arab mile = 1.957 8 km.

Two different methods seem to have been used. The first method was to observe the altitude of Polaris at one end of the arc and to set off the meridian direction. This could have been by quadrant since there is mention of Al Kindî who, in 9th century Baghdad freed one of the fixed limbs of a triquetrum to turn it into what was essentially a quadrant capable of measuring the angular distance between two stars. However the measurers do not appear to have been very adept at this part of the work.

When a suitable terminal was reached a second altitude was taken. From the measured distance, the value for 1° was calculated.

In the second method, used in the Sinjar Plain, after the initial altitude of Polaris distances were measured until the star was either higher or lower than the starting point by one degree. The method used to measure the distances is uncertain. One reference says that ropes were used and another that horses were used. A further reference quotes the use of wooden rods but this does not seem at all feasible over such distances. If it were a rope, there is no mention of its length.

There is uncertainty about the exact equivalent of an Arab mile. Records note that there were 4000 cubits to the Arab mile, but there is uncertainty about the length of a cubit although it is generally quoted as 490 mm ± 5mm; 490 mm gives 1 Arab mile = 1.960 km and 56 2/3 Arab miles = 111.073 km per degree or almost 10,000 km per quadrant (acutally 9996.59 km).

References

See in particular:

3, 6, 39, 76, 80, 124, 173, 180, 186, 195, and 260.

Chapter Seven

ASTRONOMERS AND MATHEMATICIANS
I-Hsing to Ulugh Beg

Arab astronomy began in the 8th century A.D. and reached its peak in the 9th century. It is said to have been the Golden Age of Islamic culture under the Abbasid Caliphs. The astronomy reached the Arabs from the scientific works of the Greeks which were translated into Arabic possibly through more than one language on the way.

Before discussing the Arab influence mention should be made of one Englishman at the turn of the 7th century. The Venerable Bede was one of the learned men of his time and wrote extensively on both mathematics and astronomy. He developed finger reckoning and studied the representation of numbers.

Baghdad was founded in 762 A.D. by Abbasid Caliph Al-Mansûr (= The Victorious) who reigned 754-775 A.D. Unfortunately he was very mean but under his successor Haroun-al-Rashîd (reigned 786-809 A.D.) matters were different. He was a patron of science, art and literature. Under him Baghdad took on the mantle that had for so long been worn by Alexandria. His nickname was Aaron The Just.

One of the early Arab astronomers was Yá-qub ibn Tariq who wrote on mathematical astronomy and divisions of the kardaja. The Hindus and Muslims divided the circle into 96 parts as originally done by the Archimedians. The arc, or sine, of each of these parts was called 'kardaja'. Of the same period was Al-Sûfi who wrote on the astrolabe and was the only author of the period to write on uranometry.

The first Muslim to construct an astrolabe was Al-Fazari who also wrote on mathematical instruments. Under a command from Al-Mansûr he translated into Arabic an Indian work on the calculus of the stars and equations based on sines calculated for every half degree. His work was called the great Sindhind.

By the early 9th century the centenarian astronomer Al-Hasib introduced the idea of the shadow or umbra equivalent to our tangent function and compiled a table of them. The idea of a shadow obviously stems from the use of gnomons where the height of the gnomon and the shadow length give the tangent.

Al-Rashîd was succeeded on his death by Caliph Al-Mamun (809-833 A.D.) who was his son. His nickname was 'The Trustworthy' and under him Islamic culture came to full flower but at the same time he was intolerant and tried to enforce his ideas by violence. His patronage of the sciences led to the founding of an observatory in Baghdad in 829 to emphasize its growth as the centre of Muslim science as well as culture. He occurs also in relation to the arc measurement by the Arabs.

Further Muslim observatories were built at Damascus, Fatimid Egypt in Cairo, Jandi-Shapur and Nishapur in Persia.

Al-Kindî (786-873 A.D.), known as the philosopher of the Arabs, wrote on many aspects of astronomy, optics and numbers. He was also credited with the building of a quadrant.

A notable mathematician at the court of Al-Mamun was Al-Khowârizmi (800-847 A.D.). He was an author of several mathematical books particularly on algebra and arithmetic, calculating the areas of triangles, the parallelogram and circle, where he used 3 1/7 for π. Astronomical tables prepared by him contained both sine and tangent. The book on algebra gave rise to that word, where the original title was Al-jabr wa'l muqabalah.

In addition he wrote on geography and included lists of the latitudes and longitudes of various cities. A further work was devoted to the construction of the astrolabe. He is thought to have helped in the Arabian arc measure.

An early astrolabe maker was the Muslim astronomer Al-Marwarrudhi and about the same time Al-Farghânî wrote an important work on the lines inscribed on an astrolabe, on sundials, and on astronomy in general. He was another of those who assisted in the Arabian arc measurement.

One of the foremost 9th cenutry astronomers was Al-Battani or Albategnius (858-929 A.D.) It was through him that the Arab work on trigonometry and sine functions came to Europe. He did much work on spherical trigonometry and produced a table of cotangents for every degree. Among the formulae he used was the equivalent to

$$b = a.\cot A = a.\sin(90 - A)/\sin A$$

The word 'Sine' is said to originate from his work 'De scientia stellarum' which used 'sinus'. The Arabic for the function, 'jiba' came from the Sanskrit 'jiva' and resembled the Arab 'jaib' meaning indentation or gulf. In this context he introduced the Indian 'sine' or half chord in place of the whole chord used by Ptolemy.

Credit is given to him for adding the shadow square to the astrolabe. This allowed direct readings of the length of shadows of angles less than 45°.

A notable 10th century writer on Euclid was Al-Nairizi who also wrote an elaborate treatise on the spherical astrolabe.

In the mid-tenth century Hero wrote a treatise on surveying based on the work of Heron, especially his dioptra.

In 988 Sharaf al Daula built an observatory in the palace grounds and among his astronomers was Al-Wafa (940-998 A.D.). His particular interest was in the production of sine tables correct to 8 decimal places at 15 minute intervals. He specially studied tangents and proved theorems for double and half angles.

Among the later Arab writers was Al-Biruni (973-1048 A.D.) who introduced Hindu mathematics to the Arabs. Other writings included gnomon lengths, Hindi shadow reckoning and the astrolabe. He was one of the earliest to set the sights of an astrolabe in a tube. He also produced a simplified version of the stereographic projection similar to that produced much later (1660 A.D.) by G. B. Nicolosi di Paterno. Although he is said to have made geodetic measurements there are no details.

The Abbâsid dynasty lasted from 750-1258 A.D. but by 940 A.D. it's importance was waning. When the Arabs conquered Spain in the 8th century they brought astronomy to Western Europe. By the Middle Ages mathematicians and astronomers of Muslim Spain were world leaders and what had been the role of Baghdad had moved to Toledo. The Arab settlement in Spain had started in 755 A.D. when Abd al-Sûfî escaped the massacre of all the rest of his family and fled to Spain. Baghdad city was finally sacked by Mongol II Khan Hulagu in 1258 A.D.

The Hindu mathematician and astronomer Bhâskara (1114-1185 A.D.) produced elaborate sine tables based on the 1/24 th of a quadrant although he wrote on many subjects his astronomical publication 'Siddhanta S'iromani' of 1150 added little to the knowledge of 500 years earlier. He derived a value for the earth diameter of 1600 yojans (about 12.2 km each) but no details are available. He used a horizontal staff and mirror to measure the height of an object. His value for π was 3927/1250.

By 1145 A.D. one reads of Robert of Chester translating the Algebra of Al-Khowarizmi. Robert was the first to actually use the term 'sine' when translating from the Arabic.

The Hispano-Muslim astronomer Al-Bitrûji of the late 12th century is one of those considered as possible inventor of the turquet or torquetum.

Al-Muzaffar from Tus in Khurasan invented the Tusi staff or linear astrolabe around 1200 A.D. A kinsman of his, Al-Tûsî (1201-1274) was the last great Arab astronomer. He wrote numerous treatises on trigonometry, astronomy, computations, geometry and the astrolabe. In 1259 he constructed an observatory at Maragha and was its director until just before his death. He is credited, as was Al-Bitrûjî, with being the possible inventor of the torquetum, and is known to have had a 12 ft. radius mural quadrant with sights, an 11 ft. meridian circle and azimuth circle. Unfortunately the observatory did not long survive his death.

A contemporary of Al-Tûsî was the Syrian astronomer and engineer Al-Urdi who also built a mural quadrant. He gave detailed instructions for checking the metal rings on his astronomical instruments and refined the method of observing the meridian by use of several concentric circles.

Moving briefly to China, the early 13th century (1221) saw a party under Chhiu Chhang Chhun observing at the summer solstice on the banks of the river Kerulen in north Mongolia at a latitude of about 48° N. This was effectively an extension of the earlier gnomon arc of Hsing.

In the mid-13th century Al-Marrakushi, a Moroccan astronomer and mathematician, produced a table of sines for every 1/2° and other tables for versed sine, arc sine and arc cotangent. He described two distinct types of plumb bob level and gave full details of a quadrant graduated for measuring heights, inclinations and with a shadow square.

In 1276 Robert The Englishman wrote a popular treatise on a quadrant in addition to work on the astrolabe. Such a quadrant could have been invented in the second half of the 12th century as an adaption of an Arabic instrument to Christian and Western needs. Certainly there are details of a muslim quadrant in 1231.

Of particular interest is the contribution of Levi Ben Gershon (1228-1344 A. D.). He was a Jewish Rabbi working in France. He used chords, sines and cosines but not tangents, although they had been known in Europe for a century or so. He formulated the sine theorem although it had been in use in the orient for about 300 years. He invented the Jacob staff or baculum although it did not come into wide use until much later. In connection with the staff he developed the use of transversals and the diagonal scale.

Shortly after Gershon is the first detailed reference to the use of the magnetic needle, both floating and pivoted, by Peregrinus. His invention revolved round a graduated disc.

The last astronomer of this period was Ulugh Beg (1394-1449 A.D.). He made many observations in Samarkand including some with a sextant said to have been 180 ft. radius. In 1420 he founded an institution of higher learning (masdrasa) there and three years later built a 3 storey observatory where the main instrument was a Fakhri sextant of 120 ft. (36.6m) radius. On this it was possible to read to 5 seconds of arc. Such a sextant was set in a trench that was oriented in the meridian and subsequent excavations tend to prove this. On the arc 64 cm corresponded to 1°, hence 11 mm equalled 1 minute and 1 mm equalled 5 seconds.

In addition he had an armillary sphere, triquetrum, astrolabe and shamila, a combined astrolabe and quadrant. He produced tables of sine and tangent at 1′ intervals to 9 decimal places. His assassination in 1449 led to the end of Islamic astronomy.

References
See in particular:
2, 3, 6, 28, 33, 40, 80, 90, 91, 123, 143, 173, 183, 186, 195, 223, 232, and 260

Chapter Eight

CIRCLE SUBDIVISION AND TRIGONOMETRIC FUNCTIONS, TRANSVERSALS, VERNIER, MICROMETER AND LOGARITHMS

Circle subdivision

At the dawn of philosophy or astronomy the earliest calendars were founded on a year of 360 days but this was soon found to be an incorrect assumption. Nevertheless there were those who would have taken this value and related it to a circle — the most perfect shape of the time.

It soon became apparent that some 5 days had to be added if the calendar were to keep in step with recurring annual events. By the year "0" however it was appreciated that even this was giving an error of a day every 4 years and Julius Ceasar was advised by the Egyptian astronomers to adopt a year of 365 1/4 days. Why he was advised on 365 1/4 is not clear since Hipparchus gave a value of 1/300 day less than this some 150 years before.

As computational and observational techniques improved over the centuries so it was further realized that 365 1/4 was not exact. By the time of Copernicus it was a further 10 days out of order and as a result of his computations it was recommended to Pope Gregory XIII, in 1582, that some days had to be lost. The instigator of this was Christoph Clavius whose suggestion that October 15th should follow October 4th in 1582 did not find favour with his French protagonist Franscois Viéte. Prior to this Regiomontanus had been summoned to Rome in 1475 to advise Pope Sixtus IV on calendar reform but he died the next year before completing his task. Pope Gregory XIII had also had Clavius working on improvements before he engaged Copernicus.

The figure 360 had however already been accepted as an appropriate one for divisions of a circle. Even by 2000 B.C. there was a complete sexagesimal system both for whole numbers and fractions. The Sumerians divided each

of their 360 days into 12 sets of 30 gesh each (i.e. 1 gesh = 4 minutes).

Hypsicles, of the early 2nd century B.C., divided the ecliptic into 360 parts and Ptolemy divided the circle into 360 and the hour into 60. The earliest division of the ecliptic would have been in terms of the 12 signs of the zodiac each of which was subdivided into 30 dargatu.

The Chaldeans divided their circles into 180 parts, each called an ell which derived from division of a radius into 30 ell and using the early approximation of 3 for π would give 180 ell to the circle.

The use of 360 parts was however not universal. The Chinese subdivided their circles into 365 1/4 parts or tu. See the secton on I-Hsing.

Oenipedes is credited with quantifying the year as 365 22/59 days but this is a strange fraction for such an early period. With the sexagesimal system came the arrangement of large or small numbers as functions of 60, such as 60^2, and 60^3. The Greeks adopted such a system together with various symbols, in their calculations. The diameter of a circle was divided into 120 parts each sub-divided into 60 and 60 again with the symbols $^p\ '\ ''$. Thus a length might be given as $73^p\ 28'\ 43''$ or 73.4786 in modern decimal notation. In fact they did not stop at the equivalent of seconds but are known to have gone several further subdivisions each in multiples of 1/60.

Although in today's notation it would not be possible to multiply $5°\ 11'\ 14''$ by $8°\ 19'\ 32''$ since the symbolism only refers to angles; in the beginning when it also referred to distances, or any sub-division of a number, then multiplication was applicable. The symbols for the subdivisions have taken many forms and that used at present is probably only a few centuries old.

At this stage mathematicians such as Hipparchus were aware of simple relationships between chords, arcs and radii such as:

$$(crd\ 2\alpha)^2 + [crd\ (180 - 2\alpha)]^2 = 4r^2$$

Where crd 2α = chord subtended by an arc of 2α
and r = radius

Figure 5

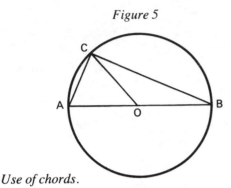

Use of chords.

Angle COB = 2α
Thus Angle COA = $180 - 2\alpha$
AB = $2r$
Then $CB^2 + AC^2 = AB^2$

Substituting gives the above chord formula.

It was readily possible at that time to sub-divide a circle into 96 parts by inscribing a figure of 96 sides. Thus it became the practice to describe an angle in terms of 96ths of a circle. They were also adept at inscribing other regular figures and relating subsequent chord lengths to the diameter of the circle.

Ptolemy studied in particular the relation of chords to central subtended angles while the Chinese studied half chord and half angle subtended by a whole chord.

As 1/96 circle = 3° 45' it will then be obvious why some of the early tabulations relating to circular functions were in units of this angle.

Trigonometric functions

Although the circle was divided into 360 parts in early times for practical purposes most early trigonometry was in terms of chord lengths instead of angles and with a diameter of 120 parts. This arose from the ability to inscribe various polygons in a circle. Thus a 6 sided figure. or hexagon, must divide the centre angle into units of 60° with the chord length equal to the radius or, at 120 parts to a diameter, equal to 60^p.

Thus –

Sides	Subtended angle	Chord Length	
3	120°	$(Crd\ 120)^2 + (Crd\ 60)^2 = 4r^2$	Crd = 103^P 55' 23"
4	90	$r\sqrt{2} = 60\sqrt{2}$	84^P 51' 10"
5	72	see below	70^P 32' 03"
6	60	$r = 60^P$	60^P 00' 00"
8	45	see below	45^P 55' 19"
9	40		41^P 02' 33"
10	36	see below	37^P 04' 55"
12	30	see below	31^P 03' 30"

Some of these are easily calculated, others require more formulae. For the pentagon and decagon the following construction is necessary

If B bisects the radius OD then

$$AB = \left[r^2 + \frac{r^2}{4}\right]^{\frac{1}{2}}$$

$$= \frac{r\sqrt{5}}{2}$$

Mark off BC = AB; then AC = side of inscribed pentagon and OC that of a decagon.

Figure 6

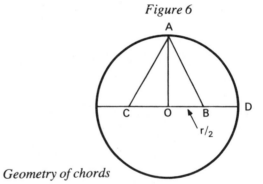

Geometry of chords

Then in triangle ACO:

$$CO = AB - \frac{r}{2} = \frac{r}{2}\left[\sqrt{5} - 1\right]$$

$$AC = \left[r^2 + \left(\frac{r}{2}[\sqrt{5} - 1]\right)^2\right]^{\frac{1}{2}}$$

Or
in terms of parts r/2 = 30

$$AB = (60^2 + 30^2)^{\frac{1}{2}}$$

$$= 67^P\ 04'\ 55"$$

25

$$CO = AB - 30$$

$$= 37^p\ 04'\ 55''$$

Whence $\quad AC^2 = CO^2 + A0^2$

$$= (37^p\ 04'\ 55'')^2 + (60)^2$$

$$= 1375^p\ 04'\ 14'' + 3600$$

$$\therefore AC = 70^p\ 32'\ 03''$$

In decimal notation

$$37^p\ 04'\ 55'' = 37 + \frac{4}{60} + \frac{55}{60^2}$$

$$= 37.0819$$

and $1375^p\ 04'\ 14'' = 1375 + \dfrac{4}{60} + \dfrac{14}{60^2}$

$$= 1375.0706$$

To extract the value of AC assumes knowledge of square roots which was a complex operation. The method used stemmed from knowledge of the relation—

$$\text{if}\quad a^{\frac12} = \left(b + \frac{\chi}{60} + \frac{y}{60^2} + \cdots\right)$$

To the term in χ^2

$$a = b^2 + \frac{2b\chi}{60} + \left(\frac{\chi}{60}\right)^2$$

When χ is found

$$a = \left[\left(b + \frac{\chi}{60}\right) + \frac{y}{60^2} + \cdots\right]^2$$

$$= \left(b + \frac{\chi}{60}\right)^2 + 2\left(b + \frac{\chi}{60}\right)\frac{y}{60^2} + \left(\frac{y}{60^2}\right)^2 + \cdots$$

Then consider $\sqrt{4975^p\ 04'\ 14''}$ from above.

The integer has to be obtained by guesswork as $70^p = b$

As $70^2 = 4\ 900$ there is a remainder of $75^p\ 04'\ 14''$, the square root of which will be expressed as

$$\frac{\chi}{60} + \frac{y}{60^2} + \frac{z}{60^3}$$

For χ. $\qquad \dfrac{2(70)\chi}{60} < 75^p\ 04'\ 14''$

$$\chi < 32^p\ 10'\ 23''$$

Whence $\qquad \left(70 + \dfrac{32}{60}\right)^2 < 4975^p\ 04'\ 14''$

Then $75^p\ 04'\ 14'' - \dfrac{2\times70\times32}{60} - \left(\dfrac{32}{60}\right)^2$

$$= 75^p\ 04'\ 14'' - 74^p\ 57'\ 04''$$

$$= 0^p\ 07'\ 10''\ \text{remainder}$$

Then for y.

$$2\left(70 + \frac{32}{60}\right)\frac{y}{60^2} < 07'\ 10''$$

$$0.04\ y < 07'\ 10''$$

$$y < 03''$$

Thus $\sqrt{4975^p\ 04'\ 14''} = 70^p\ 32'\ 03''$

If further accuracy is required, i.e. a value for z.

Then

$$2\left(70 + \frac{32}{60}\right)\frac{3}{60^2} + \left(\frac{3}{60^2}\right)^2 = \frac{25392}{60^3} + \frac{9}{60^4}$$

So

$$\frac{7}{60} + \frac{10}{60^2} - \frac{25392}{60^3} - \frac{9}{60^4} = \frac{7}{60} + \frac{10}{60^2} - \frac{7}{60} - \frac{3}{60^2}$$
$$- \frac{12}{60^3} - \frac{9}{60^4}$$
$$= \frac{7}{60^2} - \frac{12}{60^3}$$

$$2\left(70 + \frac{32}{60} + \frac{3}{60^2}\right)\frac{z}{60^3} < 7''\ 12'''$$

$$141.0683\ z < 60^3\left(\frac{7}{60^2} + \frac{12}{60^3}\right)$$

$$z < 3'''$$

Or a square root value of $70^p\ 32'\ 03''\ 03'''$

Such an approach is in order for selected angles but there is an obvious requirement to get down to single degrees or even half degrees and this requires different formulae.

Values such as $\sqrt{2}$, $\sqrt{3}$, $\sqrt{5}$ etc. can be found in the above manner but there is a need to develop formulae for sums and differences if a complete table is to result.

The following relationship can be found from a quadrilateral inscribed in a semi-circle.

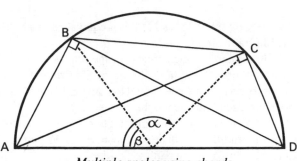

Multiple angles using chords.
Figure 7

Where AB = chord of β
AC = chord of α
Then BC = chord of $(\alpha-\beta)$
Whence CD = chord of $(180-\alpha)$
and BD = chord of $(180-\beta)$
crd $(\alpha-\beta)$ crd 180 = crd α.crd$(180-\beta)$ − crdβ. crd $(180-\alpha)$

Thus if α and β are known it is possible to determine $(\alpha-\beta)$ since crd 180 = 120

A further relationship is:

$$\left(\text{crd }\frac{x}{2}\right)^2 = \text{½ crd 180 } [\text{crd 180 } - \text{crd } (180 - x)]$$

If three chords AB, BC and CA form a triangle then—

crd 180 × crd$(180-AC)$ = crd$(180-AB)$
× crd$(180-BC)$ −crd AB × crd BC

Then crd AC can be found if crd AB and crd BC are known.   For some small angles it is possible to use inequalities to get the required crd values. To subdivide within 1/2° Ptolemy used a proportional parts technique. Alternatively it was possible to use

$$\text{crd } 3x = 3 \text{ crd } x - \frac{\text{crd}^3 x}{60^2}$$

Using a combination of these techniques Ptolemy was able to compile a table of chords from 1/2° to 180° in intervals of 1/2° that were generally correct to 5 decimal places. In addition to the chord for every 1/2° Ptolemy listed differences for units of 30ths of each entry i.e. corresponding to single minutes. Interpolation was linear.

For example, to find crd 45 knowing crd 90

Using:

$$\left(\text{crd }\frac{x}{2}\right)^2 = \text{½ crd 180 } [\text{crd 180 } - \text{crd } (180 - x)]$$

Substitute 90 for x

$$\begin{aligned}
(\text{crd } 45)^2 &= 60(120 - \text{crd } 90)\\
&= 7200 - 60(84^p\ 51'\ 10'')\\
&= 7200 - 5091^p\ 10'\ 00''\\
\text{crd } 45 &= 45^p\ 55'\ 19''
\end{aligned}$$

To find crd 30 if crd 90 and crd 60 are known.

$120\,\text{crd } 30 = \text{crd } 90\,\text{crd }(180-60) - \text{crd } 60\,\text{crd }(180-90)$

$120\,\text{crd } 30 = 84^p\ 51'\ 10'' \times 103^p\ 55'\ 23'' - 60^p \times 84^p\ 51'\ 10''$

$= 8818^p\ 09'\ 36'' - 5091^p\ 10'\ 00''$

$= 3726^p\ 59'\ 36''$

$\text{crd } 30 = \quad 31^p\ 03'\ 30''$

To find crd 108 given crd 36

$$\text{crd } 108 = 3 \text{ crd } 36 - \frac{\text{crd}^3 36}{60^2}$$

$$\text{crd } 108 = 3 \times 37^p\ 04'\ 55'' - \frac{(37^p\ 04'\ 55'')^3}{60^2}$$

$$= 111^p\ 14'\ 45'' - 14^p\ 09'\ 50''$$

$$= 97^p\ 04'\ 55''$$

Then to find crd 36 given crd 72 (which requires crd 108)

$$(\text{crd}\,36)^2 = 60(120 - \text{crd}\,(180 - 72))$$

$$= 7200 - 5824^p\,55'\,00''$$

$$\text{crd}\,36 = 37^p\,04'\,55''$$

The idea of sines appeared in the early 5th century A.D. but it was around 500 A.D. when Āryabhata first gave special names to each function and compiled a table of sines at 1° intervals. A contemporary of his, Varahamihira derived formulae that can be equated to the present expressions

$$\sin \frac{\alpha}{2} = \left[\frac{1 - \cos \alpha}{2} \right]^{1/2}$$

and $\sin^2\alpha + \text{versin}^2\alpha = 4\sin^2\frac{\alpha}{2}$

Prior to this the eastern work *Paulisá Siddhãnta* tabulated sines at an increment of 3° 45' (1/96th of a circle).

Accepting the value for the arc as equal to its angle in minutes (= 225') the other multiples were found as —

$$\sin(n + 1) = 2\sin n\alpha - \sin(n-1)\alpha - \frac{\sin n\alpha}{\sin \alpha}$$

An alternative form of this expression is

$$s_{n+1} = s_n + s_1 - S_n/s_1$$

Where s_n = nth sine

and S_n = sum of first n sines.

For example.

With $\alpha = 225' = \sin\alpha$ and $n = 1, 2 \ldots 5$

To find sin 22° 30'

7° 30'	$\sin 2\alpha = 2\sin\alpha - \sin 0.\alpha - \sin\alpha/\sin\alpha = 450 - 1 =$	449
11 15	$\sin 3\alpha = 2\sin 2\alpha - \sin\alpha - \sin 2\alpha/\sin\alpha \quad\quad =$	671
15 00	$\sin 4\alpha = 2\sin 3\alpha - \sin 2\alpha - \sin 3\alpha/\sin\alpha \quad\quad =$	890
18 45	$\sin 5\alpha = 2\sin 4\alpha - \sin 3\alpha - \sin 4\alpha/\sin\alpha \quad\quad =$	1105
22 30	$\sin 6\alpha = 2\sin 5\alpha - \sin 4\alpha - \sin 5\alpha/\sin\alpha \quad\quad =$	1316

If these values are compared with their modern equivalents the agreement is exceedingly good;

	Old	New
3° 45'	225	225
7 30	449	449
11 15	671	671
15 00	890	890
18 45	1105	1106
22 30	1316	1317

In selecting sin 3° 45' as 225 this is equivalent to a radius of 3438 for the circle.

Whilst Ptolemy concentrated on the angles subtended by whole chords the writers of *Paulisá Siddhãnta* devoted their attentions to half chords and half angles subtended by whole chords and this led to the present-day sine.

The Hindus and Muslims of the 8th century A.D. continued the tradition of dividing the circle to 96 parts and called the 'sine' of each part the 'kardaja'.

In the 9th century A.D. both Al-Khowãrizmi and Al-Hasib computed tables of sine and tangent. Shortly afterwards Al-Battâni became the first to produce cotangent tables and introduced the term "sinus" derived from the Arabic jiba and Sanskrit jiva.

In the late 10th century Abu-l-Wafa produced sine tables correct to 8 decimal places at increments of 15' of arc and he also made a study of tangents.

The Hindu mathematician Bhâskara produced elaborate sine tables in the 12th century based on the same unit of 1/24th of a right angle. He used an approximate chord formula of

$$\text{crd} = \frac{4d.a(p-a)}{\frac{5}{4}p^2 - a(p-a)}$$

Where a = arc, d = diameter and p = perimeter

Al-Marrakushi of the mid 13th century produced a table of sines at 1/2° increments and also tables for versed sine, arc sine and arc cot. Of the same vintage was Nasir al din al Tûsi who was the first to consider plane and spherical trigonometry as subjects in their own right independent of astronomy.

The Chinese at that time were still dividing the circle into 365¼ tu, each of which was sub-divided by 100 and 100 again.

By the early 15th century in Samarkand, Muhammed Taragay was producing trigonometric tables tabulated at 1' intervals from 0° to 45° and by 5' from 45° to 90°. Cotangents were derived for every degree and values were to 9 places.

The change from using chords to sines was established in Western countries by Georg Peurbach in the mid 15th century who proceeded to compute a set of sines tabulated at 10' intervals in 1460. For that era the University of Vienna, where he was Professor, was the centre of mathematical progress. One of his pupils was Johannes Königsberg (Regiomontanus) who obtained a degree by the age of 15 and became a very influential mathematician. He suggested replacing the sexagesimal form of trigonometric tables with a base in powers of 10. In doing this he wanted to avoid the use of fractions so he took sine 90° = 100,000 instead of the previously used 60,000. For tangents he took tan 45° = 100,000 and the tables were at intervals of 1 degree. Instead of the term *tangent* he used *numerus*. His tangent value for 89° was 5,729,796 compared with the present day equivalent of 5,728,996.

In 1468 he extended his accuracy by computing a table of sines with sine 90° = 10^7, at increments of 1' and with proportional parts for seconds. His early death at age 40 delayed publication of the tables until well into the 16th century.

When Gemma Frisius wrote the first work on triangulation in 1533 he mentioned the possibility of using a table of sines for the calculation but considered it to be too complicated for the normal person.

The next notable progress was by Georg Rheticus (1516-1576) who was a student of Copernicus. Decimal fractions had still not appeared so he used initially, for his tables of functions, a radius of 10^7 for sine and cosine and a base of 10^7 for the other four functions. The increment was 10 seconds. Later however he began sets based on radii of 10^{10} and 10^{15}, again with increments of 10 seconds. Not only were there first differences in these tables but second and third differences as well.

He was the first to define trigonometric functions as ratios of the sides of a triangle. Unfortunately, just as Regiomontanus had died before completion of his work, so did Rheticus. He had employed several people to do the calculations over a period of 12 years. A true labour of love! A pupil of his, Valentin Otho, completed the task in 1596 and published it as *Opus Palatinum de Triangulis*.

Franscois Viète pressed hard for the adoption of decimal rather than sexagesimal fractions, particularly in his *Canon-mathematicus seu ad triangula cum appendicibus* of 1579. He explained how to use the six functions for the computation of elements in plane and spherical triangles. In this context he developed such relations as:

$$\sin \alpha = \sin (60 + \alpha) - \sin (60 - \alpha)$$

and

$$\operatorname{cosec} \alpha + \cot \alpha = \cot \alpha/2$$

His decimal fractions he wrote in the form $123,\dfrac{456}{}$ or $123,\dfrac{456}{1000}$ or 123/456.

The tables he prepared gave the 6 functions to increments of 1' with 10^5 for base and radii values.

Rules for solving plane and spherical triangles with the aid of trigonometric functions were published by Christoph Clavius in the late 16th century whilst in the Jesuit order in Rome. He introduced use of the point as a decimal separator in sine tables published in 1593 but it did not become popular until the time of Napier (1550 - 1617). The use of decimal fractions were popularized by Simon Stevin in the late 16th century by his publication *De Thiende* in 1585 but they had been devised much earlier. Giovanni Magini used a comma or virgula as a decimal separator in his *De planis triangulis* in 1592.

By now trigonometric tables had come of age and took on a form similar to those of the 20th century.

References
See in particular:
2, 5, 28, 40, 76, 80, 102, 107, 123, 128, 129, 171, 186, 197, 223, 224, 226, 227 and 232

As instruments improved, a need was found for means of reading circles more accurately. Observation and reading of the circle by an unaided eye, even on instruments of 10 or more feet in radius, could not expect to be better than several minutes of arc.

The first movement towards a method of subdivision is uncertain. Some references give the credit to personalities such as Peurbach and Regiomontanus of the 15th century or Richard Chanzler (Kantzler) of the 16th century and yet others put it in the 14th century.

Transversals

The earliest attempts at subdivision were akin to the modern diagonal scale and the introduction was often credited to Levi Ben Gershon. Originally the idea was almost certainly one applied to straight scales where it is simple yet accurate.

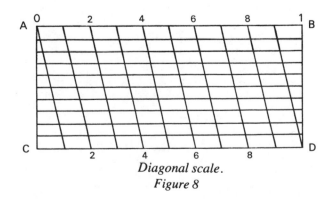

Diagonal scale.
Figure 8

29

The problem is to divide each unit of line AB into a further 10 parts so that section 0–1 is divided into 100. A second line CD is drawn parallel to AB and with similar main divisions. A further 9, equidistant parallel lines are drawn between AB and CD. Divisions are then joined diagonally such that 0 on AB joins 1 on CD; 1 on AB to 2 on CD etc.

When such a technique was applied to a curved scale it was more difficult. A series of concentric arcs replaced the parallel lines and the subdivisions were from a diagonal AG of a length that could be easily divided into ten. The problem with this approach was that the concentric circles should not be equally spaced and the transversal should not be divided linearly.

Figure 9

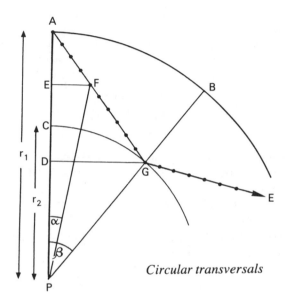

Circular transversals

If $AF/AG = k$

Then $\tan \alpha = \dfrac{r_2 . k . \sin \beta}{r_1(1 - k) + r_2 . k . \cos \beta}$

or

$= \dfrac{k . \sin \beta}{\dfrac{r_1}{r_2}(1 - k) + k . \cos \beta}$

Generally $r_1/r_2 \to 51/50$

Then, for example, if $k = 0.4$

$\alpha'' \to \dfrac{20.0}{50.6}\beta'' \quad < 0.4\beta''$

In general

$\alpha'' \to \dfrac{50k}{51 - k} \beta''$

It transpires from this that each diagonal mark on AG will be to the left of its correct position such that each value recorded will be too high by 2″ or 3″.

Where the diagonal is rising from left to right – as GE, the opposite applies and each value recorded would be too low by 2″ or 3″.

As Galilei and the introduction of telescopes came later than Brahe (who was a prime user of transversals) sights were still taken with the unaided eye and errors of 2″ or 3″ were negligible on a 10′ graduation divided into tenths.

Small modifications were provided by John Hommel with a zigzag arrangement and then Richard Chanzler in 1552 introduced a line of dots in the graduations.

In 1542 Pedro Núñez, a professor of mathematics in Coimbra, Portugal, produced a variation on the transversal on a quadrant. He had a series of concentric arcs such that the outer one was divided into 90 parts, the next into 89 and so on down to an inner arc divided into 46 parts. The number of divisions in any one arc was marked at the end. To determine the angle corresponding to any particular position of the alidade required a table. Thus for the 14th mark on the 77 arc

the angle $= \dfrac{14}{77}.90° = 16° 21' 49''$

This arrangement came to be known as a nonius but the particular difficulty with it was the division of the arcs into the various parts.

Several modifications of this idea were to appear after trials of the nonius by Tycho Brahe were abandoned. In 1611 Clavius reported the work of Johannes Curtius in his *Operum Mathematicorum*. Again there were a series of concentric arcs. This time however the arcs were of different angular lengths. The outer one was of 90° divided to 1°. The next arc covered 91°, then 92° up to an inner arc of 128°. Each arc was subdivided into 90 parts and then terminated at the bounds of the quadrant. Whereas with the nonius graduations on successive arcs became smaller; here they increased in size. The method of reading was the same. Thus if the alidade cut the 37th mark on the 115° arc,

the angle $= \dfrac{37}{90} . 115° = 47° 16' 40''$

Later Clavius reported a further modification of this idea. He used the same range of arcs from 90° to 128° but this time each was subdivided to 128 parts prior to termination at the bounds of the quadrant. In this instance if the alidade cut the 42nd mark of the 105° arc,

the angle $= \dfrac{42}{128} . 105° = 34° 27' 11''$

1. The use transversals on a mural quadrant made by Bleau for Snellius.

A further variation on a theme allowed the direct reading of minutes. This time a series of 59 arcs were each divided into 60 parts. The first had a span of 61° so that 1 division = 1° 01'; the second of 62° had 1 division = 1° 02' etc. down to 119° where 1 division = 1° 59'. Considering the 62° arc as an example its first graduation was at 1° 02' then a series at 1° until a final one of 58 minutes. With whole degrees on the outer scale, if the alidade cut a division on the 83° arc then this was equivalent to 83 − 60 = 23 minutes.

By the mid 17th century William Leybourn was using the following method of forming transversals. For example, let it be required to divide angle BAC into 6 parts. To determine where arcs E, F, G, H and J ahould be so that lines from A through E, F, G, H and J divide arc BC equally.

Draw arc of circle through BDA. Divide that part between B and D into 6 equal parts at E, F, G, H, and J. Now draw the concentric arcs through E, F, G, H and J. The separation is smaller the futher the arc is from A.

A marginal note in the same reference [146] disputes this method and result and states that distances should be greater the farther they are from the centre.
To achieve this it suggests: Use compasses set with point at E and set to nearest distance to arc KD. Set this distance from B to a.   Set compass point at F, span to line KD and set off from B to b.   Continue in the same manner for the other points.

In 1673 Dr. Wallis calculated the required separation of the circles to give equal angles between each subdivision. To do this he derived values for a, b, c, ... h, i.

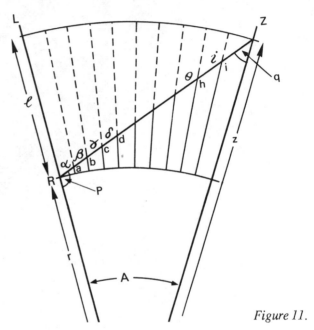

Figure 11.

Construction of Circular transversals. (b)

Let angle ARZ = p and AZR = q with angle A subdivided equally into n parts

$$RA\ \alpha\ =\ A/n$$

$$RA\ \beta\ =\ 2A/n\ \text{etc.}$$

Whence $$\frac{z\ +\ r}{z\ -\ r}\ :\ :\ \frac{\tan(p+q)/2}{\tan(p-q)/2}$$

As p and A/n are known so α can be calculated. Hence:

$$\frac{\sin \alpha}{r} = \frac{\sin p}{a}$$

or

$$a\ =\ r.\sin p/\sin\alpha$$

similarly for b,   c. . . . i

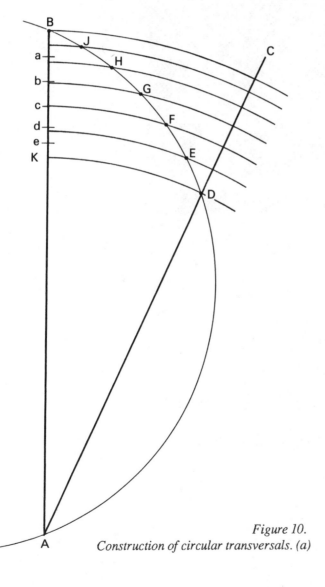

Figure 10.
Construction of circular transversals. (a)

32

For example, if r = 1, ℓ = 0.2   A = 10'

Then   p + q = 179° 50'
                          ½(p+q) =   89° 55'

                              z + r = 2.2
                              z − r = 0.2

Whence $\tan (p - q)/2 = \dfrac{z - r}{z + r} \tan (p+q)/2$

$= \dfrac{0.2}{2.2} \times 687.548\ 841$

tan (p−q)/2 = 62.504 440

(p−q)/2 =   89° 05' 00" 17'''

Then   p = 179° 00' 00" 17'''

or   180 −p =   0° 59' 59" 43'''

Then to evaluate   a = r.sinp/sinα etc.

α 0° 58' 59" 43'''	sin α 0.017 1602	sin p 0.017 4510	a 1.016 94
β 0° 57' 59" 43'''	sin β 0.016 8693		b 1.034 48
. . . .	. . .	. . .	. .
. . . .	. . .	. . .	. .
. . . .	. . .	. . .	. .
θ 0° 51' 59" 43'''	sin θ 0.015 1242		h 1.153 84
ι 0° 50' 59" 43'''	sin ι 0.014 8334		i 1.176 47

If successive differences are taken out they will be found to increase from 1694 to 2353

Wallis found this approach tedious because of the seconds and thirds and felt it would be more convenient to try and work in whole minutes and a modified value of z, for example:

If   r = 1.   A = 10'   and p = 179° 00'

Then 180 −p = 1° 00'

α 0° 59'	sin α 0.017 1616	sin p 0.017 4524	a 1.016 94
β 0° 58'	sin β 0.016 8707		b 1.034 48
. . .	. . .	. . .	. .
. . .	. . .	. . .	. .
ι 0° 51'	sin ι 0.014 8348		i 1.176 45
κ 0° 50'	sin κ 0.014 5439		k 1.199 98

Thus z is now 1.199 98 rather than 1.2.

The smaller the value of ℓ in relation to r so the more the differences become insignificant.

Either of these approaches was obviously very tedious and it was near the end of the 1600s when La Hire devised a method of subdivision that allowed the transversals to be drawn as curves. It proved possible to derive the parameters of the circles, parts of which would be the curved transversal for the successive graduation intervals. Each interval had part of a different circle but their centres were all themselves on a single circle. The centre of this circle was the centre of the instrument. Thus each transversal had the same radius. By first determining and plotting this circle on a base board the drawing of each transversal was considerably simplified. For the intermediate concentric arcs the interval between the two containing arcs was subdivided into the required number of equal parts and the arcs drawn accordingly.

The vernier

After the many variations of transversal the next major step forward came from Pierre Vernier who was a military engineer for the Spanish Hapsbergs.

Instead of having a series of stationary, concentric arcs to achieve subdivision he hit upon the idea of using just one movable scale to achieve the same result. His method was a development of the variation by Clavius whereby he divided arcs of different length into 60 parts each.

In 1631 Vernier described a quadrant to which he had attached a vernier such that a main scale of 30 minute divisions could be read directly to 1 minute with a vernier of 30 divisions which spanned 31 minutes of arc i.e. each vernier division was $1 + 1/30$ minutes. Thus if the two scales are put together a displacement of the vernier 1/30 minute to the right will make the last but one division of it coincide with the main scale division. Similarly a further movement of 1/30 minute makes the last but two division coincident. Thus if the vernier is numbered from right to left there is a direct indication of single minutes.

A similar effect is achieved if 29 minutes is divided into 30 but the numbering would be from left to right. With the still relatively poor quality of graduating circles at that time, the degree of subdivision that was feasible was very much a function of the radius of the arc in question.

Despite the introduction of the vernier in 1631 it was not widely adopted at that time probably because of the continued use of open sights. As this was improved with telescope and cross hairs so the vernier began to be accepted in the late 17th and early 18th centuries.

It was the second half of the 18th century before Ramsden and his dividing engine revolutionized the graduation of circles although the first dividing engine could well have been as early as 1739. Until then the instrument makers resorted to various combinations of successive halving and of proportional scales and the use of compasses.

The problem of mechanically dividing a circle was a difficult one. In 1793 it was said that from the time of Ptolemy to that of Copernicus the best that could be expected was 5 to 10 minutes. Tycho Brahe reduced this to 1 minute; Hevelius to 15 or 20 seconds. Flamsteed achieved 10 seconds and the work of Graham by 1740 made 7 or 8 seconds possible.

The appearance of the telescope to replace open sights considerably improved pointing and bisection without a similar improvement in the reliability of circle graduation.

Probably the first step towards improvement came from Robert Hooke around 1670. His idea was to rotate a screw thread round the edge of the arc and to use the marks that would be left as graduations. Such graduations could not be at predetermined separation (or at least not with any accuracy) and the intervals were to be calculated after marking. This approach was not a success since later checks suggested errors of about 1 minute. In effect the errors accumulated the longer the arc.

In the early years of the 18th century Olaus Roemer was using dividers set at 10 minutes to step off divisions round an arc. This had similar disadvantages to those of Hooke's method but at least Roemer was using a series of silk threads in the optical path to subdivide into single minutes.

The advance of particular importance was that of George Graham. To subdivide an 8 ft. quadrant he first of all set off a 60° angle. This he bisected to get 30°, the dimension of which he added to the 60° to get a right angle. The 30 degree segment could be bisected to 15°. Trial and error was then used to obtain $3 \times 5°$ and then division into 5 single degrees. Five minute intervals then came from 2 bisections followed by a trisection. Later comparisons put the reliability of such divisions at 5 seconds.

This approach was further developed during the 18th century by John Bird who also introduced allowance for the effects of temperature and endeavoured to get all his equipment at the same temperature before commencing work.

The micrometer

Although the telescope was introduced about 1610 it had several drawbacks that required solution before its use could advance on a wide front. Not least among these was the need for cross hairs for accurate bisection and for a micrometer to quantify small movements.

These two aspects owed their full development to Auzout and Picard in the 1660s but William Gascoigne had made a working micrometer by 1639. There was, in fact, considerable controversy as to who first used telescopic sights and a micrometer. This generated much correspondence in the *Philosophical Transactions of the Royal Society* when, upon hearing that Auzout and the French were claiming the invention, Richard Townley sprang to the defence of Gascoigne.

In the Philosophical Transactions No. 25 of 1667 [255] Townley wrote "Seeing in No. 21 of Philosophical Transactions how much M. Auzout esteems his invention of dividing a foot into near 30,000 parts and thence taking angles to great exactness I think it right to inform the world of some scattered papers of Mr. Gascoigne. The very instrument he first made I have now by me and two others. The instrument is small, not exceeding in weight, nor much in bigness, an ordinary pocket watch exactly marking about 40,000 divisions in a foot, by the help of two indices, the one showing hundreds of divisions, the other divisions of the hundred, every last division in my small one containing 1/10 inch and that so precisely that as I use it there goes above 2 1/2 divisions of a second. Yet I have taken land angles several times to 1 division, though it be very hard to come to that exactness in the heavens."

In construction, his eyepiece micrometer had two knife edges of metal parallel to one another and such that they could be moved in opposite directions about the optical axis by a fine-pitch screw. This was in turn connected to a scale and a micrometer drum of 100 subdivisions. Initially the idea of the invention was for the measurement of the diameter of heavenly bodies.

If it had not been for his untimely death, with that of many of his friends, at the Battle of Marston Moor in 1644, it is certain that his work would have become widely known at a much earlier date than Townley in 1667.

Various scientists then entered the arena with variations on a theme. Huygens made mention of an eyepiece micrometer in a publication of 1659; Hooke replaced metal edges with hairs around 1668; Divini and Malvasia introduced grids of lines with thin silver wires. Auzout and Picard had one fixed and one moveable thread such that the movement of the thread across an object could be recorded on a drum. Later they introduced more threads across the field of view.

To begin with the micrometer was almost exclusively used for diameters and movements of heavenly bodies but before long it found a place with the telescope on quadrants and sectors.

A variation on the micrometer, called a heliometer, is worthy of mention here, although it dates from the mid 18th century, because its invention is credited to Pierre Bouguer. It was a device that could be attached to the objective end of a telescope. It consisted of two lenses where one was fixed and one moveable, the latter connected to a graduated drum. As with early verniers, the heliometer, as the name might suggest, was particularly applied to instruments for astronomical observations.

References
See in particular:
2, 61, 102, 123, 124, 146, 186, 247, 255 and 257.

Logarithms

The word *'logarithm'* derives from the Greek *logos* (ratio) and *arithmos* (number). The development of a means of multiplying and dividing numbers must be credited to John Napier although others such as Jobst Bürgi were working on the same problem at the same time.

John Napier was a Scottish laird and Baron of Murchiston who treated mathematics as a hobby as respite from running his estate. He published the results of his work in 1614 as *'Miri fici logarithmorum canonis descriptio'* although he was said to have been active with the system for some 20 years.

What came to be known as Napier's bones, rods or sticks were an aid to multiplication by virtue of the method of marking the multiplication tables on them. It seems quite likely that they originated several centuries earlier in India and then made their way to Europe via Arabia. Digits are summed between the diagonal lines to give those circled. Read in order anti-clockwise they give the result.

For example, $342 = 54 = 18\ 468$.

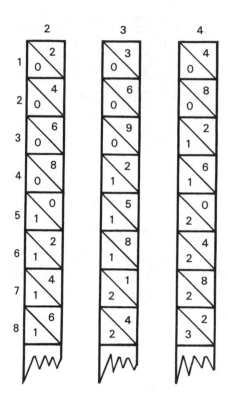

Digits are summed between the diagonal lines to give those circled. Read in order anti-clockwise they give the result.

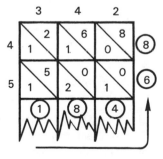

Napier's bones
Figure 12

For 342 × 36 the calculation has to be in two parts: 30 times plus 6 times.

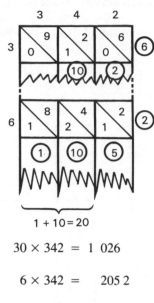

$$1 + 10 = 20$$

$$30 \times 342 = 1\ 026$$

$$6 \times 342 = \quad 205\ 2$$

Then $\quad\quad 36 \times 342 = 1\ 231\ 2$

When developing logarithms Napier was initially only interested in the logs of sines. Why this should have been so is only conjecture although there was knowledge of such a formula as:

$$2\ \text{Sin}\ A\ .\ \text{Sin}\ B = \text{Cos}(A - B) - \text{Cos}(A + B)$$

and its relations.

He defined logarithms in terms of lines thus:

Figure 13

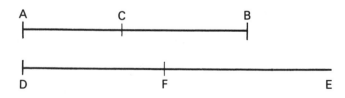

Basis of logarithms

Consider a finite line AB and infinite line DE. Let two points C and F start simultaneously from A and D at the same initial rate. The velocity along AB then to remain proportional to the length CB while F continues uniformly. If the points reach C and F simultaneously then Napier defined DF as the log of BC.

i.e. $\quad$ DF = Nap·log CB

In terms of sines he took AB as sine 90° but equal to 10^7 instead of 1, in keeping with the sine tables of the day: BC was the sine of the required angle and DF became its log. From the definition, then DF increases arithmetically as CB decreases geometrically.

By applying calculus to this relationship it could be shown that

$$\text{Nap.log}\ y = 10^7 \log_{1/e}(y/10^7) \quad \text{where } y = \text{CB}$$

from which, unlike natural logs, the Nap.log decreases as the number increases. Thus

$$\text{sine } 90° = 10^7$$

Log sine 90° = 0

In his publication of 1614 he gave a table of log sines at 1 minute intervals and in a later work described how to calculate the logs. Rearranging the above formula:

If $\quad y = (1 - 1/10^7)^L, \quad$ then $L = \text{Nap.log of } y$

Once Napier had published details of his logarithmic system it attracted interest and in particular it roused Henry Briggs who was the first Savilian professor of geometry at Oxford. So interested was he in Napier's invention that he travelled to Edinburgh to exchange some of his own ideas with Napier. In particular he suggested that it would be more convenient if log 1 = 0 and log 10 = 10^{10} but they subsequently agreed that log 10 should be 1.

Although they agreed on a system it was left to Briggs to compute the tables since Napier was in failing health. Thus were born Briggsian or common logs.

For their computation he started from various roots. Thus

$$\sqrt{10} = x = 10^{\frac{1}{2}} \quad \text{therefore log } x = \frac{1}{2}$$

or log 3.162 28 = 0.500 000

If this is done successively thus

$$(3.162\ 28)^{\frac{1}{2}} = x = 10^{\frac{1}{4}} \quad \log x = \frac{1}{4}$$

Then it proved possible to compile tables for all numbers. This Briggs did for 1 to 20 000 and 90 000 to 100 000 all to 14 decimal places. He died before filling in the gap from 20 000 to 90 000 but this was done by his successor Adriaen Vlacq who published the results in 1628.

The terms 'mantissa' and 'characteristic' are also said to have been introduced by Briggs although at least one reference credits 'mantissa' to John Wallis, some 60 years after the death of Briggs.

The first tables of Briggsian log sines and tangents were by Edmund Gunter in 1620. These were to 7 places and at 1′ intervals. At the same time the terms cosine and cotangent were introduced.

In parallel with the work of Gunter was that of John Speidell who tabulated the logs of sine, tan and sec to 6 places in a form akin to, but not exactly the same as, natural logs. These he published in 1619 in 'New logarithmes'. The difference was that whereas the digits were similar Speidell preferred all values to be positive which required the subtraction of Napier's values from 10^8 and retention of the first 6 digits.

In discussing the development of log tables one must not overlook the contribution of Joost (Jŏbst) Bürgi. If he had published his investigations at the same time as they were formalized he could well have been credited with the invention of logs. as it was he did not publish until 1620 in Prague as *Arithmetische und geometrische Progress-Tabulen* by which time the tables of Napier were widely known.

Whilst Bürgi had worked along the same lines as Napier he started from

$$y = 10^8(1 + 1/10^4)^L$$

In his tabulation Bürgi used so called black numbers and red numbers. Thus for a black number y its red equivalent was 10^L. The tables had the red numbers down the side and the black ones forming the body of the table.

In his book of 1653 William Leybourn reproduced tables of log sines (actually headed *The Table of Sines*) at 10 minute intervals up to 90° and to 6 decimal places. These were set out in the 9 system of the present day.

For example   log sin 14° 00′   9,383 675

Logs of numbers were at unit intervals and also to 6 decimal places in the present form. For example

log 231 = 2,363 612

The solution of an oblique triangle was described thus:
To find the side QR

As sine of angle QRS 45 deg 10 min	9,850 745
Is to the logarithm of the side QS 303 feet	2,481 443
So is the sine of angle QSR 24 deg 20 min	9,614 944
the sum of the second and third terms	12,096 387
the first term subtracted	9,850 745
To the logarithm of the side QR	2,245 642

The nearest absolute number answering to this logarithm is 176 and so many feet is the side QR.

It was towards the end of the 17th century before the introduction of series expansions to simplify the labours of computing tables. The expansion of

$$\ln(1 + x) \text{ as } x - \frac{x^2}{2} + \frac{x^3}{3} - \cdots$$

while often called a Mercator series after Nicolaus Mercator (not to be confused with the Mercator of the map projection) was really developed for the production of tables by John Wallis.

References

See in particular:

28, 40, 80, 146, and 266

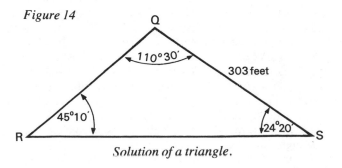

Figure 14

Solution of a triangle.

Chapter Nine

ASTRONOMERS AND MATHEMATICIANS
Peurbach to Pascal

In the first half of the 15th century George Peurbach was Professor at Vienna University and Court astronomer to the King of Bohemia. Although primarily interested in astronomy and trigonometry he also wrote about arithmetic. He saw advantage in using sines instead of chords and computed a table of sines for every 10 minutes. He also constructed a canones gnomonis or quadratum geometricum which took the form of an open square with two graduated sides, a pointer, and a sight arranged to rotate on the corner opposite the graduated sides. It could be used for measuring altitudes or for determining the distances to inaccessible points from angular observations at different stations.

Among his students was Johannes Müller von Königsberg — who was usually known as Regiomontanus. An infant prodigy, he obtained a degree by the age of 15 and at 16 was a pupil of Peurbach. They collaborated closely, particularly on trigonometric tables. Up to that time the tables had a sexagesimal form with sine 90 ° = 60,000. Königsberg replaced this with sine 90° = 100,000. For a table of tangents up to 90° at 1° increment he took tan 45° = 100,000.

After spending six years in Italy he settled in Nurnberg and built an observatory there and installed a printing press. In 1498 he produced sine tables with sine 90° = 10^7 and an increment of 1 minute with proportional parts available for seconds, but they were not published until 1541. He used an astrolabe, geometric square and quadrant, developed the torquetum, and re-introduced the Jacob staff for astronomical observations. The staff was 2 to 3 metres long with 1300 divisions. As Peurbach had done, so Königsberg adopted the Hindi "sine" in place of the Greek "chord of double arc".

Just before the death of Königsberg, Copernicus was born in Poland. Sometimes described as the founder of modern science, he studied law, medicine and astronomy, wrote on both plane and spherical trigonometry and assisted the Catholic Pope on calendar reform. He discarded consideration of functions in terms of arcs of circles and concentrated instead on the right angled triangle. His best known published work covering cosmology and trigonometry did not appear until the time of his death. Called *De Revolutionibus Orbium Coelestium Libri VI* or *Revolutions* for short, it could at that time have more than one interpretation. Possibly this was on purpose since it was not very safe to make revolutionary proposals in the religious climate at that time.

His instruments followed the designs of Ptolemy but were more crude and less reliable than their early predecessors. He used a triquetrum (later used by Tycho Brahe) that was 12 feet long, Jacob staff, a quadrant, and armillary sphere. There were no lenses or mirrors to aid observation.

Mention must be made here of the French physician Jean Fernel (ca 1485-1558 AD). Although astronomer and doctor to King Henry II of France his most notable achievement was in geodesy where he measured a meridian arc. This is detailed elsewhere.

In 1492 a volume of Francesco Pellos was published where a decimal point was used to indicate division by ten.

Pedro Nũnez, a Professor at the University of Lisbon, became prominent in the early 16th century. He is best known for his works on navigation and astronomy but in the present context the important development was his arrangement, the nonius, for reading small angles. Professor Nuñez described the mariner's astrolabe and use of the Jacob staff in navigation. He proved that a ship sailing to make equal angles with the meridians must trace a loxodrome rather than a straight line.

Gemma Frisius (1508-1555) was educated by friends

before going to the University of Louvain. At age 20 he was writing scientific papers and one of particular interest in the context of this book was *Cosmographicus Liber Petri Apiani*. In the 2nd edition of 1533 (Antwerp) he added an appendix *Libellus de locorum describendorum ratione*. It was a booklet of only 16 pages but described, for the first time, the idea of triangulation. This appendix appeared in all the 28 later editions up to 1609.

It described how to measure and compute the distance between two points without measuring directly between them. From one base line and a chain of triangles stemmed the first accurate method for computing the size and shape of the earth.

He constructed various instruments including globes, astrolabes and quadrants as well as the Jacob staff about which he wrote at length.

1534 saw the founding of the Jesuit Order (Societas Jesu) by Inigo López de Loyola, a Basque nobleman. As their activities spread across the world so they are found making contributions in many of the scientific spheres of mathematics and astronomy.

Johann Hommel (Homelius) and Richard Chanzler, in the mid 16th century were instrumental in the development of transversals—which later developed into the vernier. In the interim modifications were introduced by Christopher Clavius in 1593. He is credited with introduction of the decimal point which appeared in his table of sines of 1593 but some credit G.A. Magini with being the first with this in his 'De planis triangulis' of 1592, and others suggest Pelazzi of Nice in 1492. Pope Gregory XIII sought the advice of Clavius on reform of the calendar and was instrumental in putting October 15 after October 4 in 1582 to rectify the Julian calendar.

A leisure-time mathematician who contributed extensively to the subject was François Viète (1540-1603) who was employed throughout his life in the service of Henry III and IV of France. He introduced the practice of using vowels for unknown quantities and consonants for known ones. His "Canon mathematicus" of 1579 had tables of all six functions to 1′ interval and to avoid using decimals he adopted an hypotenuse of 10^5 for sine and cosine and a base of 10^5 for the other four functions. He derived various trigonometric identities including multiple angle formulae and pleaded for the use of decimal rather than sexagesimal fractions. He also systematically explained methods of computing plane and spherical triangles with the aid of the six trig. functions.

One of the best known of early English surveyors was Leonard Digges from Kent. He applied science to surveying and in 1556 published "Tectonicon" on mensuration. "Pantometria" was published posthumously by his son Thomas in 1571. He described an instrument topographicall as well as a theodolitus which was somewhat similar to an astrolabe. Before long the term theodolitus was applied to the instrument topographicall and subsequently developed into the theodolite. He is also said by some to have invented the telescope, a detailed account of which was written by Thomas Digges.

In 1596 Valentin Otto published the posthumous work of Georg Rheticus. The 'Opus Palatinum de Triangulis' had trigonometrical tables to 15 decimal places at 10″ intervals. He brought trigonometry of age and used all six functions. Instead of thinking in terms of arcs of a circle he concentrated on the right angled triangle. In his tables he used hypotenuses of 10^7, 10^{10} and later 10^{15} for sine and cosine and a base of 10^7 for the other functions.

In the late 16th century the Danish astronomer Tycho Brahe was making his mark. Using the tables of sine and cosine he adopted the prosthaphaeresis method of multiplication. That is the use of formulae that convert products of functions into sums or differences and hence simplify the evaluation.

From 1576 to 1597 he developed the observatory of Uraniborg on Hven, an island between Denmark and Sweden, where he had a mural quadrant of about 6 ft. radius and reading to about 5 seconds of arc. For this he developed the use of transversals. He had other quadrants up to 15 feet radius but particularly developed one of about 4 feet radius capable of measuring horizontal angles. He had a range of other equipment including sextants, torqueta and armillae. On these he developed the use of pinnules to aid sighting and adopted repetitive observations to reduce margins of error.

As his observations became more refined so they were more noticeably affected by astronomical refraction, for which he compiled a set of empirical tables.

Up to this period sexagesimal fractions were still in vogue but Simon Stevin (1548-1620), although not the inventor, was instrumental in introducing decimal fractions into mathematics in 1585 in his "De Thiende".

In 1589 Edward Wright is reported to have measured a short line (base) near Plymouth Sound and measured some angles to obtain the height of a nearby hill as 375 feet above sea level. From this and other data he is said to have found a diameter of the earth that was too small by about 15%. What information he used and how he arrived at his value is not known.

The notable British name of the late 16th century was John Napier (1550-1617), a Scottish laird and Baron of Merchiston. For some 20 years he worked on the development of logarithms and on computing rods, or bones. He published his results in 1614 and 1617, after redoubling his efforts when he heard that Brahe was using prosthaphaeresis for multiplication. His first publication explained the philosophy of logarithms and gave a table of log sines to 90° at 1′ intervals. A second publication described the method of calculating the tables.

After publishing his results he was visited by Briggs who suggested various modifications.

Giovanni Magini also made use of the decimal points after Clavius, but it was Napier who brought it into regular use.

Logarithms were invented independently of Napier by Joóst Bürgi (1552-1632). He published some six years after Napier and approached the whole problem from an algebraic angle rather than the geometric one of Napier. In addition he modified the form of transversals for the measurement of small divisions.

Henry Briggs (1561-1630) was the first Savilian Professor of Geometry at Oxford in 1619. He made various suggestions to Napier for the improvement of logarithmic tables.

From his book of 1624 the terms 'mantissa' and 'characteristic' were derived. The use of such tables rapidly became popular.

In the early 17th century Galileo Galilei (1564-1642) heard that a Dutch lens grinder Hans Lippershey of Middleberg in Zeeland had combined two lenses in a way that magnified distant objects. There were however, rival claimants in Jacob Metius of Alkmaar and Zacharias Jansen of Middleberg as having been the first to have the idea of combining lenses in this way. This quickly led in 1609 to Galilei constructing a telescope on the same design and he directed it to the skies for the first time.

Galilei was active in other areas of science as well. It was he who, as a result of his observations of water in a shaft, the forerunner of the barometer, started Toricelli on the path to a mercury barometer. In the field of pendulums he found a method of imparting impulses to a pendulum to keep it in motion. His name further appears in relation to the invention of the first thermometer to look at all like the modern equivalent.

At the time Briggs was compiling his first table of common logs so John Speidell compiled a table of logs of trigonometric functions for sine, tangent and secant, and published the result in 1619 as "New Logarithmes".

Close on the heels of log tables came the slide rule, details of which were first published in 1632 by William Oughtred. He described the circular version in "Circles of Proportion" in 1632 and the straight version in an "Addition" of 1633 to his first publication. A student of his, Richard Delamain, did however beat Oughtred to the publishers with details of the circular slide rule in 1630 and this caused some controversy.

While different authorities differ somewhat as to who should be credited with the introduction of triangulation the strongest contender in practical terms would appear to be Willibrod Snellius (1580-1626). It should be noted however that Gemma Frisius wrote about such a system in 1533 and that in 1580 Tycho Brahe had a scheme in Denmark. Snellius was a Professor of mathematics in Leyden and it was there that he executed the first large scale triangulation in Holland in 1617. He termed himself the Dutch Eratosthenes and this was reflected in the title to his publication describing the work "Eratosthenes Batavus de Terrae ambitus veru quantitate etc", Leyden, 1617. In addition he observed the first resection, introduced the term loxodrome for the rhumb line in navigation and developed the theory of refraction of light which carries his name as Snell's Law.

The name of Edmund Gunter will be very familiar to all surveyors. Professor of astronomy at Gresham College he published the first table of log sines and tangents to a common base in 1620 for every minute of arc to 7 places. At the same time he introduced the terms cosine and cotangent into the language. Among his other credits were the log slide rule which had no sliding parts as the modern forms, and the Gunter chain.

Another familiar name is that of Pierre Vernier (1584-1638) who, from his youth, was keenly interested in measuring instruments. He helped his father survey the area of Franche Comté and this led him to seek an improved method for reading angles. He first described his invention of the vernier reading system in a publication of 1631 which was principally concerned with its use on a quadrant.

Willem Blaeu the Dutch cartographer deserves mention here because of somewhat sketchy reports that he measured a meridian arc in Holland in the early 17th century. In particular it was reported by Picard as founded on a baseline along the shore line between Maas and Texel and the derived value of a degree to vary by only 60 rods from that obtained by Picard himself. An alternative reference refers to 60 Rhenish feet as the difference.

An early arc measure in England is credited to Richard Norwood (1590-1675) a reader in mathematics in London. In a publication of 1637 he described a measurement from London to York. His method of using a carriage wheel was really a backward trend from the triangulation method adopted twenty years previously by Snellius.

Around 1628 René Descartes invented cartesian geometry and in a publication of 1637 was the first to mention analytical geometry. He extended the existing idea of latitude and longitude to define any point by two coordinates.

The next measurement was in Italy around 1645 by Giovanni Riccioli who used four triangles to connect his terminals. He had a collaborator in Francesco Grimaldi who, like Riccioli, was a Jesuit from Bologna. In 1642 Riccioli determined a length for the seconds pendulum and developed a level that consisted of two vessels connected by a tube, in essence the forerunner of the modern water level.

The gap left in the Briggs logarithmic tables was filled by the Dutch bookseller Adriaen Vlacq in 1628. In 1626 a book published in the Hague by Albert Girard mentioned the use of spherical excess in finding the area of a spherical triangle. Within a few years it was also recorded in books published in Italy.

By the mid 17th century Gilles Robertval, a very disagreeable member of the Mersenne group, had developed the systematic use of telescope sights. In 1644 he wrote on the astronomy of Aristarchos. Such sights did not have a universal attraction to observers and in particular Johann Hevelius argued with Hooke that it was possible to observe as accurately with open sights as with telescopic ones. In an observatory he built in Danzig among the instruments was a telescope 150 ft. long.

A notable step forward, again of disputed invention, was the micrometer. The records however, would seem to favour the Yorkshireman William Gascoigne in 1640. The first claim was by Auzout, a French Academician but as soon as his claim was published, Richard Townley, in the Philosophical Transactions of the Royal Society, was able to produce evidence that Gascoigne had made such an instrument many years earlier. Nevertheless even in the 20th century the claim of Auzout is still described, to the exclusion of any mention of Gascoigne, in a reputable geodetic work. Gascoigne introduced stadia hairs into his device and the micrometer was divided into 100 parts.

Another Jesuit, Gabriel Mouton suggested, in 1660, that there was need for an unalterable universal measure. His choice was the length of a meridian arc of 1 minute divided decimally into 1000 milliar each of 1000 virga. He showed how the length of a seconds pendulum could be made to preserve such a unit. Mouton however was some 130 years before his time as it was not until 1790 that the idea was resurrected and the thought of a metre developed.

A particularly notable personality in the 17th century was L'Abbé Jean Picard, a French Academician and Professor of astronomy. He was the first to use a telescope for surveying when he had it as a part of a levelling instrument in 1660. He executed a considerable amount of surveying in France in the latter part of the century. Among his equipment was a quadrant of 10 ft. radius with telescopic sights and cross hairs. He also made measurements with stadia hairs.

In collaboration with Auzout he made improvements to the micrometer for incorporation in the quadrant and it was with such equipment that he measured a meridian arc in France 1668-1670. It was as a result of Picard's meridian arc value that Newton was able to complete his work on the shape of the earth.

Mention has already been made of Adrien Auzout and how his claim to inventing the micrometer was refuted by Townley. Nevertheless he was a renowned scientist and was the one who drew up plans for a French Academy of Sciences. He collaborated with Picard on developing the micrometer and use of a telescope on survey instruments. By 1668 however he had a quarrel with the Academy and departed for Italy for some 14 years. Before his departure he developed a moving wire version of the micrometer and this was later extended to six fixed wires and three movable ones.

To complete this section there is the child prodigy Blaise Pascal who was building and selling calculating machines by the age of 18. He had a superb talent for mathematics which led to his development of such a machine but his enthusiasm for its improvement waned and by the age of 25 he had given up the sciences for religion.

The calculator was only capable of addition but it was the first step in a new era of computing that was to be capitalized on 50 years later by Leibniz.

References

See in particular: 2, 6, 8, 28, 29, 30, 40, 41, 55, 60, 71, 76, 80, 92, 98, 102, 110, 112, 123, 124, 127, 128, 129, 143, 180, 186, 191, 224, 227, 232, 255, 260, 266 and 267.

Chapter Ten

EARLY INSTRUMENTATION

Crude instrumentation for angular measure goes back at least 2000 years starting with items such as the square which still survives today. The items of particular relevance are the cross staff, astrolabe, dioptra, torquetum, sector and quadrant. The last two belong to the 17th and 18th centuries while the others are of earlier development.

As well as angle measuring devices it is pertinent here to have a brief look at mechanical computing aids although neither expedition used them. Then there were the ancillary items such as thermometer, barometer and pendulum. For the distance measurement the prime item was the standard toise against which measuring rods were constructed.

A book could be devoted to such items but here there is only space for a few paragraphs on each to illustrate the background against which the expeditions operated. Not for them, the pressing of a few buttons and getting all they required to cm accuracy. They had to fight hard for every degree and every toise. Because of this, they should not be in any way despised for misclosures that by today's standards could be termed gross; or the non-application of corrections which at that time were either not known at all, or at best, only partially understood.

Angle measuring devices.

The groma.

Earliest known measuring devices in the realms of surveying date from the times of the Pharoahs, more than 1500 B.C., when knotted cords were used in field layouts. Such is apparent from carvings on tombs, yet a method still adopted on the Lapland expedition for some of the less accurate measurements. Later from similar sources one comes across the groma which was probably the first widely used instrument for setting out angles. Or perhaps one should say 'angle', since all it could achieve was a right angle.

The best evidence of the groma was the carving on the tomb of Lucius Aebutius Faustus found near Turin. It would appear that he was a mensor or agrimensor (land surveyor) around the first century A.D.

The exact form of the groma is open to doubt but in essence it was a cross formed of two rods set at right angles. From the ends of the rods hung strings with plumb bobs. Efforts at reconstruction vary as to whether or not the ends of the two rods were joined to provide stiffening and also in the form of the support. One reconstruction suggests that the whole was supported on a quadripod; another that it was supported eccentrically on a single iron shod pole that could be stuck into the ground.

The eccentric suspension point was to allow free sight from one string to its partner for setting a straight line. Similarly the right angle would be obstructed if the cross were set centrally.

Heron was not favourably inclined to the groma or star, which was probably one and the same instrument. As any inventor, he was particularly interested in pushing the advantages of his own device, the dioptra. However his assertion that the wind could cause difficulties with the plumb strings was no doubt well founded.

In 1912 a well-preserved example of the groma was found at Pompeii where the cross pieces were of iron and the plumb bobs of bronze. The eccentricity of the cross in relation to the support pole was about 10 inches. It is not known how this eccentricity was allowed for at the ground point. In addition it would have been necessary to ensure that the cross was horizontal if errors were not to be introduced by virtue of an inclined cross.

The Merkhet.

Early astronomical measurements for survey purposes included such tasks as the orientation of the pyramids. It is thought that this was done by use of a merkhet which was little more than a notched stick and plumb line.

The dioptra.

Probably the earliest form of instrument that could be likened to the theodolite was called the dioptra and was described in detail by Heron of Alexandria. Various sources place his dates anywhere between 150 B.C. and 250 A.D. It is even said by some that the actual inventor could have been Hipparchus and that Heron was simply describing the instrument.

The first books that might be referred to as survey textbooks were by Heron; in particular, *Treatise on the Dioptra*. For its time the dioptra was by far the best instrument available among the Romans and Egyptians. Heron reported how it was used in all kinds of construction work, from harbours to buildings and ramparts, and astronomical observations of heavenly bodies.

Mounted on a column was a circular plate and sleeve that could be rotated by means of a toothed wheel and worm. This could also be used as a clamp. Mounted on the plate were two upright standards between which a further toothed wheel was held in the vertical plane and could be moved or held in the same way as with the horizontal motion.

On top of this arrangement it would appear that there were two possible forms of unit. One of these was essentially a levelling instrument, the other an angle measuring device.

For the level, there was a long rod, some six feet, fixed centrally above the standard, and into which fitted a metal tube in a horizontal position. At the ends of this tube were two short uprights supporting glass tubes. With water in these it was effectively the modern water level and was used with suitably graduated staves.

The alternative unit was a circular disc, of some six feet diameter, to which were fixed sighting devices at right angles to each other. It was also said to have graduations in degrees. Such an instrument had obvious uses for setting out points as well as for various inaccessible distances, heights of objects, tunnelling and setting water courses.

Triquetrum.

This was a form of angle measuring instrument used by Ptolemy as well as by such illustrious successors as Copernicus and Brahe.

It consisted essentially of three graduated rods joined by two hinges so that they could form a triangle. In the form that Ptolemy described it, one rod was planted in the ground in a vertical attitude while the other two could rotate in the vertical plane to form varying sizes of triangle.

The rod hinged to the top of the upright one carried two sighting holes or pinnules and the observer moved this until he was looking at the required heavenly body. The vertical section between the hinges was of the same length as the rod carrying the sights so that they formed an isosceles triangle. The position of the sighting rod was then found in terms of the graduation it cut on the third rod to give a measure of

the zenith angle. This third rod was hinged towards the bottom of the vertical rod.

As this type of instrument was in use for so many years there were numerous variations on the theme. Among these were versions designed for use in the horizontal plane.

Astrolabe.

One of the best known of early angle measuring devices was the astrolabe. In a myriad of designs they date back at least to the time of Ptolemy.

In its simplest form the astrolabe was a circular disc around the edge of which were marks dividing it into a series of equal parts. Originally this would have been in terms of signs of the zodiac at 12 to the circle, and subdivisions of it.

Their principal use was in navigation for determining the elevations of heavenly bodies. Graduations were normally in terms of a particular latitude but by means of a series of adaptor plates an instrument could be used over a range of latitudes.

While the astrolabe probably originated in the Mediterranean countries it spread through Babylon and to the Far West. Many of the early astronomers wrote about their use and had specific models made for themselves. Almost invariably however their use was related to the heavens and navigation rather than the measurement of horizontal angles.

Cross staff to quadrant.

The angle measuring devices used on the Peru and Lapland expeditions were quadrants. As hand held, or when large and tripod supported, these quarter circles might well be considered to have developed from the cross staff and backstaff.

The cross staff was first described in the early 14th century by Levi Ben Gershon, a Jewish mathematician from Catalonia. Developed as the baculum or baculus Jacob, it was dedicated in 1342 by Gershon to Pope Clement VI but did not come into wide use for many years.

By virtue of its early name of baculus Jacob, or Jacob staff, some attribute it to Jacob ben Mahir ibn Tibbon who was born in Marseilles about 1236. On the other hand the principle on which it was based was certainly known in the time of Hipparchus.

It had many other names including virga visoria, radius astronomicus, baculus geometricus, balestilha, arbalete, cross de Saint Jacques and Jacobsstab.

Simple in construction and use it basically consisted of two graduated rods in the form of a cross where the longer was up to a yard long and the shorter usually around a foot long. Some had 3 or 4 interchangeable cross pieces of various lengths.

Initially devised for the angular measurement of the elevations of heavenly bodies, it was later used by Frisius and Brahe for recording horizontal angles. Earliest forms bore no graduations but by the time of Tycho Brahe both parts were graduated. For observing the elevation of a heavenly body the line AD would be sighted to the horizon at the same time that CD was slid along AB until AC was aligned on the body. The vertical angle could then be determined from the position of CD along AB. For a given length of CD a scale of degrees could be marked along AB.

43

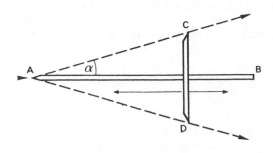

The Cross Staff

Figure 15

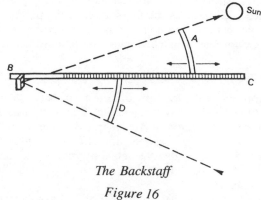

The Backstaff

Figure 16

For example, if CD was 12 inches long then if α is the half angle at A the lengths along AB for graduations would be:

$\alpha°$	**Along AB**
10	34.03 inches
15	22.39
20	16.48
25	12.87
30	10.39
35	8.57
40	7.15

Thus with a 36 inch bar and 12 inch cross piece it would have been quite feasible for the whole angle, 2α, to be divided down to single degrees and parts thereof.

With interchangeable cross pieces each of the four faces of the bar AB could be suitably inscribed for a particular combination.

When the staff was used for horizontal angles it was first levelled by use of plumb bobs. The cross piece was then moved along AB until both targets were apparently just touching the ends of CD. The better value for any given angle would be obtained by use of the largest possible crosspiece. On the other hand reliable values could only be obtained for angles less than about 60°, otherwise the crosspiece was too close to the eye.

Such an instrument was subject to other errors as well. For example, geometrically the point of intersection of CA and DA is at A whereas it is actually sighted from the position of the eye a little way from A. See note [102] where this aspect is investigated. It indicates that if the eye is 6mm from A and the cross piece is 1.20m from A then the error incurred would be 17.19 sin 2α minutes of arc. This obviously has the least effect on the smaller angle.

In some models this was overcome by modifying the eye end A or in a more cumbersome manner could be eliminated by using two cross pieces, of different lengths, simultaneously.

Tycho Brahe observed the horizontal angles in his network to single minutes·although later work suggests that many of the values were in error by up to 20 minutes of arc.

One obvious drawback with the cross staff when used for sighting the sun was that it did the observer's eyes no good at all. A development to overcome this came in 1607 from John Davis. Initially called the backstaff it later developed into the Davis quadrant. Lloyd Brown [33] discusses the possible arrangement of the early backstaff. The main staff BC had two moveable transoms A and D and a unit at B which allowed both sight of the horizon and the shadow of A as projected by the sun. A could be moved to get the shadow of the tip directed onto B. The main advantage of such an arrangement was that the observer had his back to the sun.

In this form it was not particularly sensitive and was soon developed by Davis into the Davis quadrant. The sights on A and D were now moveable such that a rough setting was first achieved on A and then the fine adjustment made on CD. The two readings together gave the required angle. The two arcs A and CD totalled 90° as in a straightforward quarter circle.

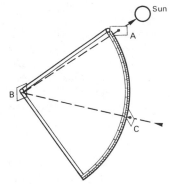

Bouguer's quadrant
Figure 17

Bouguer (the same one who went to Peru) queried why there should be two arcs on the Davis quadrant. Why not combine them? In particular, graduation is simplified and, like Brahe, he subdivided into minutes by using transversals. Sight C was moveable until it was possible to see the horizon through B at the same time as a spot projected from A.

In the early years of the 18th century George Graham was constructing mural quadrants such as the one with an 8 ft. radius at Greenwich Observatory. The graduations were on two concentric arcs with the inside one divided to degrees and twelfths while the outer one was divided to 96 and each of these to 16.

For expedition use, however, and in particular the measurement of horizontal angles as opposed to star observations, there was requirement for quadrants that were both portable and that could be mounted on a stand or tripod to allow positioning over a point.

Craftsman of sufficient ability were not numerous and it fell to Graham and his French counterpart Claude Langlois to supply the expeditions.

Quadrant.

In contrast to the astrolabe, the quadrant was graduated only for a quarter of a circle. In fact, the instrument was often referred to as a 'quart de cercle'.

It can be traced back to the time of Ptolemy when a graduated arc was marked on the upright side of a block of stone. A horizontal stylus cast a shadow to allow a determination of the elevation of the sun at different times of the year. Such a device would obviously not be portable but on the other hand its solid permanency allowed long term, continuous observations of the sun's elevation.

Among the earliest discoveries from such an instrument would have been the greatest and least elevations or the values at the summer and winter solstices. This difference is met also in the work of Eratosthenes where it was valued at 11/83 of a circle (47° 42′ 40″). Half of this value became known as the obliquity of the ecliptic.

Following from this they would not have taken long to discover that even when set up at different latitudes the difference between the extreme positions remained the same. The mid-point indicated the latitude.

When this arrangement became portable there were several related devices such as the already mentioned cross staff and back staff.

When the quadrant evolved in the form seen on the expeditions it did not produce strictly horizontal angles. Unlike a theodolite circle the quadrant had to be held in a tilted plane containing the two sighted stations and the observer's eye. Hence, there was a requirement for all angles to be reduced to the horizontal before use.

The use of quadrants had particular prominence from the time of Tycho Brahe to the mid 1700s.

The Torquetum

The Torquetum is particularly associated with Regiomontanus of the late 15th century. More for astronomical observations than survey angles, part of its base was tilted in a manner similar to that of the style of a sundial. The whole could be set along a meridian line.

The Polimetrum.

This was far more akin to the modern theodolite than the torquetum and was a development of the dioptra of Heron. It was current in the early 1500s and usually linked with the name of the cartographer Waldseemuller. It was capable of both horizontal and vertical angles. In addition to graduations round the edge of the circles to single degrees, it carried graduated squares. The vertical angles were recorded against a plumb line and the horizontal ones against a pointer.

The Theodolite.

The term 'theodolite' fist appeared as 'theodolitus' in the work of Leonard Digges. In his 'Pantometria' of 1571 he described both an 'instrument topographicall' and a 'theodolitus' where the latter was very similar to an astrolabe. Before long 'theodolitus' came to be applied to the 'instrument topographicall'.

In principle the instrument had a graduated horizontal circle from the centre of which was a vertical column to which a graduated semi-circle was fixed in a vertical plane. Readings were related to a movable bar that carried a pair of sights — the telescope was still many years away. It was not unlike the polimetrum.

In 1653, some 80 years before the expedition to Peru and Lapland, William Leybourn described an instrument called a theodolite as one of only three necessary instruments to do all survey. The other two were the circumferentor (early form of portable magnetic compass) and Plain table, where he considered the last to be the most useful.

He described the theodolite as an instrument of four main parts. The first was a circle divided to 360 degrees and subdivided as much as the size of the instrument would allow. It was usually made of brass and was from 12 to 14 inches in diameter. The subdivision was by diagonal lines between concentric circles.

The second part was an optional geometrical square drawn within a circle with the sides equally divided and used for heights and distances.

The third part was a magnetic needle at the centre of the circle, where also there was an index.

The fourth part was a socket so that the whole could be screwed to a pole.

It was not until the time of Jonathan Sisson, in 1725, that an instrument recognizably the same as the modern theodolite was developed. It had a 4½ inch circle with three verniers to read to 6′. The vertical circle could give readings up to 70° in relation to a level set on the theodolite. The whole was mounted over four foot screws.

Zenith sector. (Fig. 18)

This instrument was commonly used to determine the difference in latitude between two places. When set up in the meridian plane it was possible to determine zenith distances of stars as they crossed the meridian. If the same stars were observed at each end of an arc the difference of zenith distances was equal to the difference in latitude of the two observation positions.

The idea of observing zenith distances by use of a long, almost vertical telescope and scale dates from around 1669. It was at that time that one finds mention of Hooke making observations with a telescope 36 feet long.

A long bar of iron AB (often 8 to 10 feet long), suspended at one end had a limb CD at the other end at right angles to the main arm and bearing graduations. The graduated section was sometimes straight or sometimes curved with centre P. A telescope RS was rigidly attached to bar AB. By means of the screw T, the bar and telescope could be inclined through a few degrees to effect bisection of the chosen star. As CD moved sideways so a plumb bob suspended from P cut across the graduations and an accurate reading made of the inclination.

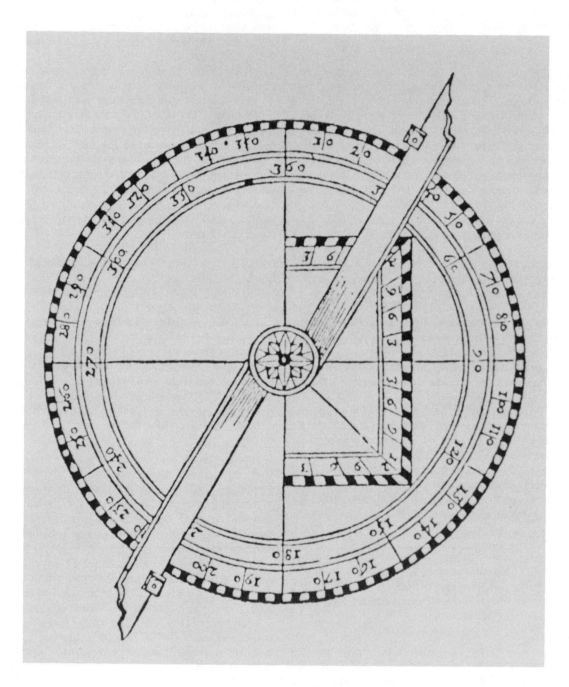

2.   *The Digges theodolitus.*

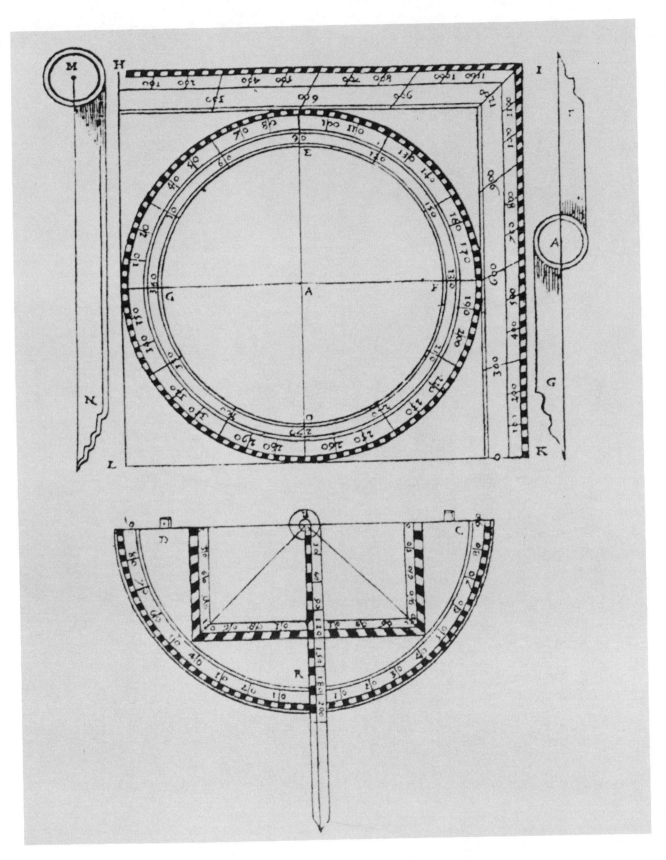

3. Digges instrument topographicall.

47

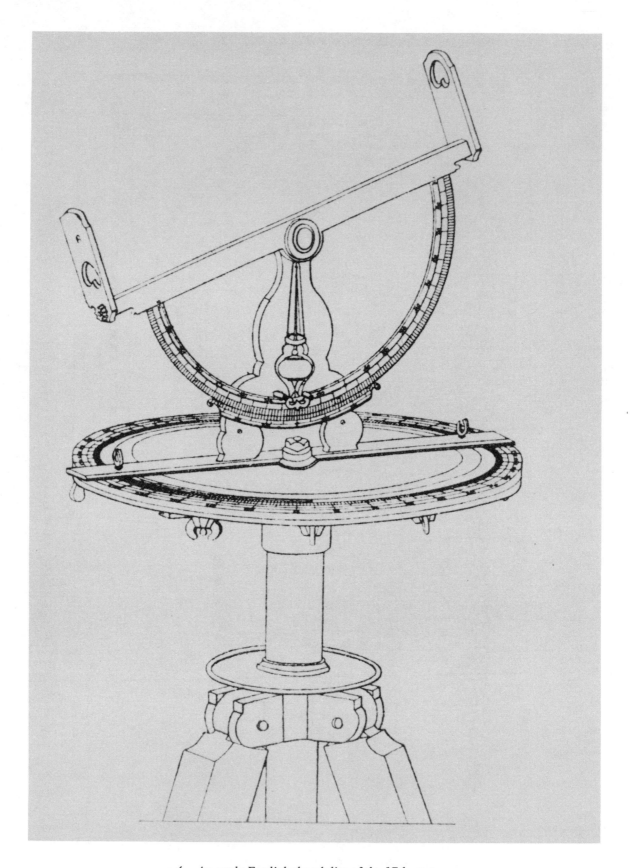

4. An early English theodolite of the 17th century.

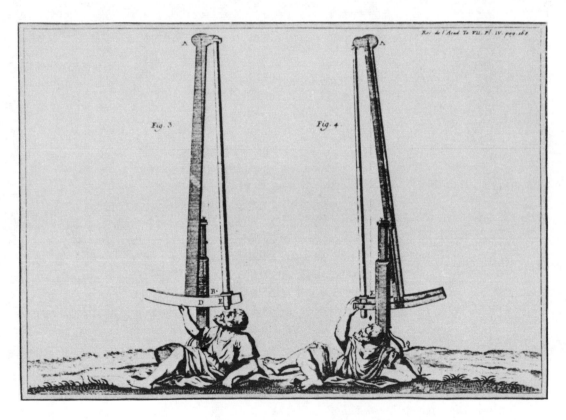

5. *A zenith sector used by Picard.   Not the most comfortable of instrument reading positions.*

6. Zenith sector as used in an observatory of the 17th and 18th centuries. Note the form of
illumination and early pendulum clock.

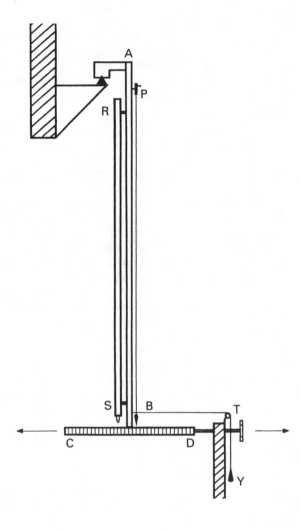

The zenith sector.

Figure 18

The support for the whole system could be either a pivot or a hinge arrangement but in either case it was made possible to effectively change face i.e. the whole would rotate through 180°. Near the point B of the bar would be attached a thread and weight BY to ensure tight contact between the limb and the screw.

The mode of operation was such that the meridian line had to be determined first and this was done by use of transits or equal altitudes together with a clock regulated to solar or sidereal time. A cord was then stretched across the room in such a way that the sector was restrained to moving only in the meridian.

When all was ready for observation the screw was turned until the telescope sighted in the vicinity of the star to be used. The position of the plumb bob over the scale was noted, together with the fractional reading from the drum at T. When the star appeared it was followed such that it moved along the cross wire and the reading again noted. This would give an apparent zenith distance except that there could be errors of various sorts inherent in the system. As it was possible to rotate the sector through 180° and a second reading taken, so the change of face would mean out some of the errors and give an improved result.

In doing so it was necessary for the telescope to retain its same position relative to the limb in both faces. In some this was achieved by attaching the graduated limb CD to the telescope rather than to the bar AB. A further refinement could be made by moving a micrometer driven wire in the telescope to bisect the star.

For good quality instruments one should look to the examples by Graham of the early 18th century; one that he constructed for Bradley measured 12½ feet radius but did allow the scientist to discover aberration and nutation. The arc length was of 12½° and was read by micrometer.

In the late 18th century the zenith sector was replaced by the repeating circle but not without controversy over the accuracy of the results obtained with it.

For the Peruvian expedition the group took a 12 ft. radius sector where the arc CD was of 30°. This arc in particular was found to be too large for convenience so they devised an ingenious method of redividing. Rather than continuously divide the arc they worked backwards from knowledge of the star they intended to use.

At the Tarqui observation point the approximate zenith distance was known to be 1° 41'. Now considering the idea of changing face, or reversion, the two positions would be about 3° 22' apart on the graduated limb. Thus if two points could be accurately marked at about this displacement any differences could be recorded by the micrometer screw. It just happened that 3° 22' 15" had a chord of 1/18th of the radius. There would then be the problem of dividing a radius into 18 parts which was not easy. On the other hand it was very straightforward to decide on a chord length and multiply by 18 to get the radius. Then sweeping an arc of 18 x chord as radius the terminals of the given chord could be marked. A chord of 1/17th radius was also used. See page 230.

At the Cotchesqui observation point the same technique was employed but with a chord of 1/20th of the radius.

In their original sector the telescope was connected to the arm by supports several inches long which were somewhat unstable. They set to work and shortened the supports but still had cause to fear flexure of some form in the telescope. As a last resort the telescope and bar were tightly wrapped together with wire!

With a sector divided in this way there was then a need to know the various parameters of proper motion of the star if the same instrument were to be used at each observatory since the observations would not be simultaneous. The alternative was to construct a second sector and so make simultaneous observations feasible. At that time the corrections for nutation and aberration were not well known as they had not long been appreciated as existing.

For the Lapland expedition a Graham sector of 9 feet radius was used. The telescope formed the radius and had the limb attached to it. Unlike that used in Peru, this one was not reversible. It was inclined by use of a micrometer screw held in firm contact by a suspended weight. By using the same star at both stations, the difference of latitude equalled the difference of zenith distances. This was, however, dependent upon care to avoid disturbance of the whole instrument during transportation from one end of the arc to the other. Later tests confirmed that in its many travels the sector had indeed maintained its line of collimation correct.

Mechanical computing devices

The abacus.

There are various theories as to the origin of the term "abacus" and the most probable of these are "tablet" and "dust". The former because of the tablet shape and the latter from the knowledge that in its earliest form the abacus was a tablet covered with sand or dust in which the numbers or symbols were drawn. Somewhat akin to the slate of the 19th and early 20th century schoolchildren.

Various forms of abacus have existed over the years and all are well documented. No date can be given to its invention other than that it has well over 2000 years of history.

Besides a dust tablet the abacus also came in the form of grooves in a tablet with loose beads as well as the presently recognized form. Its use was by no means restricted to those countries with which it is now associated. The origins and early use are certainly in Greece, Egypt, Babylon and the Roman Empire. Its association and widespread use in its present form in China and Japan probably only dates from about the 12th century. Obviously however trade with the Middle East considerably predates this and it could have been introduced much earlier in a simplified form. As well as the Middle and Far East the abacus has had wide use in Russia and Europe.

All four basic arithmetic operations were possible.

Not until the early 17th century and the computing systems of Napier was there a competitor to the abacus.

Although there were several forms of the abacus they all endeavoured to portray lines for each digit position; successive lines were often in powers of 10. Some had a cross line representing half the value of the line, others had no cross line. Some also had lines representing fractions.

The slide rule.

At the same time as Napier was developing logarithms and only a few years before Pascal was working on a calculating machine, Edmund Gunter produced a forerunner to the slide rule which he described in his 'Canon Triangulorum' of 1620. On this, distances were marked in such a way as to be proportional to the logarithms of the numbers they represented. Originally the sliding component was missing and instead a pair of dividers were needed to add and subtract parts of the scale representing the required numbers. Within a few years however William Oughtred (1622) invented both the straight and circular slide rules but did not publish details until 1632. For the circular slide rule there was a rival inventor in Richard Delamain when he and Oughtred disputed each other's claims.

The first type with a sliding section as known today dated from around 1654. Unlike various other inventions however, its use did not become widespread until well into the 18th century when the cursor also appeared. Within Europe its acceptance was even more sluggish and little development occurred there until the 19th century.

Calculating machines.

While there is no knowledge of the use of calculating machines on the expeditions they were in existence, albeit in somewhat crude form. The first recognized machine, by Blaise Pascal, was almost a century before the two expeditions, but developments were very slow.

Pascal's machine of 1642 was a rectangular box, on top of which were six numbered wheels and a six digit display. Designed only for addition and subtraction, one of the initial problems was to devise a reliable tens transfer.

In 1666 Samuel Morland produced, independently of Pascal's work, a much smaller machine, but again aimed at addition and subtraction only. As the prime intention was the manipulation of sums of money so not all transfers were in tens, for instance, there were farthings to pennies that required a four transfer.

Morland developed other machines but it was really left to Gottfried Leibniz to complete a machine in 1694 that was able to execute multiplication. Nevertheless the development had a very long way to go before it could be considered either reliable, easy to use or readily portable to the field.

Thus it is quite certain that it was a case of logarithmic tables or long hand for all calculations relating to the expeditions.

Ancillary equipment.

Any survey party uses various items of ancillary equipment but in the case of the expeditions they were unique in that several of the items they made use of were still in their infancy. This particularly applied to the thermometer and barometer, both of which were of prime importance.

The thermometer.

Throughout the descriptions of the expeditions there are references to the extremes of temperature suffered and how it changed diurnally. What sort of reliability might be placed on such values? Certainly not parts of degrees, probably not single degrees and most likely the best that could be expected was a few degrees.

In the late 16th and early 17th centuries Galilei made experiments on the variation of degrees of heat such as the difference between air and the warmth of hands. A bulb, with long tube inserted in a container of liquid was given changes in warmth and as a result the water level varied in the tube. Although it is thought that Galilei had some form of graduation on the tube it would not have borne any resemblance to those of today.

The term *thermometre* first appeared [266] in 1624 which was very early in the development of such arrangements. With friends such as Sagredo and Sanctorius, Galilei made observations at different places and at different times of the year and endeavoured to determine changes in body heat.

Various improvements came slowly during the 17th century with the first appearance of the use of alcohol in 1641 by the Grand Duke Ferdinand II of Tuscany. Instruments were beginning to take on the appearance of today although there were some spiralled as a spring in order to achieve a greater length of tube and hence greater accuracy.

Over the years a large number of alternatives were tried as fixed points on the scale. For example, at any given locality it could be the temperatures at mid-winter day and at mid-summer day. As is well known however, this would not give very stable values.

The first suggestion of the freezing or boiling point of water as a given mark was by Huygens in 1665 but it was almost the end of the century before the modern fixed points were decided upon.

There still remained however the problems of what liquid to use and what sort of scale divisions to adopt. Scientists belonging to the various National Academies experimented with many substances before alcohol and mercury became the main contenders.

The well known names of Fahrenheit, Celsius and Réaumur were contemporaries working respectively in Amsterdam, Sweden and Paris.

Fahrenheit adopted mercury as an indicator in the early 18th century, around 1721, and using melting ice and blood temperature as fixed points he gave them the somewhat unusual values of 7½° and 22½° respectively, as shown in column a in the following diagram.

	(a)	(b)	(c)
Blood	22½	90	96
ice	7½	30	32
zero	0	0	0

Each of the degrees he then divided into 4 to give 60 intervals between the two fixed positions (column b). Not satisfied with 90 as the upper limit he changed this to 96 which in its turn, by direct proportion, moved the value for ice to 32 (column c). When these had been finally settled it was subsequently found that the boiling point of water on this scale was 212° although the thermometers were only graduated to 96°.

The French scientist Réaumur worked on a completely different premise and concluded in 1730 with 80 divisions between the freezing point and the boiling point. It was this form of thermometer that was used in Peru. The indicator was alcohol rather than mercury and the scale was devised as a direct function of the amount by which alcohol expanded in the change from freezing to boiling of water. Since this was found to be from 1000 parts to 1080 parts his scale initially took these values so that among the reports of the expedition are many references such as 1012 degrees Réaumur. Full details of the construction of this and other thermometers are given in [267] page 293 et seq. Before long the first two digits were dropped and the scale became 0 to 80.

Anders Celsius took part in the Lapland expedition but it was after his return that he developed the scale of his name. Initially, in 1742, he used a division of 100 but such that it decreased with increased warmth. The thinking behind this was to avoid negative values as it became colder below the freezing point which he experienced in the extreme on the expedition as well as no doubt being a common occurrence in Uppsala where he lived.

By the mid 1740s the scale numbering had been reversed to its present form, not by Celsius who had died, but by Linné. Later this new version was incorrectly credited to Celsius in a textbook and the name was perpetuated.

Considerable advances came a little later through De Luc who made notable investigations in to the various sources of uncertainty in the construction fo both thermometers and barometers. Unfortunately this did not come until the mid 18th century. Of particular importance in relation to the Peru expedition was his finding that mercury was much more stable in its expansion than alcohol which is affected by its concentration. Thus the temperatures quoted by Bouguer and his colleagues could have been noticeably in error.

The barometer

As with the thermometer, one of the earliest names to be associated with the barometer (or at least the related phenomena) was that of Galilei. In 1638 he noticed how water would rise no more than 32 feet above its base level. This soon led to Toricelli to make experiments with the much denser mercury and achieving results around 30 inches as he had already forecast should be the result.

Early in its development it was thought that the barometer could be used to determine heights and it was in this way that Bouguer and La Condamine experimented in Peru.

The first to experiment along these lines was Pascal, who in 1648, took a mercury barometer up the Puy-de-Dôme, noting the reading in relation to the known height at various points on the way. He effectively adopted a single base approach to this barometric levelling since a second instrument was left near the foot of the mountain and indicated little change during the same period. The field instrument proved his theory and led to further experiments at different localities.

As with many other instruments, the barometer went through a number of development stages that gave it a variety of shapes but as early as 1700 Leibniz was considering the idea of an aneroid, (or non-liquid) arrangement.

Of particular interest were the activities of De Luc who was keen to develop its application to height measurement. The idea was to supplant the need for a base line, angle measuring instruments and the vagaries of refraction as were the ingredients of previous height measuring techniques. The difficulty was to find an acceptable relationship between the parameters. If he had but known, he might have been better guided to have pursued knowledge of these other parameters instead.

Among the suggested relationships was a fall of one line for the 61 feet next to sea level, another line for the next 62 feet and so on. Alternatively there was the experiment in 1709 that deduced a fall of ten lines over a height of 714 feet in Switzerland. Using the former approach would have given a minimum value of 655 feet but would have been dependent on the base height above sea level.

A further alternative was that the fall rate was harmonic, whereas Bouguer later suggested a logarithmic relationship. In effect his idea could be expressed thus:

If h_b and h_t = height of mercury in lines at base and top of a mountain respectively

and H = difference of height between the same two points

then H = $(\log h_b - \log h_t)^{29}/_{30}$ in toise

Note that the log values are only to four decimal places. An example of this is given in [137] for the difference between Carabourou and Pitchincha. The readings were 254¾ lines and 191 lines:

Log 254.75	=	2.4061
Log 191	=	2.2810
10^4 x difference		1251
1/30th		42
Difference in height		1209 toise

The modern equivalent formula would be:

$$9442 \log_{10}\left(\frac{P_b}{P_t}\right)\frac{T}{T_s}$$

Where T/T_s is the correction for temperature difference between the two points. Its value varies as:

Mean temp. °C	T/T_s
0	1.000
10	1.037
20	1.073
30	1.110

Thus the same example would become 1181.2 T/T_s and in the circumstances of the Andes the mean temperature would probably be, say 10 °C depending on the time of year of the measurement, whence 1181.2 x 1.037 = 1225 toise.

When compared with sea level at 28 inch. 1 line = 337 lines and the top of Pitchincha = 191 lines, then 9442 x 0.2466 x 1.037 = 2414 toise as against the value of 2432 toise given in the main references and 2385 toise if the original formula is used. These are very reasonable agreements in the circumstances though even Bouguer found that it was not so reliable away from the mountains.

De Luc's experiments were very wide and detailed but mostly after the period of interest here. [267] page 183 et seq gives many details.

The pendulum.

During the 17th and 18th centuries many experiments were made with a simple pendulum. In particular, the aim was to determine the length of such a pendulum that beat seconds. The first milestone came in 1672 when Jean Richer went to Cayenne and found that the seconds pendulum there needed to be shorter than in Paris. Why should this be? Was it poor observations? Was it perhaps that the earth was not the sphere it had been thought to be?

The basic ground work of noticing the independence of the period from the length of the arc through which the pendulum swung had already been laid by Galilei. Much of the early investigation was by Christiaan Huygens while he was in Paris 1666-1681. Besides direct investigations into the simple pendulum his other interests helped to emphasize his conviction that the earth was flattened.

Picard in 1669 had studied the seconds pendulum and determined its length in the direct manner of adjusting the length until such time as it beat exactly a second and then measured the length against a scale. Not the most accurate of methods but the most obvious.

It was during this period, 1670, that L'Abbé Mouton, who had a deep interest in the acceptance of a universal unit of length, initially suggested the length of 1 minute of meridian arc and then additionally that this should be established in terms of the length of a seconds pendulum. Both these however, could only be constants if the earth were a sphere and it was very soon after that Richer cast the doubts.

Huygens sent Richer to Cayenne in French Guiana, at nearly 5°.N latitude. He left La Rochelle on 8 February 1672, arrived Cayenne 22 April 1672 and stayed for just over a year until ill health forced him to return to France.

He found that whereas the seconds pendulum in Paris was 3 feet 8.6 lines, at Cayenne it was 1¼ lines less. The clock lost 2 minutes 28 seconds a day, and he observed the seconds pendulum over a period of 10 months.

Supporting results were soon forthcoming from Varin and Des Hayes with their work on the coast of Africa and in America. They had been fully briefed at the Paris Observatory before their departure by Cassini. They took with them a 2½ foot quadrant, one 19 ft. and one 18 ft. telescope, 2 pendulum clocks, a zenith sector and all the necessary ancillary items.

The procedure at each longitude station was to first set up the pendulums and clocks at the approximately correct time. The meridian was established and checked by several methods such as by use of equal altitudes of the sun. Once the telescope was set in the meridian it was possible to observe the transits of various stars over a period of time. From these observations a pendulum was adjusted to equate to a sidereal day of $23^h 56^m 4^s$.

Latitude was determined by the sun or pole star and the eclipses of the satellites of Jupiter utilized to allow calculation of the difference in time between their base in Paris and each observation point. These and other results can be summarized:

54

Date	Observer	Location	Length
1672	Richer	Cayenne	1 ¼ line shorter than Paris
1677	Halley	St. Helena	1 ½ line shorter than London
1682	Verin & Des Hayes	Goreé	2 line shorter than Paris
1682	Verin & Des Hayes	Guadaloupe	2.1 line shorter than Paris
1697	Couplet Jnr.	Lisbon	2 ½ line shorter than Paris
1698	Couplet Jnr.	Paraiba	3.7 line shorter than Paris

(Newton queried these last two saying they should be 1.3 and 2.6 lines if computed properly.)

1699-1700	Des Hayes	Cayenne	2.1 line shorter than Paris
		Granada	2.1 line shorter than Paris
		St. Christopher	1.8 line shorter than Paris
		St. Domingo	1.6 line shorter than Paris

The 1 ¼ lines by which the pendulum at Cayenne was shorter than that in Paris was at variance with the calculated difference suggested by Huygens before Richer departed, of 5/6 lines.

Nevertheless there was now sufficient information to reinforce Newton's theory of an oblate earth with flattening of 1/229. (Today's accepted value is 1/297.4).

Problems with pendulums were numerous, the construction, the suspension system, the effects of various atmospheric parameters and how to correct for them, as well as problems of accurate comparison with acceptable clocks. Remembering that 1 line = 1/12 inch then it is additionally difficult to make reliable measurements of the length of the pendulum to decimals of a line when the exact point of suspension was in some doubt.

Initially suspended from a thread, this gave way in the 18th century to use of a knife edge although even this had its inherent doubts since the edge can never be a line, and is, in fact some form of curve.

Temperature effects were allowed for by the introduction of the grid iron bimetallic pendulum of Harrison in the 1720s. During the Peru expedition however La Condamine put his knowledge of this to good use when determining the expansion of his metal standard bar.

Other small corrections were also appreciated by this time and Bouguer was able to make allowances in his results for the buoyancy of the air and the variation of gravity with altitude. By this time the method for determining the length was rather different from that used by Picard in that the use of beats had been thought of by De Mairan. Observation was made of the instants when the bob, at the extremity of its swing, was in time with a nearby clock. Subsequent coincidences were noted and from this knowledge of the exact time interval the period was simply the number of seconds divided by the number of oscillations.

Initial efforts at deriving a mathematical relation between the length and the latitude were from the pen of Newton who gave:

$$\ell_\phi = \ell \ (1 + m \cdot \sin \phi)$$

Where ℓ_0 = length at Equator
ℓ_ϕ = length at latitude $\phi°$
m = $\frac{1}{229}$; which was his figure for the flattening based on the scant information as to the earth's dimensions available to him.

This might be rearranged as

$$\ell_{\phi_1} = \ell_{\phi_2} \frac{(1 + m.\sin^2 \phi_1)}{(1 + m. \ \sin^2 \phi_2)}$$

Where ϕ_1 and ϕ_2 are two positions at which measures have been taken. Then if Paris, as ϕ_2, is taken as a standard

$$\ell_{\phi_1} = 3.05216838(1 + m \cdot \sin^2 \phi_1)$$

The Plain table (Now spelled plane table).

Of a rectangular shape some 14 inches by 11 inches it was then composed of 3 boards that could be taken apart for ease of carriage. They were held together by a frame which also held the paper in place. The frame had scales upon it to aid the drawing of parallel lines on the sheet and also to help the matching of sheets.

The ruler had to be as long as the diagonal of the table and 2 inches wide. It had two sights, one about 6 inches long, the other 12 inches. The smaller bore a pin and plummet, the larger, a long slit (c.f. an alidade). The long sight bore graduations numbered up and down from the level line to indicate the difference between an hypotenuse and its horizontal equivalent. (c.f. the Indian clinometer).

It was used extensively on the expeditions for mapping as the parties moved about what was at that time, virtually unknown territory.

Standard toise.

At the time of the expeditions the standard measure in France was an iron toise kept safely at the Châtelet since 1668 but it was a relatively crude measure set in the stairs—or as some refer to it, sealed in the wall at the foot of the stairs at the Grand Châtelet. As it was roughly produced no attempt had been made to ensure parallel faces nor was it protected from rust and other forms of deterioration.

As early as 1610 there was reference to this toise but by the 1670s scientists were considering the need for a new, readily reproduced, standard.

Various suggestions were made including the length of a geodetic minute of arc, and a third of the length of a seconds pendulum. In 1720 Cassini proposed a geodetic foot as 1/6000 of a terrestrial minute of arc but such deliberations were not finalized until the end of the century and birth of the metre. Too late by far for Peru and Lapland.

For Peru it required a specially constructed copy of the toise that came to be known as the 'Toise du Perou'. It was 864.0000 Paris lines at 16.25 °R and constructed by Langlois in the form of a flat iron bar, 'half' of which was an end measure and the other longer 'half' had the length defined by points on its surface. It was 17 lines wide and 4 ½ lines thick.

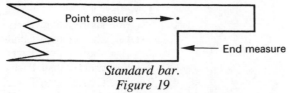

Standard bar.
Figure 19

At the same time that this was constructed there were two other versions of note — the Toise du Nord and the Toise du Mairan. What became known as the Toise du Nord should have stayed in Paris but was taken north by Maupertuis. That of Mairan was sent to Boscovitch. The three were said to be of equal length in 1735. However, when the Peru expedition returned to France, on the suggestion of Louis Godin, the three were recompared and on 26th June 1756 the relation was found to be:

$$t_P = t_N + 1/25 \text{ Paris line} = t_M + 8/75 \text{ Paris line}$$

It was noted that as the length marked on the Peru toise by dots or points was about 1/20 to 1/30 Paris lines longer than its end measure it was considered to be unchanged to within 1/25 line.

On 16 May 1766 King Louis XV made the Peru toise a national standard by Royal Decree. Where:

> 1 toise du
> Perou   = 6 Paris feet (Pied du Roi of 12.79 inches)
>   = 72 Paris inches
>   = 864 Paris lines

Some 80 copies were made by Canivet for the various French authorities.

The *Toise du Perou* was standardized again in 1887 and 1890 by Benoit at the BIPM. Although the marks were defective he gave his results as:

> between
> the points   1.949 001 m
> between
> the ends   1.949 090 m

or a discrepancy of 1:20,000 that would have affected the results of the Peru expedition.

A novel method of determining the expansion of the toise bar was used by La Condamine in Peru and is described in that section.

References
See in particular: 6, 40, 42, 55, 61, 79, 80, 87, 98, 102, 107, 123, 124, 135, 146, 152, 172, 174, 186, 223, 232, 233, 238, 252, 258, 266 and 267.

Chapter Eleven

THE FIRST TRIANGULATION

While it is generally considered that the first triangulation worthy of the name was carried out in Holland by Willebrord Snellius, certainly two earlier efforts should be noted.

Much of this section is based upon *'Gemma Frisius, Tycho Brahe and Snellius and their triangulations'* N.D. Haasbroek, Netherlands Geodetic Commission, 1968.

Gemma Frisius, in an appendix to a publication of 1533 put forward the principles of triangulation in teaching how to measure and to compute the distance between two places. He described the need for a complete circle—as opposed to a quadrant, for measuring the angles; that each new point requires fixing from at least two others; that if the triangle so formed is ill-conditioned a third line should be used; and that in order to get a map to a known scale it was necessary to know one of the distances. He concluded with mention of the problem of distortion necessary in order to depict the curved earth on a flat sheet of paper. The values he gave for the angles (or directions since that is the way he described them) are only approximate and two of the towers—Middelburg and Bergen op Zoom are not even visible from Brussels.

Unfortunately it is not known if these ideas were ever put into practice by Frisius or whether his sketch of Northern Belgium bore only assumed values.

It is known however that his work came to the notice of Tycho Brahe, the eminent Danish astronomer. Brahe had a scheme centered on Uraniborg where he had an observatory built in 1580, on the Danish island of Hven and was to form the base of a map of Denmark. (since the Peace of Roskilde in 1658 Denmark lost the territory it had in S.Sweden and this included Hven).

He observed a net of 11 points using a mixture of astronomical azimuths and observed angles together with a base line of 830 passus geometricus (1,287.9 m). The relation between the units was not determined until 1943 in work by N.E. Nørlund.

With his angle measuring instruments he favoured subdivision by use of transversals and he appreciated various instrumental errors and endeavoured to detect and reduce them by the care of construction and observation. His were the first empirical refraction tables, although, as might have been expected at such an early date, they were not very accurate.

Despite more than sufficient observations to compute the whole as a triangulation net there is no evidence to indicate that Brahe did compute the whole figure. In fact there is little other than the observations on which to base any conclusions.

Haasbroek [102] thoroughly investigated all available material and performed an adjustment by condition equations. He made various assumptions and corrections prior to the adjustment and concluded with a standard error of ± 5.9 minutes for observations probably made with a cross staff, a very good figure for such an instrument.

No indication is given of how the base line was measured. With regard to the angular observations all were made between towers or spires and hence needed reduction to centres. This Brahe omitted to do as its magnitude would generally have been less than 1′ and at most 2′, this would have been negligible compared with other errors. Similarly no account was taken of spherical excess but this would only have ac-

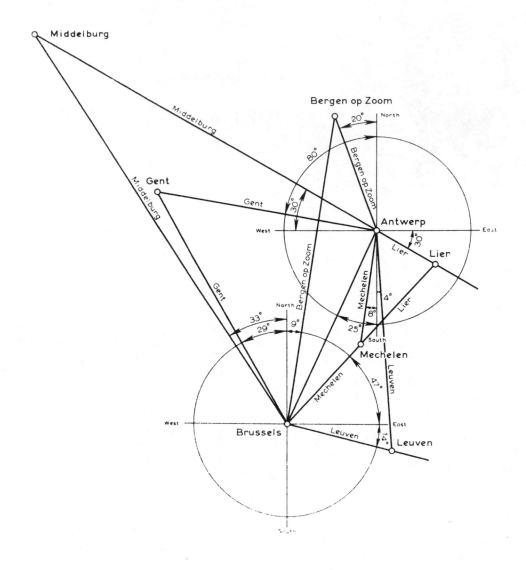

Figure 20

Diagram of the scheme of Gemma Frisius.

58

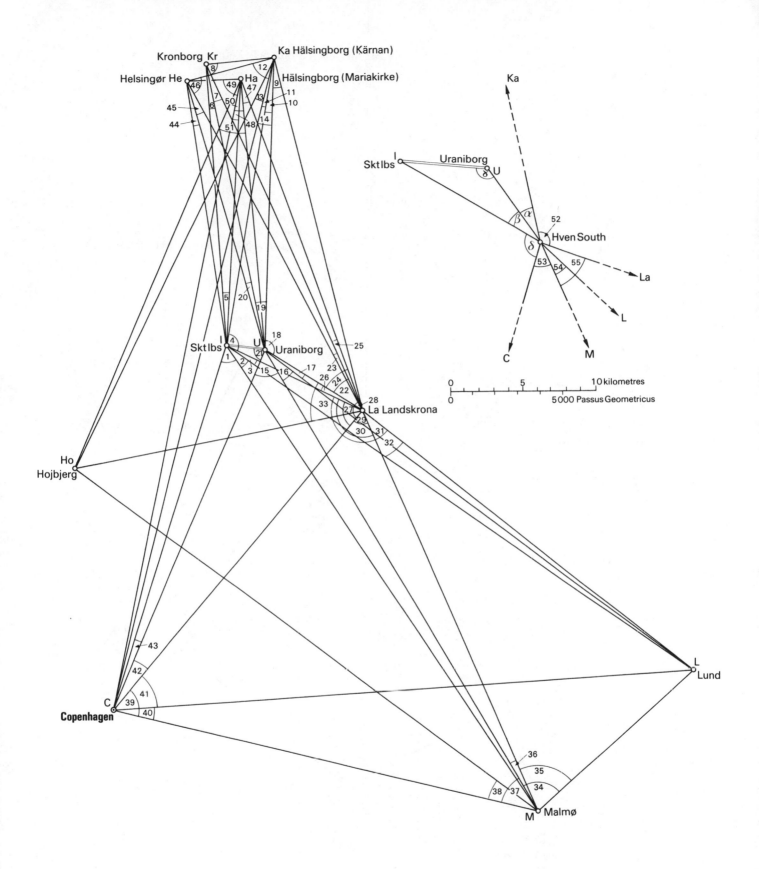

Figure 21

Diagram of the scheme of Tycho Brahe.

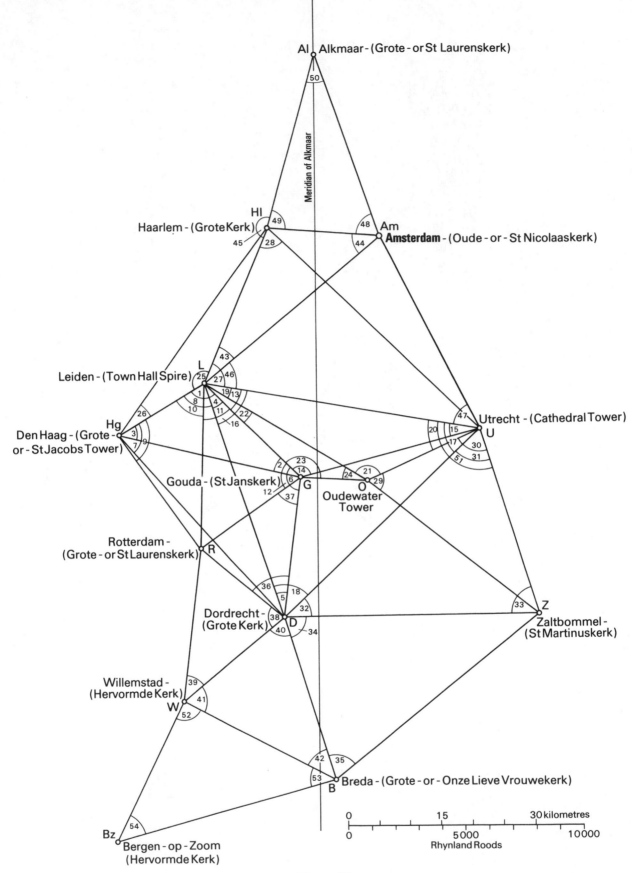

Figure 22

Diagram of the scheme of Snellius.

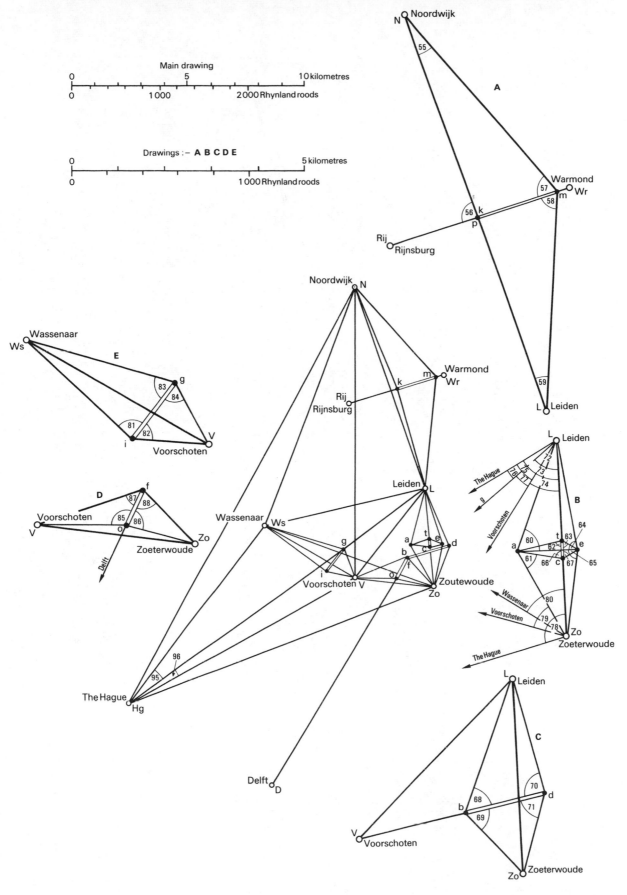

Figure 23
Diagram of the baselines used by Snellius.

61

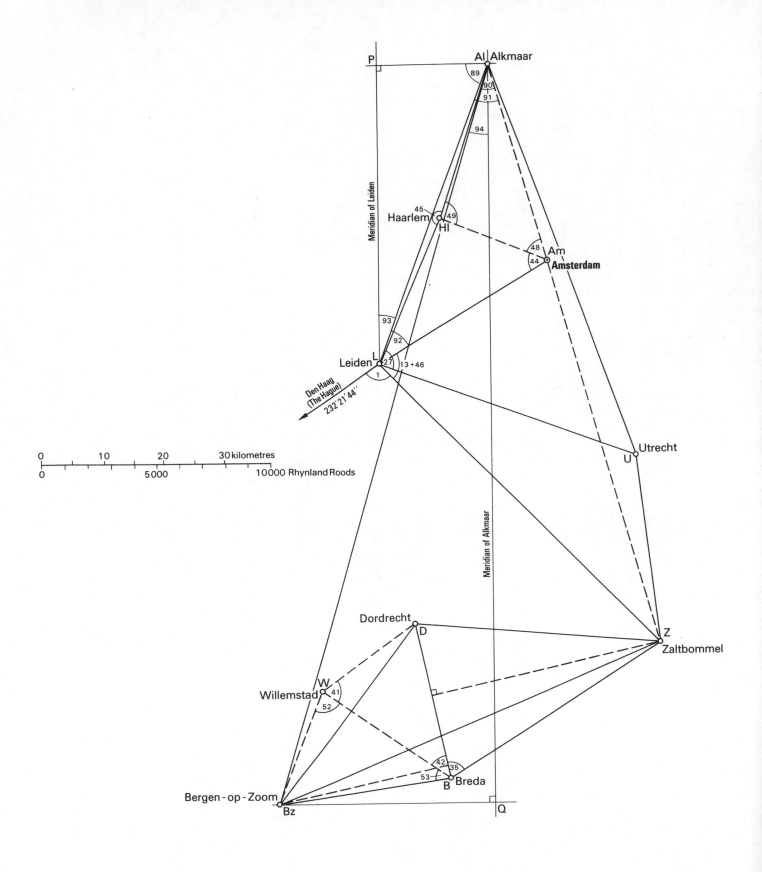

Figure 24
Diagram of the computational scheme of Snellius.

62

counted for seconds. The angular observations were:

Station I. = Skt. Ibs.	Station U = Uraniborg	41. 48°49′	25. 11°46′
1. 45°34′	(See note 1)	42. 20°08′	26. 1°27′
2. 23°27′	15. 47°03.5°	43. 2°44′	27. 78°16′
3. 11°31′	16. 24°05′		28. 106°24′
4. 109°13′	17. 10°52′	Station He = Helsingór	29. 57°24′
5. 18°20′	18. 115°00.5′	44. 4°18′	30. 135°32′
	19. 17°46.5′	45. 16°50′	31. 165°41′
Station Kr = Kronborg	20. 2°08′	46. 68°31′	32. 30°23′
6. 4°26′	21. 86°25′		33. 36°51′
7. 17°21′		Station Ha = Halsingborg	
8. 67°27′	Station M = Malmo	Maria Kirke	Station Hven South
	34. 77°01′	47. 23°31′	(See note 2)
Station Ka = Karnan	35. 67°17′	48. 6°08′	52. 108°51′ (114°58.7′)
9. 23°27′	36. 9°44′	49. 67°14′	53. 50°20′ (47°01.0′)
10. 4°40′	37. 43°36′	50. 10°50′	54. 25°10′ (23°53.8′)
11. 6°03′	38. 21°21′	51. 17°51′	55. 43°21′ (34°51.8′)
12. 66°41′			α. 26°56′
13. 17°50′	Station C = Copenhagen	Station La = Landskrona	β. 17°46′
14. 10°41′	39. 69°45′	22. 28°08′	γ. 133°37′
	40. 20°41′	23. 1°23′	δ. 111°31′
		24. 41°18′	

Note 1. All angles at this station were derived from differences of observed azimuths.
Note 2. No observations were taken to Hven South but angles from it were included and were later reduced as if the station were eccentric to Uraniborg. The adjusted values are in brackets.

From the detailed adjustment of [102] the coordinates for the 11 points of Brahe's system, based on present coordinates for Copenhagen and Landskrona, were found as:

		E(m)	N(m)
Copenhagen, Frue Kirke	=C	138 130.75	77 816.56
Landskrona, Skt.Joh.Bapt.	=La	153 638	99 858
Uraniborg, Obs. centre	=U	145 101.7	103 541.8
Hven Skt.Ibs(Gamle Kirke)	=I	143 850.0	103 798.4
Malmo Skt.Petri Kirke	=M	165 532.0	70 622.8
Lund Donkirke S.Tower	=Lu	176 921.2	81 939.6
Helsingor Skt.Olai Kirke	=He	139 504.4	117 623.8
Kronborg s.e. tower	=Kr	139 952.0	117 984.9
Halsingborg Maria Kirke	=Ha	144 605.3	118 880.0
Halsingborg Karnan	=Ka	144 632.2	119 190.7
Hojbjerg	=Ho	132 364.3	94 753.8

In one of his publications Brahe assumes the earth to have a circumference of 5400 German miles (or 1° = 15 miles) where 1 German mile = 4.611 English miles and 1 English mile = 1.609 km. Whence the length of 1° was 69.16 English miles or 111.28 km and the radius 6382 km. These values are NOT however derived from Brahe's triangulation.

From the values of Brahe it would seem, on a sample line from Copenhagen to Karnan where the difference of latitude was given by Brahe as 20′, that 1° was approximately 124 km (77.5 miles). However if the 'correct' latitude difference of 22′ is used this reduces to 1° = 113 km (70.5 miles) which is reasonably in keeping with the other figures quoted above.

Although Willem Blaeu is reported to have measured along the coast from the mouth of the R.Maas to Texel, his results were never published.

The first mention of such a measure was by a family friend, Vossius, who expected Blaeu's son to publish his father's results but there is no record of this happening.

Jean Picard made reference to it in his *Voyage* *d'Uranibourg* when he visited the ruins of Tycho Brahe's former observatory. Even if Blaeu did keep records of his work they would have been lost in 1672 when the family home was destroyed by fire.

From the comments of Picard it would appear that the value by Blaeu for the length of a degree differed from that of Picard by only 60 rods (or was it Rhenish feet?). The references vary in that one quotes 60 rods, the others, including Delambre in [69] 60 Rhenish feet. But there are 12 Rhenish feet to the rod or rood i.e. a variation of 12 times! How easy it is with the passage of time for facts to become distorted.

No mention is made of how the distance was measured but there is reference to the observation of latitude at the terminals with a sector of 14 Rhenish feet in radius (one reference simply says 'of great diameter').

The date of the work is also unknown although thought to be prior to the work of Snellius of 1615. What is certain is that as a youth Blaeu spent two years working for Tycho

Brahe at Uraniborg. Since Brahe had an interest in arc measurement it is quite feasible that this would have been passed on to Blaeu.

As Picard's accepted value was 57,060 toise and 1 rod was approximately 1.93 toise, so 60 rods = 115.8 (say 116 toise). Hence the length of 1° according to Blaeu would be either 56,944 or 57,176 toise since no sign for the difference was given. If the correct unit for the difference was Rhenish feet rather than rods then the variation would be reduced by a factor of 12. i.e. 57,050 or 57,070 toise.

For what might be described as the real starting point of triangulation one has to refer to Willibrord Snellius and his work in Holland. He is also credited with the first resection.

Somewhat of a prodigy, he was lecturing at Leiden University on mathematics and astronomy when he was only 19 years old.

He set out to measure an arc of the meridian between Alkmaar and Bergen op Zoom very much along the same principles as used by his predecessors. In other words determining both the angular and linear distances between two points on the same meridian. The basic difference was that while the predecessors had measured the complete distances directly, Snellius was the first to use a short base and extension through a chain of triangles.

He described all his work on this in his celebrated book *Eratosthenes Batavus de Terrae ambitus vera quantitate* Leiden 1617. Here he obviously considered himself to be the Dutch successor to Eratosthenes.

With his introduction of triangulation he also developed the recognised approach to base extension and gave considerable thought to the unit of length.

Reference [108] states that the Leyden or Rhynland (Rhineland) rood (Rod) commonly accepted in the 18th century as the length of an iron bar in the wall of the courthouse in Leiden was supposed to be a copy of the rood used by Snellius in 1621. Mention can also be found of the 'Rhynland standard measure' preserved in the Archives of Leiden Observatory and with which it is thought Snellius measured his arc.

In 1752, when Lulofs was experimenting on the length of the pendulum in Leiden he used 1/12th part of the Snellius rood and termed it the Rhynland foot. When these standards were compared in 1807 to the French legal metre the result was:

1 Rhynland rood = 3.767 358 m

A year later, 8 February 1808, Royal decree by the King of Holland declared the true ratio as:

1 metre = 3.185 256 0866 Rhynland feet (Which agrees with the above value of the rood).

Over the centuries various other roods can be found varying from one another by as much as 0.4m.

Various authorities do, however, differ slightly in their estimates of the value in the time of Snellius:

Van der Plaats (1889) gave 3.7635m
Jordan (1920)   3.766 2420m
Haasbroek [102] accepts 3.766m

While the rood was normally subdivided into 12 feet each of 12 inches Snellius worked in terms of two decimals of a rood but not written as a continuous number. An alternative name was the Rhynland perch.

Snellius had 5 base lines all in the vicinity of Leiden.

Line	Date	Length (roods)
tc	1615	87.05
gi	1616	348.10
of	3.2.1622	250.00
bd	1.1622	475.00
km	1622	471.50

There was some suggestion that the last line was really 475.00 roods. These are discussed in detail in [102] and their relative positions are shown on figure 22. A further base between Oudewater and Woerden closed badly and was rejected by Snellius.

Some 100 years after the death of Snellius, a professor at Utrecht, Petrus van Musschenbroek, made an in-depth study of the work of Snellius. He made many corrections and additions but not all could be justified and he introduced some additional mistakes and apparent falsifications. Additional because it would appear that Snellius was somewhat careless with his computations and Haasbroek highlights, and corrects, all that he found. His observed angles were:

At Alkmaar
50. 34°22'
89. 74°32'
90. 35°02'
91. 39°02¼'
94. 11°26'
At Amsterdam
44. 43°18'
48. 67°45'
At Bergen op Zoom
54. 47°15'
At Breda
35. 70°14'
42. 45°59'
53. 43°24'
At Den Haag
 3. 50°23'
 7. 85°51'
 9. 90°18'
26. 20°45'
95. 17°09'
96. 15°10'
At Dordrecht
 5. 25°50'
18. 62°28'
32. 44°20'
34. 72°15'
36. 54°12'
38. 86°19'
40. 66°11'
At Gouda
 2. 32°25'
 6. 128°22'
12. 80°00'
14. 114°48'
23. 125°42'

37.　48°15'
At Noordwijk
55.　21°08'02*
At Oudewater
21.　125°43'
24.　36°53'
29.　65°25'
At Haarlem
28.　68°04'
45.　172°11'
49.　77°55'
At Leiden
　1.　97°11'
　4.　25°49'
　8.　71°31'
10.　53°40'
11.　43°36'
13.　37°40'
16.　63°26'
19.　20°26'
22.　17°23'
25.　147°19'
27.　77°50'
43.　27°11'
46.　50°38'
59.　22°57'58*
72.　60°32'
73.　84°05'
74.　45°21'
75.　38°45'
76.　23°36'
77.　　6°12'
92.　31°21¾'
93.　15°28'
At Utrecht
15.　27°32'
17.　54°08'
20.　33°53'
30.　82°31'
31.　62°13'
47.　54°00'
51.　116°23'
At Willemstad
39.　41°10'
41.　67°51'
52.　89°25'
At Zaltbommel
33.　73°29'
At Zoeterwoude
78.　104°32'
79.　77°12'
80.　63°57'
At "a"
60.　67°44'
61.　61°38'
At "b"
68.　54°36'
69.　61°53'

At "c"
66.　82°08½'
67.　63°52'
At "d"
70.　89°31'
71.　57°42'
At "e"
64.　83°20'
65.　81°29'
At "f"
87.　40°02'35"
88.　78°38'
At "g"
83.　66°05'
84.　59°20'
At "i"
81.　92°10'
82.　60°11'
At "m"
57.　67°04*
58.　68°50*
At "o"
85.　119°44'
86.　70°34'

At "p"
56.　88°12'02"
At "t"
62.　78°30'
63.　54°00'

*[102] derived a correction of +v to 57 and 59 and —v to 55 and 58 with v = 6.3'

The base lines were measured with a surveyor's chain and the angles with a 2.2 ft radius quadrant or a semi-circle of 1.75 ft radius. It must be remembered however that there was still no optical assistance in sighting or reading and that minutes were the best that could be expected. Both instruments had been constructed by Willem Janszoon Blaeu.

The lines of 1622 were measured on the ice covering the low lying flooded fields.

In addition to having no optics to aid observation, Snellius had no assistance in his computations - logarithms had only just arrived and were yet to be accepted and calculating machines were yet to be invented.

When computing his triangulation Snellius was in the position of only being able to apply the simple triangle closure conditions for angles. Side conditions were yet to be appreciated. On the other hand Haasbroek was in a position to use a computer and modern adjustment techniques. This allowed computation of coordinates in both the system of Snellius and in the Netherlands national system. Interest here is only in the Snellius system.

Station	E (Metres)	N (Metres)
Alkmaar	−43 647.69	+53 421.30
Haarlem	−51 120.24	+25 298.04
Amsterdam	−33 410.34	+24 413.65
Leiden	−61 342.49	+ 725.08
Utrecht	−18 222.58	− 7 145.44
Gouda	−46 433.49	−15 850.98
Oudewater	−35 466.23	−14 654.19
Den Haag	−74 090.59	− 8 115.78
Rotterdam	−62 083.10	−25 641.58
Zaltbommel	− 9 327.31	−38 127.55
Breda	−42 375.99	−62 951.68
Willemstad	−65 625.18	−51 162.39
Dordrecht	−50 157.16	−37 743.55
Bergen op Zoom	−76 425.06	−73 030.31

In order to determine the meridian length Snellius first derived the length from Alkmaar to Bergen op Zoom as 34 710.6 roods (as against 34,635.9 roods in the Netherlands system) or a meridian distance of 33,930 roods.

For the determination of latitudes and azimuths an iron quadrant of more then 5½ ft radius was used where 0.5 mm on the limb represented 1′ of arc.

The derived latitudes (since observations were not at the same points as the horizontal angles) were:

Alkmaar	52° 40½′		Leiden	52° 10½′
Bergen op Zoom	51 29			51 29
Difference	1 11½			0 41½

The observed azimuth was Leiden to Den Haag 232° 21′ 44″

Which with various of the observed angles can be turned into:

Alkmaar - Bergen op Zoom	191° 26′
Alkmaar - Leiden	195 28

From these it was possible to determine two values for the length of 1° Using:

Alkmaar - Bergen op Zoom	28 473 Rhynland perches
Alkmaar - Leiden	28 510 Rhynland perches

From which Snellius accepted a mean value of 28 500 Rhynland perches or 55 100 toise [38]

[98] gives 1 rood = 1.9324 toise whence	1° = 55 072 toise
[116] gives a conversion such that	1° = 55 021 toise

Recalculation by Musschenbroek in 1729 gave [116], [163] 29,514 perches 2 ft 3 inches or 57,033 toise but the doubts about this have already been expressed.

Reference [98] concludes that there is probably a 2 000 toise error, which is indeed indicated by Haasbroek. Four possible reasons are analyzed for this:

1. All triangles were considered plane although the effect of this would not be great.
2. All measured angles were by quadrant and hence oblique. The vertical angles involved were unlikely to be great so the effect of treating the angles as horizontal should have had little effect.
3. Were the base measures suspect? Early ones were measured by chain, the later ones by wooden rod.
4. By the time Snellius discovered some errors in his angles he was too ill to recompute the results.

It would appear that the last two were as much to blame as anything else.

References
See in particular: 8, 38, 98, 102, 108, 116, 124, 163 and 266

Chapter Twelve

THE ARCS OF FERNEL, NORWOOD, RICCIOLI, AND GRIMALDI

Fernel

Jean François Fernel (1485-1558) was the son of a prosperous publican from Montdidier who became astronomer and court physician to Henry II of France. Before he was in his teens the family moved to Clermont near Paris.

In 1527 he published details of an arc measurement he had made between Paris and Amiens but the information available is sketchy. The observations were probably made in 1525, when, on August 25th he observed the meridian altitude of the sun in Paris. He then travelled northwards towards Amiens until he judged he had covered the equivalent of one degree. This took him until August 29th when he again observed the altitude of the sun.

This observation however indicated an arc of less than 1° so he moved on further until he was at an indicated 1° from Paris. One reference [6] implies that he made allowance for the change in declination of the sun between the dates of observation. Another reference however [98] discusses neglect of the sun's change of declination as amounting to 2 or 3 minutes of arc. This second reference does not however state definitely that the change was not applied.

For the sun's altitude he used a triquetrum 8 ft. high. This was the instrument first used by Ptolemy and, contemporary with Fernel, by Copernicus. This consisted of three rods hinged to form a triangle. One 8 ft. side was placed vertical. From its top, another side of similar length was hinged and carried open sights. The observer sighted along this to objects of interest. The third side was hinged lower down the vertical and was graduated such that the inclination of the sighting rod could be read off.

The distance between the points was derived from the number of revolutions of a carriage wheel and said to be 17,024 revolutions each of 20 ft. 4 pas since the diameter was 6 ft. 6 doigts. No indication is given as to whether or not allowance was made for the twists and turns of the route.

The length found for the arc is variously given as:

[38], [98] 68 Italian miles 95¼ paces=68 095¼ paces
[143]   57 099 toise
[252], [6] 56 746 toise
[67]   62.12 miles = 57 746 toise by Delambre.

Certainly Fernel gave the result in Italian miles so that any doubt is in the different conversions.

In [252] however, Lalande is quoted from [140] as suggesting that allowance should be made for a change in the length of the toise. The result would then be 57,070 toise.

Todhunter [252] expresses serious doubt as to the actual length of the foot used by Fernel.

If the two figures 68,096 Italian miles and 56,746 toise are considered, then the conversion used was

$$1 \text{ mile} = 833.32 \text{ toise}$$
$$\text{or } 6 \text{ feet} = 1.000 \text{ toise}$$

On turning this round and reading 5,000 French feet to the mile, then

68.096 Italian miles = 340,480 French feet
                     = 56,746 toise at 1 toise = 6 feet.

To complicate this [38] gives 1 Italian mile as about 1628 yards (English?) or a result of almost 63 English miles.

However it would seem feasible that even although one reference uses Italian miles and paces as the original unit, whoever did the conversion to toise (and it may not have been Fernel) took the equality of 1 Italian mile = 5,000 French feet and 6 French feet = 1 toise.

For Lalande to suggest a change to 57,070 toise means that he would be doing the equivalent of changing the relation of

1 toise = 864 lines to 1 toise = 859 lines

A change of 5 lines where there are 144 lines to the foot is an appreciable amount to explain away.

References See in particular: 2, 6, 38, 67, 98, 140, 143 and 252

Norwood

The first English arc was that of 1633-5 by Richard Norwood, and published in 1637 in *The Seaman's Practice*.... His measurement was from London to York but was really a backward step in surveying since no account was made of the method of triangulation introduced by Snellius. As he mentioned most previous attempts at arc measure it is inconceivable that he had not heard of the Dutch operation.

As it was he reverted to the direct measurement of the distance between two terminals at each of which astronomical observations for latitude had been made.

On 11 June 1633, he observed the apparent meridian altitude of the sun near the Tower of London as 62° 01′. On 11 June 1635, with a sextant of more than 5 ft. radius, he found the apparent meridian altitude of the sun near the middle of York as 59° 33′. There is some discrepancy here since [6] and [38] give 6th June, while [252] and [98] give 11 June. Since all four say that great stress was laid on the fact that both observations were on the same day of the month it would appear to be a mistake in the first two. On the other hand, [191] quotes that he observed, as exactly as he could, the summer solsticial altitude at both places. On the two days the declination of the sun was noted as being sensibly the same. Hence, he made no corrections for declination, refraction or parallax. Thus, he found the difference of latitude as 2° 28′ with a mean value of 52°49′ N.

A further discrepancy arises with regard to the instrument used in London. Reference [6] says that he used a sextant of more than 5 ft. radius...in London. References [252], [38] and [98] specifically say that he made no mention of the instrument used in London but quote the use of a sextant of that size in York. Since [252] and [38] are quoting directly from Norwood's book it would suggest that they should be the more reliable if it were not for the fact that the same logic would not seem to work with regard to the dates!

The distance was measured with a chain of 99 feet and some by pacing, as 9149 chains each of 6 poles and every pole of 16½ English feet. All main deviations of the route were measured by circumferentor with due allowance made for minor deviations and slopes. These corrections were derived from a table. Where he paced he was of the opinion that he normally came very near the truth.

From his results he found:

$$1° = 61,199 \text{ fathoms} = 367,196 \text{ English ft.}$$
$$= 57,300 \text{ or } 57,424 \text{ toise}$$

depending on the conversion factor used.

The result is also quoted as 69½ English miles + 14 poles.

References See in particular: 6, 38, 42, 98, 112, 143, 163, 191, 252 and 260

Riccioli and Grimaldi

An Italian attempt at arc measurement was made in 1645 but was doomed to failure by lack of appreciation of the effects of refraction. The Jesuits Riccioli and Grimaldi took a novel approach but a very difficult one. Their reasoning was:

a. Measure the distance AB
b. By star observations determine the angles S_1AB and S_2BA; hence angles BAP and ABP
c. It is then possible to calculate θ and hence solve triangle ABP for R.

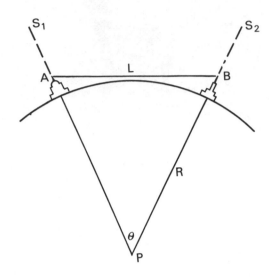

Figure 25
Riccioli's method of arc measure

The terminals A and B were the Tower of Modena Cathedral and the Jesuit Residence on Mount Serra near Bologna, at latitude 44°34′ N. From a base line of 1094 Bolognese paces + 2¼ feet (one reference quotes 1088.4 Bologna paces) and a series of four triangles the distance from Bologna to Modena was found as 20,439 paces. The base had been measured with wooden rods.

Their value for θ by the above method was 19′25″ but no account was taken of refraction. Against this value Picard in *La mesure de la Terre* says that Riccioli and Grimaldi had zenith angles of 89° 26′ 13″ and 90° 15′ 07″ which give a value of θ as 18′ 40″.

[38] gives a range from 56 130 to 61 797 toise and 62 300 toise.

[98] and [116] give 64 363 Bologna paces=62 650 toise

[6]   63 159 Bologna paces=61 478 toise (this agrees with an arc of 20 439 paces)

[144]   =62 900 toise

From these figures the conversion used was:

1 Bologna pace=0.973 385 toise=5.840 French feet

These results are normally rejected in any analysis of the figure of the earth.

References See in particular: 6, 38, 98, 116, 144 and 260

Chapter Thirteen

INTERNATIONAL SCIENTIFIC BODIES

The Royal Society

The 28th of November 1660 saw a suggestion made for the founding of a college for promoting 'Physico-mathematical Experimental Learning'. This idea had grown over some 15 years out of informal gatherings of men of science. Science is here used in a very loose sense to embrace all who might be interested in experimentation on almost any topic except theology or politics. Such meetings initially took place in London, often at Gresham College or failing that, the chambers of various kinds of the interested persons.

Around 1648 however, some of the participants moved to Oxford and continued their meetings there in parallel with those in London. In Oxford the centre of activity was Wadham College and the group became the Philosophical Society of Oxford but it lasted only until 1690.

On the Wednesday following the suggestion of 28 November 1660, a message was brought to the gathering that King Charles II had approved of their ideas and a formal document was signed by the founding members. The meeting placc was to remain Gresham College. It was a further year and a half before a Royal Charter was granted on 15 July 1662 and they became 'The Royal Society of London, for the improving of Natural Knowledge'. Further privileges were granted on 22nd April 1663 and the Council of the Royal Society met for the first time on 13 May 1663.

Its first President was Lord Brouncker and Robert Hooke was appointed Curator of the Society. One of the two appointed Secretaries, Henry Oldenberg, maintained international correspondence on a prodigious scale with eminent scientists and took a close interest in all relevant publications. From these activities developed *The Philosophical Transactions* as a monthly publication begun in March 1665 by Oldenberg rather than as the work of the Society. 'The Transactions' and other similar publications from Societies in Europe became the accepted organ for the exchange of ideas, agreements and disagreements between the developers of scientific thought.

One of the first activities of the Society was to create a laboratory where the members could develop their new ideas but money was one of the main difficulties as it was not endowed with funds. Lectures were initiated in June 1664 and in 1673 established regular weekly meetings where the public could listen to erudite papers.

When adventurers decided to mount overseas expeditions they turned to such Societies for background information gleaned from previous travellers. In addition they were often the source of up-to-date instrumentation and advice as to the areas for which observations might be usefully accumulated.

Many of those who feature in aspects of this publication were at some time members of the Royal Society and/or the French Academy of Sciences.

Of particular relevance here was the involvement of Robert Hooke. In 1669 Hooke was requested by the King to determine the length of a degree of a meridian. The committee that was formed to investigate this, heard Hooke suggest that such a measure could be obtained from a combination of the measure of stellar postions with a geodetic survey. Although a site was chosen and various instruments designed, the study came to nothing. All was, in effect, left to the French to accomplish.

L'Académie Royale des Sciences.

As with the Royal Society so L'Académie arose from informal gatherings of scientists and philosophers. Such famous names as Descartes, Pascal and Fermat used to meet in the cell of Père Mersenne to discuss a variety of current scientific problems. Although Mersenne died in 1648 the meetings continued at the houses of Hubert de Montfort. the Master of Requests, and of Melchisédech Thevoner. Experimentation would have been preferable to argument but only the wealthy would be in a position to build such facilities.

The gatherings attracted also range of foreign scholars and finally Charles Perrault suggested to Colbert that a society should be formed. The proposal was put to King Louis XIV for the establishment of a permanent Académie and this had its first meeting on 22 December 1666. Subsequently there were meetings twice a week.

Members were given a Royal pension of 1500 livres a year as well as funding of 1200 livres for their researches. They were however, set specific problems to solove and tasks to complete. The Royal Society on the other hand had no planning programme and relied heavily on the interests and enthusiasms ot the members.

The Royal patronage was such that besides the French scientists it also attracted from overseas people of the calibre of Huygens and Römer.

Among the tasks it had was that of precisely determining the length of a degree of the meridian. The Royal Society had made an attempt to get such a work off the ground but both Hooke and Halley declined since they considered it to be beyond the efforts of a single person.

The French however were more persistent. Under the direction of Jean Baptiste Colbert their first effort was the work of Picard in 1669-70. This was followed by various overseas expeditions such as that by Picard in 1671 when he went to the ruined site of Brahe's observatory at Uraniborg to effect an accurate positioning. The following year the Academy sent Richer to Cayenne to make observations of the opposition of Mars as Cassini did simultaneously in Paris. While there Richer made his celebrated measurements of the seconds pendulum which heralded the beginning of doubts as to the true shape of the earth.

It was also Picard, in collaboration with Auzout, who pursued the use on quadrants of optical devices for sighting and micrometers for reading.

In 1672 an observatory was completed in the Faubourg s.Jacques near Paris, some 3 years before that at Greenwich.

Although everything from the weight of air to the force of gunpowder was investigated one the prime interests was astronomy where the Cassini family were in control for many years — son followed father as Astronomer Royal for four generations. It was the first of these, Gian Domenico Cassini, who at the end of the 17th century, began the controversy as to whether the earth was oblate or prolate and came down on the side of the latter.

This in turn led, in the 1730s, to the suggestion for and execution of the expeditions that are the subject of this volume. Although the Academy did not have the means to provide equipment and other needs for the expeditions it was, through its influence, able to obtain funding from King Louis XV.

From 1683 there was a period quiet of activity until Jean Paul Bignon completely reorganized the Academy in 1699. The Academy then had its first set of rules detailing its membership, business and relation to the Crown. The first meeting after the reorganization was 29 April 1699. In 1713 its etablishment was confirmed on the understanding that it was firmly based and it became a Royal Academy.

Other Societies

Although the personalities pertinent to this volume were mostly members of either the Royal Society or the Academy of Sciences these were by no means the only such organizations. In fact the first formal society was in Italy where the Accademia del Cimento (= Experiment) of Florence was founded in 1657. This grew out of various small groups throughout Italy and in particular the Accademia dei Lincei of Rome. Begun in 1600 this group contained Galilei and Toricelli among its members as well as Domenico Cassini, who was to go to the Paris Observatory. In particular they did many experiments on air pressure and temperature measurement although their researches spread across a wide canvas.

The Accademia del Cimento centered around the brothers, the Grand Duke Ferdinand II and Prince Leopold of Tuscany. Meetings took place at the residence of Prince Leopold but unfortunately were discontinued in 1667 when he was elevated to the position of cardinal. Could it have been a way of the Pope obtaining the closure?

As far as Germany was concerned there was a number of small societies from around 1622 including The Collegium Naturae Curiosorum began in 1651 but it was not formalized in the same way as the Royal Society or the Academy of Sciences although it did have the Emperor as patron. It specialised in medical research.

The famous Berlin Academy was not founded until 1700. The leading figure here was Gottfried Leibniz who also had a hand in similar societies in Dresden, St. Petersburg and Vienna. In this the Berlin Academy was different to those of London and Paris in that its formation was a one man effort rather than growing from group meetings. It stemmed essentially from the calendar reform of 1699 and Leibniz based many of his plans on the styles of the Royal Society and the Academy of Sciences. In addition he planned an observatory and other accommodation and the acquisition of instruments. While 1700 can be said to be the founding of the Academy it was not until 19 January 1711 that Leibniz had established it to his liking. Before it really blossomed however it went through an unhappy period of some years with the change of Monarch and death of Leibniz.

References See in particular: 3, 98, 104, 105, 192, 245 and 266

Chapter Fourteen

PICARD'S ARC MEASURE

L'Abbé Jean Picard was born at La Fléche in Anjou on 21 July 1620 and died in Paris on 12 October 1682. He began as a gardener for the Duke of Crequi, became Prior at Rille and later Professor of astronomy in Paris. He was one of the first astronomers at the Paris Observatory where he made many refinements to telescopes including the introduction of cross hairs and development of the micrometer.

In 1666 he was one of the founding members of the Academy of Sciences. It was in this environment that he was charged with improving the value for the length of a meridian degree. This he did over the period 1668-1670 using the same method as Snellius introduced 50 years earlier.

His site was in France between Malvoisin, South of Paris, and Amiens. The large plain between Clermont and Fontainbleau had only one hill at Mareuil-en-France, to the north of Paris. Malvoisin was a farm to the south of Corbeil.

There were 17 triangles, a southern base on the road between Villejuive and Juvisy near Malvoisin and a north check base near Santerre and Montdidier between the mill of Méry and Past. The southern base had one terminal at the middle of a mill at Villejuive and the other at a Pavillon at Juvisy.

The measurement was with 4 wooden rods each 2 toise long (one reference says they were iron rods). These were joined together in pairs by a screw to give two measuring pieces (pikes) each 4 toise long. There was also a cylindrical copper tube at each joint to aid rigidity. The pikes were round, light and very liable to slip on a smooth surface. The road between the southern terminals was relatively level and straight for the whole 6½ miles, and only 80 feet above sea level. To keep the measure in a straight line a long cord was stretched along the road to be in contact with the rods. Each rod was laid against the cord in turn so that the forward end of the first abutted against the rear end of the second. Each

rod was leap-frogged forward in turn. This obviously required very careful placement to avoid disturbance of any rod.

The method of keeping count of the number of rods laid was similar to that used in modern chain surveying. Each rod man had 10 arrows (stakes) and each time he placed a pike he left an arrow. Thus when each man was empty handed they had covered 80 toise.

Difference references give slightly different values for the double measure:

[207] and [105] 5662 t 5 ft and 5663 t 1 ft
[98]   5662 t 5 ft and 5662t
[113]   5662 t 5 ft and 5663 t

[2] and others give only the accepted value of 5663 t.

The second base, XY on the figure, is stated as 3902 t and it can only be assumed that the measurement procedure was the same as for the first base.

As a reference standard Picard found a seconds pendulum, measured by the same toise, to be 36 inches 8½ Paris lines. Picard suggested that this might become a universal unit of measure such that:

if pendulum of 36 inches 8½ lines	= Radius Astronomicus
⅓ Radius Astronomicus	= Pes Universalis
2 Radius Astronomicus	= Universal toise
4 Radius Astronomicus	= Universal perch
1000 perch	= Universal mile

In such a relation the universal toise to the Paris toise would have been as 881 to 864.

Instruments and method.

For his angular measurements Picard used a quadrant of 38 inch radius supported on an iron stand. The graduated

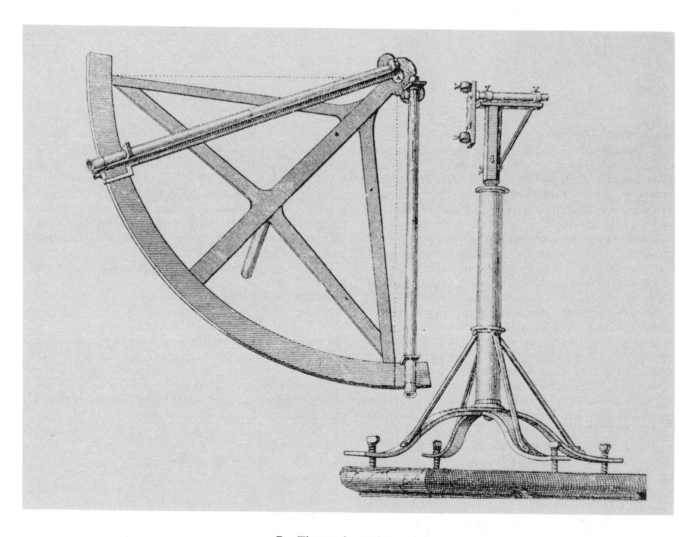

7. *The quadrant of Picard.*

8. *The field use of the quadrant.   A more relaxed posture than in plate 5 but the dress looks somewhat inappropriate.*

copper limb had a diagonal scale. Of the two attached telescopes one was fixed, the other movable. First the fixed telescope would be directed to the back station and then the movable one rotated to sight the forward station. This second telescope had an index to allow the scale to be read and the smallest scale division was 1 minute. Each triangle was said to have closed to the same order or better. There were four levelling screws below the quadrant to allow it to be tilted into the plane containing the two distant stations and the observing station.

While the angles were measured several times by different observers indication as to how, if at all, they were reduced to their horizontal equivalent is lacking.

The orientation observations were made by mounting the quadrant in the vertical plane, for which one of the telescopes had to be detached, and then observing the pole star. Next day this position was related to a station in the survey scheme. Although he had at some stage experimented with using fires as night signals he does not seem to have used the idea for his azimuths.

For determination of the angular dimension of the arc Picard used a sector of 10 ft. radius with an arc length of 1/20th of a circle (18°). In both the sector and quadrant the telescope had cross wires. The limb of the sector was divided to 4 minutes of arc and was read by determining where a long silver wire plumb line crossed the scale. To guard against draughts the line was protected by a 10 ft. long case. It was possible to reverse the sector in its mount to eliminate any error at the zero mark of the scale.

At Malvoisin, Sourdon and Amiens the star described as the knee of Cassiopeia (δ Cassiopeia) was observed on the meridian. These observations were spread over September and October 1670, so that there was no attempt at simultaneous observations from each end of an arc. His results were:

Zenith distances Sept. 1670 Malvoism 9°59'05"
Sept./Oct. 1670 Sourdon 8°47'08"
October 1670 Amiens 8°36'10"
Whence Malvoisin—Sourdon 1°1'57" 68 342 toise
Malvoisin—Amiens 1°22'55" 78 850 toise

The positions from which the astronomical observations were made were not coincident with the horizontal angle positions. The corrections to allow for these eccentricities were such as to increase the length of the Malvoisin-Sourdon arc by 83 toise and to decrease the other by 57 toise. i.e.
At Malvoisin the observation was 18 t S of the Pavillon.
Sourdon 65 t N of the church
Amiens 75 t S of the cathedral
Hence the values as quoted by Picard [98] and [105] are as given below but if one recomputes to eliminate the inconsistencies on lines HI and IK the values would be rather different.

1° = 57 064.5 toise
and 1° = 57 057.0

Picard accepted a mean value of 57,060 toise and took this to be correct to 32 toise.

[12] quotes the results to be equivalent to 69.104 English Statute miles. The stations on the scheme were as follows:
A = Middle of mill at Villejuive.
B = Nearest corner of Pavillon of Juvisy.

C = Top of steeple of Brie Compte Robert.
D = Middle of tower of Montlehéry.
E = Top of Pavillon of Malvoisin.
F = Pole on ruins of tower of Montjay with a lock of hay on it to give better visibility.
G = Middle of hummock of Mareuil. Needed a fire as beacon.
H = Middle of great oval Pavillon of Castle of Dammartin.
I = Tower of St. Sampson in Clermont.
K = Mill of Jonquières near Compeigne.
L = Tower of Coyvrel.
M = Little tree on hill of Boulogne near Montdidier.
N = Tower of Sourdon.
R = Steeple of St. Peters in Montdidier.
T = Tree on hill of Moreuil.
V = Lantern of Notredame, Amiens.

The observations of the triangles in his main scheme were:

Base line A-B 5663t 0 ft
X-Y 3902t

CAB	54°04'35"	GHI	55°58'00"
ABC	95°06'55"	GIH	27°14'00"
ACB	30°48'30"	IGH	96°48'00'
DCF	113°47'40"	LKM	28°52'30"
DFC	33°40'00"	KML	63°31'00"
FDC	32°32'20"		
		LMR	58°21'50"
FGH	39°51'00"	MRL	68°52'30"
FHG	91°46'30"		
HFG	48°22'30"	NTV	83°58'40"
		TNV	70°34'30"
LIK	58°31'50"		
IKL	58°31'00"	DEC	74°09'30"
		DCE	40°34'00"
ILN	119°32'40"	CDE	65°16'30"
NTR	72°25'40"	GDE	128°09'30"
TNR	67°21'40"		
		HIK	65°46'00"
DAC	77°25'50"	HKI	80°59'40"
ADC	55°00'10"	KHI	33°14'20"
ACD	47°34'00"		
		LMN	60°38'00"
DFG	92°05'20"	MNL	29°28'20"
DGF	57°34'00"		
GDF	30°20'40"	NRL	115°01'30"
		RNL	27°50'30"

For the meridian:
EG ϵ 00°26'00" i.e. bearing GE 180°55'00"
GI θ 01°09'00" GI 358° 51' 00"
IN ν 02°09'10" NI 177°50'50"
VN β 18°55'00" NV 341°05'00"

It should be noted that [113] gives different values of EG ϵ 0'20"; GI θ 0' 19" and IN ν 01' 09". How these values were derived is unknown since the above values from [207] should be the more authoritative.

Many subsidiary angles were recorded by use of points

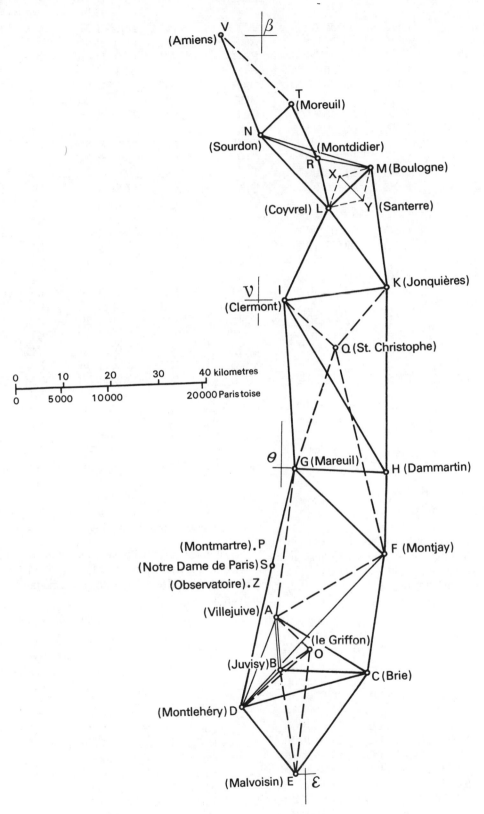

Figure 26

Diagram of the scheme of Picard

O, P, Q, S and Z. The additional angles given by these and by XY were:

AOB	62°22'00"	CDG	62°53'00"
ABO	75°08'20"	GDC	86°24'25"
BAO	42°29'40"		
		QGI	31°50'30"
ACE	88°08'00"	QIG	43°39'30"
AEC	42°27'30"		
EAC	49°24'30"	XYM	56°46'15"
		YXM	65°20'45"
PCE	102°36'40"		
PEC	43°09'30"	DOZ	82°05'10"
		DZO	51°34'00"
FAD	140°38'50"	ZDO	46°20'50"
QFG	36°50'00"	DOE	47°00'00"
QGF	104°48'30"	DEO	50°02'50"
		EDO	82°57'10"
XYL	50°37'40"		
YXL	54°10'45"	PDC	65°31'00"
		PCD	62°02'40"
DOS	88°16'40"		
DSO	46°35'00"	GAF	52°08'50"
SDO	45°08'20"	GFA	75°12'10"
		FGA	52°39'00"
AOD	76°50'00"		
ADO	37°19'20"	GCE	126°58'25"
DAO	65°50'40"		
		QIK	49°20'30"
BCE	57°19'30"	QKI	53°06'40"
BEC	44°55'45"		
EBC	77°44'45"	MYL	107°23'55"
ACF	66°13'40"		
AFC	50°33'20"		
FAC	63°13'00"		

In addition to the selected distances as used to determine the arc length Picard calculated many others and these give some insight into the major discrepancy at Station I of 5 t as well as other smaller variations.

Triangle	From		Derived	
ABC	AB	5 663t 0ft	AC	1 012t 5ft
			BC	8 954t 0ft
DAC	AC	11 012t 5ft	DC	13 121t 3ft
			AD	9 922t 2ft
DEC	DC	13 121t 3ft	DE	8 870t 3ft[1]
			CE	12 389t 3ft[2]
DCF	DC	13 121t 3ft	DF	21 658t 0ft[3]
DFG	DF	21 658t 0ft	DG	25 643t 0ft
			FG	12 963t 3ft[4]
GDE	DG	25 643t 0ft	GE	31 897t 0ft[5]
AOB	AB	5 663t 0ft	AO	6 178t 2ft
AOD	AO	6 178t 2ft	AD	9 922t 2ft
			DO	9 298t 0ft
DOE	DO	9 298t 0ft	DE	8 870t 5ft[1]
ACE	AC	11 012t 5ft	CE	12 388t 2ft[2]
BCE	BC	8 954t 0ft	CE	12 390t 0ft[2]

Triangle	From		Derived	
PDC	DC	13 121t 3ft	PC	15 064t 3ft
			DP	14 621t 3ft
PCE	PC	15 064t 3ft	CE	12 389t 0ft[2]
ACF	AC	11 012t 5ft	AF	13 051t 0ft
FAD	AF	13 051t 0ft	AD	9 922t 2ft
			DF	21 657t 3ft[3]
GAF	AF	13 051t 0ft	FG	12 963t 0ft[4]
GDC	DG	25 643t 0ft	GC	22 869t 3ft
GCE	GC	22 869t 3ft	CE	12 389t 3ft[2]
			GE	31 893t 3ft[5]
FGH	FG	12 963t 3ft	GH	9 695t 0ft
GHI	GH	9 695t 0ft	GI	17 557t 0ft[6]
			IH	21 037t 0ft[9]
QFG	GF	12 963t 3ft	QG	12 523t 0ft
QGI	QG	12 523t 0ft	GI	17 562t 0ft[6]
			QI	9 570t 0ft
HIK	HI	21 043t 0ft[9]	IK	11 678t 0ft[7]
QIK	QI	9 570t 0ft	IK	11 683t 0ft[7]
LIK	IK	11 683t 0ft	KL	11 188t 2ft
			IL	11 186t 4ft
LKM	KL	11 188t 2ft	LM	6 036t 2ft[8]
LMN	LM	6 036t 2ft	LN	10 691t 0ft
ILN	LN	10 691t 0ft	IL	11 186t 4ft
			IN	18 905t 0ft
XYL	XY	3 902t 0ft	YL	3 273t 2ft
XYM	XY	3 902t 0ft	MY	4 187t 0ft
MYL	YL	3 273t 2ft	YM	4 187t 0ft
			ML	6 037t 0ft[8]
LMR	LM	6 037t 0ft	LR	5 510t 3ft
NRL	LR	5 510t 3ft	NR	7 122t 2ft
NTR	NR	7 122t 2ft	NT	4 822t 4ft
NTV	NT	4 822t 4ft	NV	11 161t 4ft
DOS	DO	9 298t 0ft	DS	12 795t 0ft
			OS	9 073t 0ft
DOZ	DO	9 298t 0ft	DZ	11 757t 0ft
			OZ	8 588t 3ft

Compare those marked with the same numbers thus [1], [2], etc.

Picard made various conversions of his result of 57,060t. In universal measure at a ratio of 881 to 864 it became 55,959 Universal toise. He then compared existing standard units of length:

If 1 Paris foot	= 1,440 parts
1 Rhinelande or Leyden Ft	= 1,390 parts
1 London foot	= 1,350 parts
1 Bolognian foot	= 1,686 parts
1 brasse de Florence	= 2,580 parts

Then for 1°:

In toise du Châtelet	57,060
pas de Bologna	58,481
Verges du Rhin of 12 ft. each	29,556
Paris leagues of 2000t	28¼
Mean French leagues of 2282t	25
Marine leagues of 2853 t	20
English miles of 5,000 ft	73.035
Florentine miles of 3,000 brasses	63.7

Whence the circumference of the earth 20,541,600 t of Paris and diameter 6,538,594 t of Paris.

Maupertuis, Clairaut, Camus and LeMonnier made observations on the Paris-Amiens arc after their return from the north, with the same sector as had been used in Lapland. In 1739 they found the arc from the northern end of the rue de Louis-Le-Grand in Paris to the Jardin du Roy in Amiens as 1° 01' 12". To adjust this to the arc between the two cathedrals required linear corrections of +1105 t and +98½t = 1203½t, which equated to + 1'16", so that the arc became 1° 02' 28". This however neglected refraction of +1" so that their value of the arc was 1° 02' 29".

Now Picard found Malvoisin-Amiens as 1° 22' 55" but Malvoisin was 19,376t south of Notre Dame and his observation point a further 18t south, or a total of 19,394t. At Amiens the cathedral was 75t north of Picard's observation point so that the total adjustment needed was 19,319t. This would be equivalent to 20' 19".

Then Picard's equivalent value was 1° 22' 55" - 20' 19" = 1° 02' 36". But corrections omitted by Picard for aberration, precession and refraction would increase this by 11½" to 1° 02' 47½".

Compare this with the value by Maupertuis of 1° 02' 29" and there is a difference of 18½".

These differences led Cassini III and LaCaille in 1740 to remeasure 5 times, at different temperatures, the Picard base and they found a difference of nearly -6t or 1/1000 of the value by Picard. If this had been due to an error in his standard measure it would have been equivalent to about 0.9 line.

When Cassini remeasured and recomputed between some of the same stations his value for NV was 11,126.8t against that of Picard as 11,161.7t.

La Condamine in [135] said that if one were to apply those equations that were necessary yet which Picard neglected then the arc would be augmented by 11½" and the length of 1° would become 56,925t. He further recorded that the remeasure of the same degree by Maupertuis and others gave 57,183t, or if the refraction had been accounted for, 57,164t. Later still the same degree was measured by Cassini III and La Caille as 57,074½t.

La Condamine queried the possibility of changes in the length of the wooden rods with time, temperature, humidity and their linkage in pairs. Where Picard used only two combined rods, laying them end for end, La Condamine in his work always used three so that there were always two on the ground, and Cassini in the measure with La Caille used four iron rods so that three were always on the ground. In Picard's case any knocking on making contact could disturb the only rod resting in place, disturbance was far less likely with 2 or 3 rods resting. Cassini's allowance for temperature amounted to 2 feet.

A further Cassini value, for the section Montlehéry to Brie was 13,108.32t as against that of Picard of 13,121.60t i.e., short by 13,28t or very nearly 1/1000 again.

Critics also pointed out that Picard deduced, rather than measured, the two important angles NVT and NLR, and that the latter, when measured directly in 1740 differed by -1' 10" from that of Picard.

In effect Picard had made compensating errors of 7" in the arc and 96t (6") in the amplitude so that the final value by Cassini III of 57,074½ differed from Picard by only 14½t.

References See in particular: 2, 3, 12, 30, 55, 69, 98, 105, 113, 135, 144, 163, 206, 207, 245 and 266

Chapter Fifteen

THE CASSINIS AND SIR ISSAC NEWTON

THE CASSINIS

Any astronomical or survey activities in France from 1669 until the mid 1800s were influenced by the Cassini family. From father to son over four generations they were directors of the Paris Observatory and were the hub of French scientific thought. To sort them out, one requires numbers since the initials can be confusing on translation.

Cassini I	Giovanni Domenico Cassini (Jean Dominique)	1625-1712
Cassini II	Jacques Cassini	1677-1756
Cassini III	César-François Cassini De Thury III	1714-1784
Cassini IV	Jean Dominique Cassini	1748-1845

Soon after Cassini I arrived in France it was proposed that the arc measured by Picard in 1668-70 from Amiens to Malvoisin should be extended through the length of the country. Colbert gave the idea his full support and endeavoured to obtain what funds he could for it.

A two-way extension, sough from Malvoisin by Cassini I and north of Amiens by de La Hire was begun in June 1683 so as to extend the arc from 1° 21′ to 8° 30′ (Some references suggest that Cassini went north and La Hire to the south). Unfortunately this work had hardly begun before the death in 1683 of Colbert, the liberal minister of Louis XIV. Wars followed. 1689 was between Britain and France. In 1690 the English were defeated at Beachy Head and the following year several ships were lost around the Scilly Isles. The problem here was lack of knowledge of longitude.

Colbert was succeeded by the Marquis de Louvois who was more interested in territorial gains than in scientific developments. As always, the wars were expensive and economies fell on science and also upon the scientists and it was almost the end of the century before conditions eased.

The hiatus coincided with the complete reorganization of the Academy of Sciences.

The southward extension was restarted in August 1700 by Cassini II, Maraldi and La Hire II, that to the north did not recommence until 1718. It proved possible to incorporate much of Picard's original work and two further bases were added, one near Perpignan by the Pyrenees and the other near Dunkirk. These were of lengths — Perpignan 7246 t and Dunkirk 5464 t. They were measured with wooden rods of 4t each and the angles by 39 inch quadrant and 36 inch octant. Azimuth was by rising and setting of the sun and by equal altitudes and transit of Capella. Zenith distances were by a large radius sector.

By the time of the restart in 1700 Cassini I was joined by his son Cassini II, Maraldi, Couplet and Chazelles. The Gulf of Lyon was reached by the following year. From this section of the arc it was possible to derive a value for the length of a degree averaged over the whole distance. From Paris to Collioure the arc was 6° 18′ 57″ and the distance 360,614t to give 1° = 57,097t. This was some 37t longer than that found by Picard for the central section. A later value for Paris to Perpignan was 6°08′17″, 350, 142t and 1°= 57,045t.

Cassini I was by now in failing health and it was his son who devised the scheme for extending to Dunkirk. As had happened before, more wars interrupted progress, this time it was that of the Spanish Succession and this delayed operations until June, 1718. Some 28 triangles 9 of which were the same as Picard, formed this section over 2° 12′ 15″ and 125,454t to give 1° = 56,960t. References [6] and [38] give the arc as 2° 12′ 09.3″ while [252] gives 2° 12′ 09.5″ and 125,468t. A later remeasure/recomputation was 2° 11′ 55.5″ for 125,508t and 1° = 57,081.5t.

The whole arc from Dunkirk to Collioure gave the same as Picard i.e. $1° = 57,060t$.

With the central and southern parts this strongly suggested that the arc lengthened towards the Equator at a rate of about 11 or 12t per degree. In other words it was prolate, which was in contradiction to the theories put forward by Isaac Newton. The Cassinis were convinced of both the accuracy of their work and of the prolate figure even though a rate of change per degree of 12t was equivalent to less than 1″ in the accuracy of the arc. Their prolate figure had a ratio for the axes of about 95 to 96.

In an attempt to settle the dispute a degree of parallel was measured 1733-34 between Strasbourg and St Malo. The length of 1° of longitude at the latitude of St Malo was found to be 36,670t where, if it had been a sphere Cassini calculated that it should have been 37,707t. If spherical the radius of the parallel would be $\rho \cos \lambda$ where $\rho =$ radius of curvature and $\lambda =$ the latitude.

For the latitude of Strasbourg 1° of longitude was found as 37,066t against 37,745t for a spherical earth. Although this also supported a prolate shape, doubts were expressed as to the reliability of the longitudes since they were based on eclipses of the satellites of Jupiter. Further work on the parallel arc to the west of Paris gave 1° at the latitude of Brest some 300t shorter than for a sphere. Then for the latitude of Nantes there was a difference of 781t.

Modern latitude values for the places concerned are:

St Malo 48° 40′N
Strasbourg 48° 35′N
Brest 48° 24′N
Nantes 47° 12′N

The first chink in the Cassini armour came with the publication of *La Meridienne verifier...* [45] by Cassini de Thury (Cassini III) which was said by Delambre to have been entirely the work of LaCaille since Cassini was more interested in the map of France and its frontiers. The fact that it was attributed to Cassini probably stemmed from the jealously guarded preserve that existed for members of the Academy over exclusive publishing rights. This was similarly highlighted in both the Peru and Lapland expeditions. For the latter, Maupertuis as leader and senior Academician of the party had the sole right to publish details of the scientific work. Outhier circumvented this by writing on the voyage and travels rather than the science.

The verification survey originally commenced in 1733 but then stopped for a while. The aim was to measure the meridian through the Paris observatory for the length of France. It was to follow the plan of Cassini II.

For this work they constructed, in 1738, a 6 ft. sector with limb of 50° to allow the observation of many stars at varying distances from the zenith. Its bulk was such that it had to be carried throughout the length of the arc on a form of stretcher between mules. A quadrant of 2 ft. radius was constructed and carried two micrometers. When this new sector was checked for circle error there was only 2 or 3 seconds discrepancy. Similarly the checks on the quadrant indicated only 8 or 10″.

Approved by the French Academy of Sciences the scheme was started again by LaCaille and Cassini III in May 1739 with assistance from Saunac and Le Gros, with the first observation from Brie tower.

The first partial result of note was that the arc from Paris to Bourges using LaCaille's base near Bourges was noticeably different for a degree than the earlier measures from Paris to Amiens. The first chink in the French case as it supported the flattening factor.

Gradually proof was accumulating that the base of Picard must have had an error of about 1/1000. Then another piece of evidence came with the arc from Brie to Montlehéry which was again about 1/1000 less than that of Picard.

By 1744 all were agreed on the shape, if not the size of the earth.

References See in particular: 2, 6, 20, 33, 38, 45, 55, 98, 113, 144, 163, 198, 203, 237 and 252

Sir Issac Newton

As is well known, Isaac Newton played a major role in many branches of science. That makes it difficult here to concentrate solely on the aspects of his work that impinge on the figure of the earth.

It was as early as 1666 that he first thought about gravity and its effects. For some unknown reason it would appear that he was unaware at that time of the arc measures of Norwood and Snellius since he used 60 miles as an estimation of the length of 1°. Probably because of the inaccuracy of this figure he made little progress with his theories.

Within a few years the results of Picard's work made all the difference. June 1682 at the Royal Society saw the announcement of Picard's results and the realisation that they changed the earth's radius by some 10%.

Newton's theory could now progress and it was on the basis of this that it appeared in the 'Principia' of 1687. By relating the theory of centrifugal forces to the properties of fluids he was able to demonstrate that not only must the earth be spheroidal but that its equatorial radius must be longer than the polar one in the ratio of 230 to 229. The circumference increases as the radius increases and thus a degree will increase with increased radius. The longer the radius the flatter the curvature of the arc, thus the smaller the curvature the longer the length of a degree. Now for an elliptical shape the smallest curvature is where the axis is shortest and hence the degree longest i.e. at the pole.

It followed from this that gravity would be smaller at the equator than at the pole — one of the first pieces of evidence to explain the results obtained by Richer in Cayenne with his pendulum. His clock lost more than 2 minutes a day compared with Paris and until Newton there was no logical reason put forward for this difference. Soon afterwards further observers such as Varin and Des Hayes found similar variations at other locations, none of which could be explained away by, say, the effects of temperature.

How did Newton go about deriving his theory? He started from a value for 1° of 57 060t or 123 249 600 Paris feet for the circumference (derived from Picard's work) and radius 19 615 783 Paris feet. Newton actually quotes 19 615 800 feet. Later he had the Cassini I value of 19 695 539 Paris feet for the radius or 1° = 57 061t.

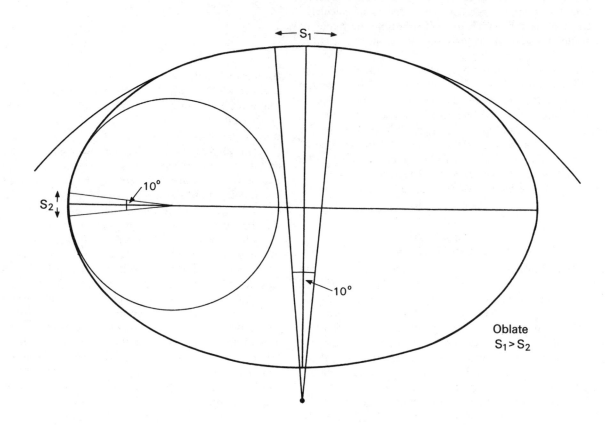

Figure 27   Oblate or prolate - The oblate situation.

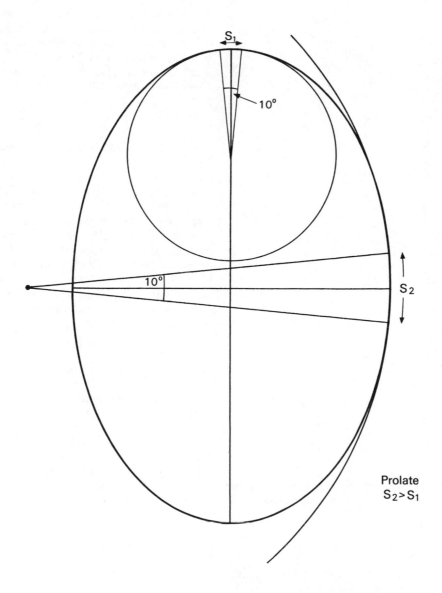

Figure 28   Oblate or prolate - The prolate situation.

Now any body on the earth's surface revolves once per sidereal day of $23^h\ 56^m\ 04^s$ (or 86 164 seconds) in a circle of radius r.

If the distance per second $\quad AC = a$

Then distance in direction $\quad A0 = n$.

$$n^2 = a^2 - x^2$$
$$= a^2 - (r^2 - (r - n)^2)$$
$$2rn = a^2$$
$$n = a^2/2r$$

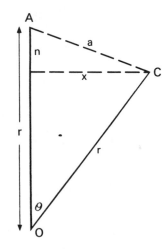

Figure 29 The basis of Newton's reasoning.

Now it would seem best to quote the figures of Newton even though they do not appear to be in complete agreement with each other

$$a = 1433.46\ ft.$$
$$so\quad n = \quad 0.052\ 365\ 61\ ft\ or$$
$$= \quad 7.540\ 64\ lines$$

But from other sources it was known that at Paris a heavy body fell 2137⅞ lines in 1 second or allowing for the effects of air, this is given as 2174 lines in vacuum.

To this must be added the centrifugal effect at any particular latitude

$$= n.cos^2\phi$$
$$= 7.540\ 64\ cos^2(48°\ 50'\ 10'')\ for\ Paris$$
$$= 3.267\ lines$$

Then the force of gravity at Paris to the centrifugal effect of rotation at the equator was as:

2177.267 to 7.540 64 in lines
or   192.50 to 0.67   in inches
or   289   to 1

Starting his analysis from a unit ratio of axes of 100 to 101 and compounding various gravity and centrifugal factors he gave the gravity at an equatorial point related to that of a polar point as 501 to 500 or relating this to the unit ratio of axes:

$$\frac{101}{100} \times \frac{500}{501} = \frac{505}{501}$$

Whence the centrifugal effect which by this ratio should be 4 in 505 was only 1 in 289 which when compounded gives the ratio of the diameter at the equator to that at the pole of 230 to 229 i.e. a compression of 1 in 230 against sent day value of 1 in 297.4

Then equatorial radius $= r + x/2$ where $x/2 =$ excess over mean r and polar radius $= r - x/2$

$$Thus\quad \frac{r + x/2}{r - x/2} = \frac{230}{229}$$

From Picard's figure for the mean radius of r = 19 615 783 Paris feet

Then $\qquad$ x = 85 472 Paris feet
Equatorial radius = 19 658 519 Paris feet
Polar radius $\qquad$ = 19 573 047 Paris feet

In deriving the weight at any particular latitude Newton showed that the increase was nearly as the versine of 2ϕ or twice the square of $sin\phi$. i.e. $1 - cos\ 2\phi = 2.sin^2\phi$. Whilst previously he had used the latitude of Paris as $48°\ 50'\ 10''$ he now used just $48°50'$ and $1 - cos\ 2\phi$ as 11334 (with radius of 10^4) Then for $\phi = 90°$ the versine of $2\phi = 20\ 000$. Since excess of gravity at the pole is 1 to 229, that at Paris would be:

$$1.\frac{11344}{20000}\ to\ 229$$

or a ratio of $\qquad$ 229.5667 to 229

Whence a seconds pendulum of 440.555 lines at Paris would, at the Equator, be:

$$\frac{229}{229.5667} \times 440.555 = 439.468\ lines$$

or 1.087 lines shorter

This compares with the measures of Richer in 1672 at Cayenne (4°55′N) where he found his pendulum shorter by 1¼ lines.

This certainly suggested that Newton's ideas were in the right area. In his 'Principia' he used the same approach to produce a table of pendulum lengths and arc lengths for various latitudes.

Latitude	Pendulum (Lines)	Meridian degree (t)
0	439.468	56 637
10	.526	659
20	,692	724
30	.948	823
40	440.261	945
50	.594	57 074
60	.907	196
70	441.162	295
80	.329	360
90	.387	382

The above values are from Cassini's observations. In comparison Newton computed similarly for Picard's work.

Latitude	Picard (t)	Cassini (t)
0	56 909	56 637
45	57 283	57 010
90	57 657	57 382

These can be compared with figures given by Maupertuis and Bouguer thus: Maupertuis [163] gives values for the difference in length of the seconds pendulum at different latitudes.

Latitude	Difference (lines)
0	0
10	0.065
20	0.254
30	0.542
40	0.896
50	1.273
60	1.626
70	1.915
80	2.103
90	2.169

Bouguer [26] gave two sets of values for the length of a degree; one in terms of variations as a function of $\sin^2\phi$ and the other as $\sin^4\phi$

Latitude	as $\sin^2\phi$	as $\sin^4\phi$
0	56 753	56 753
10	776	754
20	843	766
30	946	813
40	57 072	917
50	206	57 083
60	332	292
70	435	501
80	502	655
90	525	712

For these figures Bouguer used the basic parameters of

Equatorial radius	3 281 013 toise
Polar radius	3 262 689
Difference	18 324

References
See in particular; 6, 42, 98, and 252.

Chapter Sixteen

THE ARC OF THOMAS IN CHINA

A. THOMAS

Sometimes this name is found preceded or followed by the letters S.J. which should not be interpreted as initials since they stand for Society of Jesus i.e. Jesuit.

Born at Namur, Belgium, Antoine Thomas joined the Jesuits in 1660 when he was only 16, and took up residence in China in 1685. There he had the Chinese name An To and became vice-director of the Bureau of Astronomy (Sometimes referred to as vice-president of the Imperial Observatory).

Where was the relationship between Namur and China?

On 15 August 1678 Pére Ferdinand Verbiest, Director of the Peking Observatory (1623-1688) wrote to all Jesuit orders in Europe describing the low state of the church in China and seeking aid. He requested devout men but at the same time he required them to be good mathematicians.

It followed however that Thomas left the Jesuits in Tournai in 1678 to make his way to the Orient. He had only got as far as Portugal when he stopped off to spend a year teaching mathematics at the University of Coimbre. It was here that he met the Duchess of Aveyro who was to become his benefactress.

He finally left Lisbon on 3 April 1680 and arrived in Goa 26 September 1680. After some months there he moved on to Siam and thence to Macao where he arrived 4 July 1682 and stayed until the middle of 1685. When he covered the last leg of the journey to Peking.

During his 10 months in Siam (August 1681-May 1682) Thomas took various measurements at Juthia, the capital. These revolved around the use of the gnomon. To define a meridian line he set a horizontal plate on the top of a wall. In the centre of this was a small hole through which the sun's rays could pass and subsequently fall on a similar horizon-

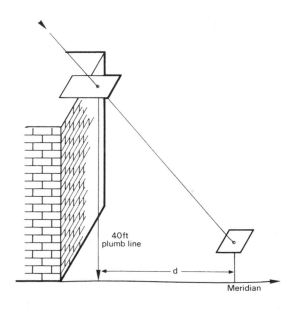

Figure 30. A Chinese gnomon.

tal plate near the base of the wall. The position of the image was recorded by measuring its position in relation to a 40 ft. long plumb line.

Thomas was none too happy with the results of this work and modified his approach considerably when, in the early years of the 18th century, he was ordered to measure the length of a meridian degree.

From the house of the Jesuits in Juthia, Thomas observed the latitude:

on 14th October 1681 as   14°17′45″
on 30th December 1681 as   14°18′27″

The difference could have been due to insufficient knowledge of the position of the sun and neglect of parallax and refraction. Later Gouye applied corrections to the observations and obtained a mean value of 14° 18′ 36″ but then even later he applied further corrections and decided on 14° 19′ 20″. Obviously there was still a long way to go in fully understanding such observations.

For the longitude of Juthia it had been suggested by Philippe de La Hire to use eclipses of the satellites of Jupiter but Thomas preferred the eclipse of the moon. On 22 February 1682 he was able to do this and used a plumb bob vibrating 3345 to the hour. When his results were compared to those for the same eclipse as seen in Paris there was a difference of 6 hours 32 minutes or 98° from Paris. As Paris is 2° 20′ E of Greenwich this gave Juthia as 100° 20′ E of Greenwich (N.B. Bangkok is 13° 45′ N, 100° 35′ E).

While in Macao he made similar observations in 1682 and 1685 to give a latitude of 22° 12′ 44″ but there was no Paris eclipse value to compare with that of Macao for longitude. The modern values are 22° 16′ N and 113° 35′ E.

When Thomas finally reached China in 1685 he found considerable interest in mapping the country. By the early 18th century there appeared a considerable need for the measurement of a meridian degree to aid this mapping.

Throughout China at that time standards for measure varied from place to place, so for a national triangulation it was necessary for the Emperor to define a Chinese foot. It was using this foot that Thomas later found 1° = 200 lys (or li) or Chinese stades, each of 180 Chinese toise of 10 feet each. Now ¹⁄₂₀th degree contained 2853 French toise and 6 feet of the Châtelet standard and this equalled 1800 Chinese toise or 10 lys.

The foot standard of the Palais was engraved on an iron scale which contained 5 feet and whose total length was defined as a 'pas géométrique' or ½ Chinese toise (perch). The Emperor then decreed that henceforth the Palais foot should be such that there was 200 stades to 1°.

Emperor Kang-Hi who had mapped much of China and Tartarie ordered Thomas to measure, as exactly as possible, the length of a terrestrial degree. This emperor was one of the most important leaders in China who tolerated the Jesuits because of his need for their scientific and technical abilities rather than the religious qualities. For his first attempt Thomas had the assistance of Father Jean François Gerbillon and they left Peking, travelling to the NNE on 24 May 1698 and on 30 June 1698 pitched tent on the banks of Lake Puyr. (Buyr Nuur latitude 47° 50′N Longitude 117° 35′ E). More or less due north of Peking across the deserts of Tartarie. This lake is to the south of the much larger lake Colon. Their value for the latitude was 48° 03′. They measured to the west some 900 Chinese stades along the route of the river Kerlon (Kerulen?) and pitched tents again. They travelled on to the rivers Tula and Urgon (Argun) which have their confluence some 300 stades from Siringa. This they

reached on 13 August and found the latitude as 49° 03′. At this point Father Gerbillon fell ill and had to return to Peking.

By 2 October they arrived near the village of Kokohoton with a latitude of 40°54′ and on 10 October passed the Great Wall at Cham-Kia-Keu and another 5 days brought them back to Peking.

On 28 November 1702 Thomas left Peking again, this time going south, to the vast Plains. Although [24] says to the south of Peking, the observed latitudes suggest that the line was north of Peking (Latitude 39° 50′ N). [260] says near Pa-Shih (Cheu) south east of Peking.

For his first terminal he chose a point near Pa-Cheu as a prominent stone monument. He commenced measuring along the meridian from here. To trace the line he set his instrument up at the start of each stade and put pegs in at every 120 pas géométrique. The instrument he used had two telescopes and a compass with a long needle. All his equipment was constructed in China.

Each stade was 180 Chinese toise (or perches) or 360 pas. When the instrument was set up it was first aligned along the direction they had come and then prolonged forward for putting more pegs in. The linear measurement was by iron wire placed against the pegs.

At the 74th stade the line went straight through a village so to avoid this a right angle offset of 1 stade was made to the east. On progressing forward again all was well until the 97th stade where a lake provided the next obstacle. This time an offset of 1 stade to the west allowed forward progress again. It was not all work and no play. On 3 December the

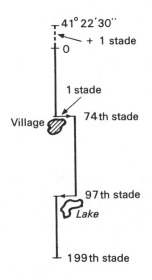

Figure 31.   The Chinese arc.

party were able to join celebrations in the village of Yuen-Kia-Keu for the feast of St François-Xavier.

Measurement continued to the 199th stade when they reached the river Cham-Ho which had to be the southern limit. Here they set up a gnomon on the bank of the river. The gnomon consisted of a vertical post designed for determination of the altitude of the sun by measuring the length of the shadow. Generally the top of such a post bore a horizontal plate with a hole in it to project a readily determined position. Thomas modified this by use of scaffolding with a vertical post at its centre. It was 30 feet high, a foot square at the base and stayed at each corner. At the top was mounted a metal plate, with hole, with its axis normal to the sun's rays. To ensure that the hole was vertically above the point marking the meridian at the base, a long plumb wire with a 2 lb. weight was suspended through the hole.

Many drawbacks to the use of such a device included problems with the hole, the uncertainty of the end of the shadow, uncertainty of the true instant of meridian passage and instability of the structure.

The meridian passage of the pole star was observed with an iron and copper quadrant of 3 ft. radius divided into minutes. As well as pinnules the quadrant had a fine telescope that allowed observations to 10″. This was set up within a tent and the star observed on the consecutive nights of 15, 16 and 17 December 1702. The latitude was found as 40° 21′.

On returning to the starting point of the arc a further 1 stade was added to make the arc 200 stade in total. A similar gnomon was erected and the latitude found to be 41° 22′ 30″. This time on the nights of 21 and 22 December.

Thus the difference of latitude was   1°01′ 30″
or allowing for varying refraction   1   01   32

Since this equated to 200 stades or 72 000 pas

Then 1° = 70 206 pas
      ≙ 195 Chinese stades + 6 pas
      = 74 886 Old Roman pas

However, the Emperor required the length of a degree to be 200 stades. Thus, neglecting the 6 pas, the new Chinese foot of the Palais was in the ratio 200 to 195

With 72 000 pas = 1°
      1 200   = 1′
         20   = 1″

i.e. the length of the standard foot was diminished by a factor of 39/40 to make it easily comparable to a 'reproducible' quantity. In essence the thinking on the metre as a particular fraction of a terrestrial quadrant was pre-dated by almost 100 years.

In 1710 the Jesuit fathers Regis and Jartoux measured 6° of meridian from 41° N to 47° N and found that each degree differed, the earth was not spherical. For their angles they used an instrument (quadrant?) of 2 ft. radius. When the 47th degree was compared to the 41st, the difference was 285 Chinese feet.

The spheroidal nature of the earth figure had been determined on the Plains of Northern China and Mongolia.

References See in particular: 14, 24, 180, 186 and 260

Chapter Seventeen

ASTRONOMERS AND MATHEMATICIANS
Cassini I to Cassini III

Many of the names in this section bear heavily on the two expeditions and maybe none more so than the first, that of Cassini; except that this can be very confusing as there were four of them. As they were father and son over four generations they are often designated by numbers. This is detailed elsewhere.

The first *Cassini* was Gian Domenico (Jean Dominique) who was brought up in Italy but who in 1669 arrived in Paris and took out French citizenship in 1673. His contribution to science was enormous. He improved instrumentation, advanced theories on atmospheric refraction, made many celestial observations and helped set up the Paris Observatory.

Suggestions by Richer and others that the earth was flattened did not impress Cassini who said more arcs were necessary. To this end he extended the work of Picard. He was also not in favour of the heliocentric ideas of Copernicus.

He maintained that the earth must be flattened at the equator, i.e. prolate, and poor observations on the extension survey only served to strengthen his ideas.

Of the same vintage as Cassini I, but working in England was William Leybourn who described many of the contemporary survey instruments in his *Compleat Surveyor* of 1653. By 1662 Cornelius Malvasia was developing an eyepiece micrometer where readings could be taken against a network of silver wires. This was to be considerably improved by Bouguer.

A distinct advancement in the use of pendulums and clocks took place around this time and one of the leading exponents was Christiaan Huygens from The Hague. He put forward ideas on centrifugal force and his publication *De Causa Gravitatis* of 1690 investigated the figure of the earth, where he found a ratio for the axes of 578 to 579. He also played an important role in analysing the changing length of a seconds pendulum with latitude.

This was a subject that was also close to the heart of Jean Richer who in 1671 was sent to Cayenne in French Guiana to investigate. He found that his clock lost $2^m 28^s$ a day compared with Paris and required a shortening of the pendulum by 1¼ lines.

The replacement of open sights by telescopic ones was introduced by Eustachio Divini and also by Jean Morin (ca 1600-1650) but Richer was among the first to put them to good use while using a quadrant in French Canada.

Telescopic sights were also of interest to Robert Hooke in 1667 but he missed a different opportunity when it was suggested that he measure an arc and he declined. He was more at home as Curator of Experiments at the newly formed Royal Society but once again England missed the chance of a famous first.

The scene moved to France and Philippe de La Hire was involved with Cassini and Maraldi with the extension of Picard's work. The results published by Cassini in 1710 only added fuel to the controversy of shape.

Contemporary with all this practical activity was the development of theories by Isaac Newton. Specific details are given elsewhere but in particular the contents of his *'Principia'* in 1687 formed the other side of the controversy raised by Cassini. Prolate versus oblate was well and truly entrenched. Alongside this Newton developed an octant, introduced the use of polar coordinates and conversion from rectangular to polar and vice versa. He was the first to develop infinite series to express trigonometric functions and explained the reasons for a seconds pendulum varying with latitude.

The Jesuit Order which had been established in 1534 was

still flourishing 150 years later and was sending members with various scientific abilities around the world. Among these was the Belgian Antoine Thomas who was sent to China in 1678 but stopped off in Portugal and Siam before finally arriving in China in 1685. He became Vice President of the Imperial Observatory and was instrumental in executing a long arc measurement.

The Greenwich Observatory was founded in 1675 and its first Astronomer Royal was John Flamsteed. During his 45 years at the observatory he spent much energy on the graduation and calibration of instrument limbs, micrometer screws and scales. In general, however, his were fixed rather than portable instruments.

A year after the founding of Greenwich was the rather different milestone of Gottfried Wilhelm Leibniz bringing his calculating machine to London. He was the first to suggest a fluidless barometer, the aneroid, and helped to develop integral and differential calculus.

Edmund Halley did much for astronomy from the 1670s onward but was also interested in developing the usefulness of log. tables.

At the turn of the 18th century A.M. Mallet was involved in the development of spirit levels but as the tubes used were not yet subject to internal curvature so the bubble movement was erratic and inaccurate. The invention of the bubble level took place as early as 1661 by Thévonet in France.

The early 18th century also saw the rise of the English instrument maker and in particular George Graham. By 1715 he was experimenting with the effects of changing temperature on the pendulum and within a few years had devised a compensating system. Various of his instruments were taken on the Lapland expedition by Maupertuis. He was also particularly active in improving micrometers and verniers.

Cassini II was Jacques who succeeded his father as astronomer royal in 1712. He continued the meridian arc started by his father and extended it to Perpignan. The further results were published in 1718 and still supported the prolate theory. In 1733 a parallel arc was measured between Strasbourg and Granville near St Malo. This yet again, suggested to Cassini that the earth was prolate.

Petrus van Musschenbroek recomputed the triangulation scheme of Snellius but founded his work on a new base of 1622 that was better than that used by Snellius. Unfortunately there is considerable doubt about the authenticity of some of his work. Haasbroek in [102] condemned all of the recomputation.

The chronology now brings one to those who were involved in the expeditions. Réginald Outhier, a canon from Bayeux and amateur scientist helped on the measure of a parallel from Caen to St Malo. As a result he was then invited to join the Lapland expedition, where he particularly assisted in the astronomical observations and also drew many maps of the area. Less than 4 years younger was Pierre Bouguer (1698-1758), a Professor of hydrography from Brittany. He made investigations into atmospheric refraction and gravitation. With the latter he attempted to measure the attraction of a 2300 m mountain on the plumb line and its effect on g. For this he used Mount Chimborazo in Peru. He played one of the major roles in the Peru expedition.

The leader of the Lapland expedition, Pierre Louis Moreau de Maupertuis (1698-1759), was the same age as Bouguer. He was a follower of Newton's ideas on the shape of the earth. He retired from the army to concentrate on mathematics. As the Lapland expedition returned long before that from Peru so Maupertuis became known as the earth flattener. He later became President of the Berlin Academy.

Charles Camus served a considerable period as administrator of the French Academy of Sciences. After the Lapland expedition he aided some of the same scientists on reobservation of part of Picard's arc.

In parallel with the English instruments of Graham the French had Claude Langlois who was their first instrument maker of any consequence. In particular he made 5 quadrants for the expeditions as well as for the Paris Observatory. In addition he made the standard toise bars that were taken on each expedition.

One of the best remembered names on the Lapland expedition was Anders Celsius the Swedish Professor of astronomy. Obviously the nearness of Uppsala to Lapland was one of the reasons for his invitation to join the group. In addition, however, he had been in France at the time of the preliminary discussions so was in the right place at the right time. It was Celsius who went to London to get some of the necessary equipment. After Lapland he was involved in the French arcs from Paris to Collioure and Paris to Dunkirk where his analyses raised some doubts. His connection with the thermometer maintains his name.

The wealthy French naturalist, mathematician and explorer Charles de La Condamine was the leading light of the Peru expedition. He was away for ten years. Most of the difficulties met in Peru fell to La Condamine to struggle with and spend years in resolution. He later took the lead in trying to establish an international unit of measure but it was not to come to fruition until some 20 years after his death. In S.America he took an interest in everything scientific and reported at extreme length on all aspects of nature.

With similar interests was Joseph de Jussieu who spent much of his time in Peru gathering plant specimens but lost all before getting them back to France. Loss of memory precluded him from much active science on his return. Longest resident in Peru was Jean Godin Odonais whose harrowing adventures are outlined elsewhere.

Pedro Maldonado was from the Jesuit College in Quito. His local knowledge, mathematics and cartographic ability proved of considerable use to the Academicians. He was later to become Governor of Esmeraldes.

The surgeon on the expedition was Dr. Jean Seniergues who met an untimely death in Cuenca towards the end of the survey. Among others in the party were Verguin—an engineer from the French Navy; Morainville an architect who also met an early death by falling from scaffolding; and Hugot the watchmaker/instrument builder.

Louis Godin was an astronomer known to have constructed various major instruments. It was he who proposed to the Academy that an expedition should be sent to the Equator. Despite this he took only a background role in the work although nominally the leader. Unfortunately he never published a separate account of his contribution although it was reported by Juan and Ulloa. He stayed on in Peru as a Professor of mathematics before returning to Europe.

Another name remembered for other than his work on the Lapland expedition was Alexis-Claude Clairaut. He was particularly interested in celestial mechanics but achieved fame in many branches of mathematics.

The Peru expedition was accompanied by two Spanish naval officers Jorge Juan and Antonio de Ulloa who, although young when they went to Peru, were nevertheless most important members of the party. Not only were their linguistic attributes useful but they were able mathematicians and observers. Juan later became a member of the Royal Society and both of them wrote accounts of their exploits. They were both interested in all branches of science and took many observations on a variety of phenomena.

Cassini III- César—Francois de Thury, succeeded his father in 1756. He was the first Cassini to pass some doubt on the convictions of his relatives on the figure of the earth.

Among the youngest of the explorers was Pierre LeMonnier who accompanied the Lapland party. He was a skilled observer and did much detailed scientific work during his lifetime. At age 20 he produced an elaborate lunar map.

A further proposal for a new measurement standard came from Mathurin Brisson in 1790 and this was finally to give birth to the Metre.

References See in particular: 1, 2, 3, 6, 14, 28, 40, 55, 61, 78, 80, 98, 102, 103, 104, 113, 123, 124, 132, 134, 135, 137, 143, 163, 186, 208, 232, 252, 258, 260, 266 and 267.

Chapter Eighteen

THE UNIVERSAL UNIT OF MEASURE

From the middle of the 17th century there was a growing realization of the need for a universal unit of measurement. The problem was how such a unit should be defined. It needed to be in a form that could be readily reproduced but at that time there was no obvious candidate.

A pioneer in this field was Gabriel Mouton, a vicar from Lyons, who, in 1670 proposed the adoption of a minute of arc of a great circle of the earth as a universal unit. Mouton, like other prominent scientists of those times was a Jesuit and treated astronomy as a hobby. He used the work of another Jesuit, Riccioli to illustrate his idea but in essence it was:

1 Milliar	=	1,000 virga
1 Centuria	=	100 virga
1 Decuria	=	10 virga
1 virga	=	0.001 minute of arc
1 decima	=	0.1 virga
1 centesima	=	0.01 virga
1 millesima	=	0.001 virga

The virga was to have the same length as a pendulum which made 3959.2 oscillations over 30 minutes at Lyons but nothing came of these efforts.

In England Sir Christopher Wren had similar aspirations and was considering the length of a half second pendulum in London as a possibly appropriate unit.

Over the Channel, in 1671, l'Abbé Picard, while occupied on the restoration of the toise du Châtelet was also working along the same lines and suggested that a universal foot should be ⅓ of the length of a seconds pendulum or 'rayon astronomique' at Paris Observatory. This was 36 pouce 8.5 lines in length where the Paris pouce (inch) was equivalent to about 1.065 present inches.

The toise had become some 5 lines too long through usage and wear. A similar pendulum suggestion came in 1673 from the Dutch astronomer Huygens.

The scent went cold for some 50 years until Cassini II in 1720 reverted to the notion of using the length of a minute of arc by suggesting that a geodetic foot should be 1/6000 of a terrestrial minute of arc. This would have been very close to the approximation used in some present-day calculations of 1 second ≙ 100 ft.

As a result of the Peru expedition La Condamine recommended in 1747 that the equatorial seconds pendulum should be used to define a standard. This he had found to be 439.15 Paris lines. He was emphatic in saying that it should be very much an international unit and that perhaps a group of academics from various countries should investigate the problem.

In 1789 a conference of delegates from numerous French towns petitioned Louis XVI for a unification of the system of measures. 1790 saw the proposal for a new unit to be called the metre and to be based yet again, on the seconds pendulum.

In 1791 the National Convention of France decided the time had come to finally settle on a standard unit of measurement. The pendulum was eliminated since it had been shown to vary from place to place. Instead they decided upon the length of a meridian quadrant through the Paris Observatory. $^1/_{10\,M}$ th of the quadrant to be metre.

It was agreed that Delambre and Mechain should measure an arc from Dunkirk to Mountjouy over some 9½° but the time for such work was rather inopportune as it was much delayed by the French Revolution.

When the results became available the assumption was made that the earth was an exact spheroid and that its ellipticity, dimensions and quadrant length should be determined from this new arc together with that of Bouguer and La Condamine in Peru. The result gave the ellipticity as $\frac{1}{304}$ and the metre as 443.296 lines.

The data they worked on to achieve this was:
For France:
551,584.7t; amp. 9° 40′ 25″; mid. latitude 46° 11′ 58″
For Peru:
176,873t;   amp. 3° 07′ 01″; mid. latitude 1° 31′ 00″
The approximate solution can be found thus:

If 2A and 2B are the sum and difference of the semi-axes then neglecting powers of B/A:
$$R = A - 3B.\cos 2\phi$$
Integrating from 0 to $\pi/2$ the length of the quadrant:
$$Q = A.\pi/2$$
From the radii at two latitudes ϕ_1 and ϕ_2 there are two equations from which B can be eliminated. Then
$$A = \Sigma + \Delta \cot \sigma \cot \delta$$
where 2Σ and 2Δ are the sum and difference of the radii and σ and δ are the sum and difference of the mean latiudes. Then from the data the curvaures of centres of the two arcs are

3 266,978 and 3 251,285

Note: The values quoted here are from [55] but recalculation produces slight differences.

Whence Σ = 3 259 131 and Δ = 7 846
From this A = 3 266 345
and multiplying by $\pi/2$ and dividing by 10M gives the metre = 0.513 0766 parts of the toise du Perou. Since the toise = 864 lines, the metre became 443.296 lines.

The metric system was legalized in France on 7 April 1795 and resulted in the expression 'Legal metre'. It must be remembered, however, that the metre was at first only defined as a certain proportion of the Toise du Perou. It was first given a material form in 1798 as a platinum bar, *Metre des Archives* by Borda. This then immediately took the earth quadrant out of consideration for the future. When manufactured in 1799 it was to about an order of precision better than that with which the French arc was measured.

For the eccentricity of the earth figure from two arcs Bouguer, [163] gave:
Let the ratio of the radii be as m: 1
Let the lengths of the arcs for 1° be E and F
Let the latitude of these be ϕ_E and ϕ_F

$$\text{Then } (1 - m^2) = \frac{2 (E - F)}{3 (E.\sin^2 \phi_E - F.\sin^2 \phi_F)}$$

Using the data for the computation of the metre

$$(1 - m^2) = \frac{2(273.89)}{3(29703.07 - 167.77)}$$

$$1 - m^2 = 0.006\ 182$$
$$m^2 = 0.993\ 818$$
$$m = 0.996\ 904 \quad \text{or 1 in 323}$$

Airy [16] was slightly different and used:-

$$e = \frac{(E - F)}{3b (\sin^2 \phi_{E_m} - \sin^2 \phi_{F_m})}$$

Where ϕ_{E_m}, ϕ_{F_m} are the mean latitudes for the arcs and b is approximated to E or F.

$$\text{Then } e = 0.003\ 078 \quad \text{or} \quad \text{1 in 325}$$

These are suitably close together except that [38] quotes the adoption of 1 in 334 and [6] quotes 1 in 304. This could well have arisen because the French arc could be divided into 4 sections and there is a suggestion that the mean result was accepted.

The 334 is in fact the mean of comparing each of the 4 French sections with that of Peru but since they were very widely spread it could not be described as a very reliable mean. 323 (or 325) is from using the overall arc rather than individual sections.

References See in particular: 2, 38, 55, 98, 163, 172, 175, 224, 233, 238, 252 and 260.

Chapter Nineteen

THE CONTROVERSY

As has been shown in the early chapters, ideas on the shape of the earth passed through many stages before there was recognition, or some measure of agreement, that it was spherical. Any refinement on such ideas was not at all feasible without improved methods, better instrumentation and universal travel.

These developments were slow and the first circumnavigation of the earth was not until Magellan in 1521. Already however lack of precise knowledge on the size and shape of the earth had caused problems. As far as Christopher Columbus knew, the earth was far smaller than it actually was and hence he thought that distant continents were much nearer than they actually were.

Magellan however paved the way for explorers to begin taking scientific measurements at an increasingly wide range of places. There was also a tendency to be very versatile with any one explorer observing and recording all manner of phenomena.

For example, when Richer was sent to Cayenne on his famous voyage of 1671 his brief was not solely the length of the seconds pendulum. In fact that might almost seem just a sideline when placed against the wide range of observations he was charged with making of various planets and stars; of the atmospheric pressure experiments and observation of tidal changes. Could such a program really be conducive to the highest standards of results?

Nevertheless he was able to take sufficient pendulum readings to return to Paris convinced that in Cayenne the seconds pendulum required to be 1¼ lines shorter than in Paris since his clock lost 2 minutes 28 seconds a day. This change was attributed to the fact that although Cayenne was at sea level, it must be further from the centre of the earth and hence the force of gravity less.

Now while this could not have pleased Cassini since it was contrary to his ideas, neither did it entirely please Huygens, who was a fan of Newton's since his calculations said it ought only to be ⅚ lines shorter. Remember that 1 line = $\frac{1}{144}$ ft. or about 2¼ mm so that the difference between the two was only of the order of 0.4 mm at a time when measurement techniques left much to be desired.

Consider that the sort of method used was for the pendulum to consist of a small weight suspended by a fibre (or pite of aloe). The length of this was measured from the lower edge of the suspension point to the upper surface of the weight. Of course the amount of change need not be measured directly but could be derived by calculation from the time loss. Very approximately this could be found from:

$$
\begin{aligned}
\ell &= 2g.t^2/\pi^2 &&= k.t^2 \\
\text{Then} \quad d\ell &= 2k.t.dt \\
2^m 28^s \text{ per day} \quad &= 148/3600 \times 24 &&= 1/600 \\
\text{Then} \quad d\ell &= 2 \times 2g/600.\,\pi^2 \\
\text{But} \quad g &= \ell.\pi^2/2 \\
d\ell &= 4 \times \ell\pi^2/(1200\,\pi^2) \\
&= \ell/300 \\
&= 440/300 \\
&= 1.47 \text{ lines}
\end{aligned}
$$

Applying further small corrections such as for temperature, would give the 1.25 lines.

War was declared between the Newtonians and the Cassinians. As was to be expected many French scientists and mathematicians sprang to the defense of Cassini on one side of the English Channel while Newton had similar defenders on the other side. For those who had a use for knowledge of the correct shape, particularly the seafarers and astronomers, the problem was not to be resolved for many years.

Maupertuis is quoted as saying of the work of the Cassini family—"Ces mesres, furent répétées par M. Cassini en différens tems, en différens lieux, avec différens instruments, et par différentes méthods, le Gouvernement y prodigua toute la depense et toute la protection imaginable pendent l'espace de 36 ans et le résultat de six operations faites en 1701, 1713, 1718, 1733, 1734 et 1735, fut toujours que la Terre étoit allongée et non applatie vers les Poles".

Roughly translated this reads, "These measurements, he said, were repeated by M.Cassini at different times, in different places, with different instruments and by different methods; the Government supplied all the expenses and all the protection imaginable during a 36 year period and the result of 6 operations made in 1701, 1713, 1718, 1733, 1734 and 1735 was always that the earth was elongated and not flattened towards the Poles".

Cassini I was not simply interested in the earth, he made numerous measurements and observations to the stars and planets and during this work he found that Jupiter appeared to have a considerable polar flattening of the order of 15 : 16. Was there not an obvious clue here to what he ought to find for the earth? If so, he did not find it.

Other evidence was being produced by yet more French scientists taking their pendulums to far parts of the globe and all returning with the same conclusion—the nearer the equator the shorter the pendulum required to beat seconds.

Newton working in the 1670s and 1680s with what information was available at the time theorized on the properties of fluids in relation to rotating bodies. Unfortunately his base data was both scant and of limited accuracy; the best earth measurement at that time, as far as he was aware, was 60 miles to 1°, whereas it should have been close to 69 miles, and he had only the early pendulum values.

He was perplexed at the discordance between his theory and these available facts. Then in 1682, at the Royal Society, the French inadvertently helped Newton by providing the results of Picard's arc. Immediately he had a completely different earth figure to work with and his jigsaw began to fall into place and his blockbuster was ready for launching at the Cassinis. The *Principia* was printed in 1687.

Before any further advance could be made towards settling the controversy methods and instrumentation had to be improved. As was to be seen much later, so small was the amount of flattening that the instrumental errors could, when expanded over a degree of arc, give a completely false impression.

In 1701 Cassini reported to the Royal Academy of Sciences that by means of triangles he went from Paris to the high Pyrenees which separate Roussillon from Catalonia. He observed that as he went south so one degree surpassed another by 1/800th part which may give great reason to doubt the exact roundness of the earth. He reported further however that Huygens and Newton held that the earth was flattened like a Dutch cheese. A mathematician from Strasbourg, Eisenschmid, held that it was elliptical like an egg, lengthened towards the poles. Cassini left the question unanswered.

In 1718 Cassini II published his tome [44] on the length of meridian arcs in France as the result of intermittent work over 35 years. Surely large enough and accurate enough to settle the argument once and for all? But no, the indication was that in France the degree still decreased away from the equator by some 11 or 12 t per degree.

Maupertuis neatly summarized the situation in [160] and it is no wonder that Cassini felt he was correct. Thus:

1701	Cassini I and Cassini II measured the arc from Paris to Collioure. The arc was 6° 18', amplitude 360,939.6 t and 1° = 57,292 t. They used a 3 ft. quadrant at Paris and 10 ft. sector at Collioure for zenith distances of Capella. It was suggested that this could account for up to 1' of error. Result that the degree of Midi de Paris was greater than that of Picard further north i.e. elongation.
1713	Cassini II reobserved part of the same arc. The same equipment was used but the results gave a smaller elongation.
1718	Cassini II extended the arc from Collioure to Dunkirk. The arc was 6° 18' 57" with amplitude 360,614 t and 1° = 57,097 t. For the extension of 2° 12' 09.5" the amplitude was 125,454 t or 1° = 56,960 t. i.e. still elongation.
1733	Cassini II measured an arc of a parallel from Paris to St Malo. The chain was of 44 triangles and the base of verification agreed to 1 t. The arc was of 4° 30' 00" in longitude, amplitude 165,015 t or 1° = 36,670 t. This proved to be smaller than it would have been for a sphere. i.e. elongation.
1734	Cassini II extended the parallel to Strasbourg. The extension was for a further 29 triangles. The base of verification agreed to 5 ft. The arc was 5°32'45" with amplitude 205. 100t or 1° = 37, 066 t. This again was smaller than if the earth had been a sphere; this time by 680 t as against 1037 t for the initial parallel. i.e. elongation.

Who, faced with such a mass of evidence that was all in agreement would consider himself to be wrong and that the earth was really flattened?

While the Cassini results predictably drew more to the French cause there were yet others who were still spherical adherents. An example of this was the English mathematician James Stirling, who wrote to Bradley in 1733 that he could prove from his astronomical observations that Cassini was very near the truth since such were surer for determining the diameter than by any actual mensuration. Yet he was still obliged to suppose the figure to be an exact spheroid. In the same letter he endeavoured to find reasons to dispute the results of the pendulum observations in Cayenne and Jamaica.

In reply Bradley stressed that he could find no reason to explain all the difference of the clock in Jamaica of $1^m 28^s$ per day, other than that it must be ascribed to the inequality of the density of different parts of the earth.

By the 1730s there was a clutch of rising mathematicians and scientists at the helm of French scientific activity and among these was one Charles-Marie de La Condamine who was widely travelled and had a deep understanding of many branches of science. In 1733 he put forward the suggestion that an expedition should be mounted to measure an arc of the meridian as near to the equator as possible. This would place it as far as practicable away from the French arcs and thus provide a sound basis for deciding once and for all the shape of the earth. However, as he was only a junior member of the Academy the idea was not able to progress until taken up by a senior Academician. The following year the younger, yet more senior, Louis Godin pushed the idea forward and it was accepted by the French Academy of Sciences under the patronage of Louis XV.

A group of Academicians and others were gathered together, suitably equipped and despatched to Peru. One unanswered question lurks. Why was it that neither of the Cassinis went on either expedition? Cassini II was admittedly 58 but his son Cassini III was 21. Even if inexperienced, one would have expected the family to wish to be represented. It would even appear that Cassini featured little in the discussions that led to the expeditions.

At about the same time as Bouguer and his colleagues were going to Peru another French scientist, Pierre Louis Maupertuis, capped the idea by suggesting that if an expedition was going as near the equator as possible then why not send another one as near to the pole as possible. Surely this was the soundest way of all to obtain convincing proof? He persuaded the authorities and led a party to Lapland the following year.

After his return, the Lapland expedition took only 18 months compared with the 9 years for Peru, he wrote on 27 September 1737 to Bradley. Initially Maupertuis outlined his results, then wrote: ''Thus, Sir, you see the earth is oblate according to the actual measurements as it has been already found by the Laws of Statics, and this flatness appears even more considerable than Sir Isaac Newton thought it. I am likewise of the opinion that gravity increases more towards the Pole, and diminishes more towards the equator than Sir Isaac supposed.''

Voltaire referred to Maupertuis as the flattener of the earth and the Cassinis and the Marquise du Châtelet called him Sir Isaac Maupertuis but the Cassinis were not yet finished.

There was acrimony between Maupertuis and Celsius on the one hand and Cassini on the other. Cassini tried to raise doubts on the accuracy of the Lapland work and the integrity of the members of the expedition. This stung Celsius into action and he indicated that while Cassini had highlighted various uncertainties in Picard's work, as well as in that of Snellius and Riccioli, he had then in his own work had 16 triangles that had one or more of the same inconsistencies. He went so far as to indicate that if Cassini's triangles were corrected the length from Paris to Collioure would be 358,980t for 6° 19′ 11″ instead of 360,614t for 6° 18′ 57″. This would change the degree from 57,097 to 56803t.

The final capitulation came in 1744 when Cassini published [45] his results in 'La meridienne verifiee' for the arc through Paris—although it was said by Delambre that it was entirely executed, calculated and written up by LaCaille. In so doing he highlighted errors in the previous French arcs. The earth was now indeed flattened and the belated return of the Peruvian expedition confirmed this and within 50 years led to the designation of the metre.

To refer to two early quotations during the controversy: ''Cassini says, that the earth is an oblong spheroid, higher at the poles than the equator, making the axis longer than a diameter of the equator about 13 French leagues, which he deduces from comparing his father's measures of the meridian from Paris to the Pyrenean Mountains with those of Picard. But having afterwards continued the meridian, which is drawn through France, from Paris to Dunkirk, he still draws consequences to prove the earth an oblong spheroid; but then he makes the axis exceed the equatorial diameter 34 leagues''.

''Sir Isaac Newton makes the earth higher at the equator, and consequently flattened towards the poles, reckoning its equatorial diameter 34 English miles longer than the axis; which he proves from the principles of gravity, and the centrifugal force arising from the diurnal rotation of the earth; and to confirm this mentions several experiments on pendulums, which have been made shorter to swing seconds near the equator than in greater latitudes''.

J.T. Desaguliers. Phil.Trans.386.[72] NB. 1 French league = 3000t = approximately 3.4 English miles.

References. See in particular:
1, 2, 6, 20, 22, 44, 45, 47, 72, 78, 98, 113, 116, 160, 252, 256.

Chapter 20

THE EXPEDITION TO PERU

Before describing the work of the expedition it will be appropriate to have some details of the main cast.

Louis Godin
Born 28 February 1704 in Paris
Died 11 September 1760 in Cadiz

His father François Godin, was a lawyer in the Parlement and his mother was Elisabeth Charron. Following an early education at the College de Beauvais he went first to humanities but later turned to philosophy and then astronomy.

By the early age of 21 he was a member of the French Academy of Sciences. There he busied himself with editing the early volumes of the Memoires of the Academy and wrote some of its history. Besides various activities in astronomy he wrote a paper in 1733 that discussed the relation between parallels of latitude and the changing relationships between them for different shapes of the earth—from elongated to flattened.

Maybe it was this interest that led him to suggest that the Academy should set out to resolve the differing opinions once and for all by organizing an expedition to the Equator. When the idea was accepted Godin was among those asked to go and as the senior Academician of the party was named its leader. As will be seen later, however, he left most of the decision making and organization to Bouguer and La Condamine.

He never published a report of his work on the expedition and stayed on in Peru until 1751 as Professor of mathematics. at the university of San Marcos. After a short stay in Paris he went to Spain as Director of the Academy of Naval Guards in Cadiz.

Before going to Peru he had married Rose Angelique de Moyne in 1728 and their son and daughter both died before Louis. He died of apoplexy while still in Cadiz.

Pierre Bouguer
Born 16 February 1698 in Croisic, Lower Brittany
Died 15 August 1758 in Paris

Pierre was the son of Jean Bouguer, the Regius Professor of hydrography at Croisic. He was an infant prodigy and by 1730 had become Professor of hydrography at Le Havre after first succeeding his father as Professor at Croisic. In 1731 he became associate geometrician of the Academy of Sciences and in 1735 a full academician.

He invented a heliometer to measure the angular diameter of the sun and constructed tables of atmospheric refraction based on his own theory. This theory connecting pressure with elevation worked well in the high Andes but was not so good elsewhere.

He determined the length of a seconds pendulum at various locations and elevations and was one of the first to use a knife edge support for improving the accuracy of locating the point of swing. Between Quito and sea level he compared his theoretical value for the change of g from the known difference of elevation with the measured difference and found a variation of 1/6983 which he thought might be due to the attractive force of the mountains. This led to attempts to measure the effect of the 20,000 ft. Mount Chimborazo on deflecting the plumb line.

He later became known as the father of photometry for the instrument he developed.

He was the first to use the symbols $\geq$ and $\leq$ and made an improved version of the early quadrant.

See the references for some of his publications.

MESURE
DES
TROIS PREMIERS DEGRÉS
DU MÉRIDIEN
DANS L'HÉMISPHERE AUSTRAL,

Tirée des Observations de M.^{rs} de l'Académie Royale des Sciences, Envoyés par le Roi sous l'Équateur:

Par M. DE LA CONDAMINE.

Fuit alter
Defcripfit radio medium *qui gentibus Orbem.* Virgil.

A PARIS,
DE L'IMPRIMERIE ROYALE.
M. DCCLI.

9. *Title page of "Mesure des Trois Premiers Degre's" as written by De la Condamine in 1751.*

10.   Charles Marie de La Condamine.

Charles Marie de La Condamine
Born 28 January 1701 in Paris
Died 4 February 1774 in Paris

He was born into a wealthy family where his father was a district collector of taxes who had married late in life to a wife half his age. He took an avid interest in all things scientific but particularly mathematics, surveying and all aspects of nature.

In 1730 he became a member of the Academy of Sciences.

The expedition to Peru kept him away from Paris from 1735 to 1745. He did much of the organization for the well being of the expedition and sorted out many of its difficulties, not the least of which were lack of finances and murder of the surgeon with the party.

To round off the expedition he took the hard route home by travelling down the Amazon, mapping and studying every possible aspect of life on the way.

In August 1756 he married his niece Charlotte Bouzier d'Estouilly. She was some 30 years younger than he, so in this he was following the precedent of his father.

In 1757 he paid a visit to Italy to follow his interest in the length of the Roman foot.

Late 1760 saw him as a member of the Academie Française. He was keen on establishing an international standard of length and thought the seconds pendulum might well be the means of achieving it. Unfortunately he died some

20 years before the appearance of the metre.

His death in 1774 was caused by an abcess of the liver.

See the references for details of his publications.

Jorge Juan y Santacilia
Born 5 January 1713 Novelda (Nobelda), Alicante
Died 21 July 1773 Madrid

His parents were Bernardo Juan y Cancia and Violante Santacilia y Soler but he was orphaned when 3 years old. An uncle took him under his wing and was his teacher for some while, then at the age of 12 he was sent to Malta for his education. While there he enrolled in the Order of Malta. By the time of his return to Spain at age 16 he was well versed in many aspects of mathematics and astronomy.

A few months after his return he joined the Guardias Marinas, an elite naval organization for young noblemen. At age 17 he enlisted as a midshipman in the Spanish navy where he served particularly in the Mediterranean. He later rose to captain. His particular interest in all aspects of science led to, or at least helped in, his appointment to the Peru expedition. Albeit, there is some suggestion that because of his youth, and that of Ulloa, there must have been some influence used in high eircles.

He was in S.America from 1735 to 1745 and then spent some while writing up his results. It was on his return that he was elected a member of the Royal Society and joined

97

11.   Jorge Juan.

12. *Antonio de Ulloa.*

Note the plane table, compass and mounted sextant.*

the Academy of Sciences, the Royal Academy of Science in Berlin, and the Spanish Academy of San Fernando.

His first assignment then was to spend 18 months in England studying shipbuilding techniques which he was later able to introduce in Spain. By now he had a wide range of specialist expertise and travelled throughout Spain on a variety of assignments. This hectic life led to a breakdown of his health in 1761 which was aggravated again in 1763 before he had fully recuperated.

In February 1767 he became Spanish Ambassador to Morocco for some 6 months. On his return to Spain he was appointed head of the Royal Seminary of Nobles.

He never married and died in Madrid of apoplexy.

See the list of references for his publications.

Antonio de Ulloa y de la Torre Giral
Born 12 January 1716 at Seville
Died 5 July 1795 at Isla de Léon, Cadiz

The son of Bernardo de Ulloa, an economist, he was not a healthy child. He was, however, given a good grounding in the science subjects while at college in Seville.

By the age of 14 he was at sea and visiting various parts of America and the Indies. As Juan, he became a member of the Guardias Marinas in 1733 prior to a spell in the Mediterranean. He rose to rank of Captain. In 1735 he was elected to go on the Peru expedition as a substitute for José del Postigo who, although initially invited, was unable to go. For this he was given the rank of lieutenant, as was Juan.

On his return journey from Peru in 1745, he was captured

by the English and spent some time as a prisoner in Louisburg and later in England. He was well treated and mixed with many eminent scientists. While in London he was made a fellow of the Royal Society. He finally had his papers returned to him and was sent back to Spain in 1746.

After completing his writings relating to the expedition he was sent to various parts of Europe on commercial and military assignments. Following his return to Madrid in 1752 he rejoined the Guardias in Cadiz until 1757. While there he was instrumental in the establishment of an observatory in 1754 and Royal Natural History collection in 1752. He became a member of the Scientific Academies of Bologna, Paris, Berlin and Stockholm. In 1757 he was appointed Governor of Huancavelica in Peru and arrived there on 2 November 1758. In particular he was charged with supervision of a mercury mine. It was not an easy task with widespread corruption and bitter feuds. By mid 1762 he had had enough but it was near the end of 1764 before he persuaded his superiors to relieve him of the post. Unfortunately for him he was soon pitched into a further unhappy situation as Governor of the newly acquired Spanish territory of Louisiana. While there he married Francisca Ramirez de Laredo in 1767 and they had nine children. He was stuck in America from 1766 to 1768 and then managed to return to the navy as an admiral.

In 1779 he was charged with loss of a Spanish ship, some say by virtue of his obsession for science taking over his official duties, although there is no proof of this. Luckily he was acquitted. He continued in various naval roles until his death.

See list of references for his publications.

Joseph de Jussieu
Born 3 September 1704 at Lyons
Died 11 April 1779 at Paris

From a scientific family, he chose medicine while his brothers Antoine and Bernard were botanists. He became Docteur-Regent de la Faculte de Paris before being asked to join the Peru expedition as the doctor and botanist.

After the rest of the party started their return to Europe Joseph became ill and was without resources. He had to apply his profession to save for his return fare. When almost in a position to leave, the outbreak of an epidemic of smallpox forced him to stay in Quito.

When the situation returned to normal he travelled to Lima where he stayed a while with Godin. From there he made various scientific expeditions into the Andes. By mid 1749 he was at Potosi after crossing and recrossing the Andes.

With failing memory he returned to Paris in 1771. While he was in S.America he was made an Academician in his absence.

The idea for an expedition to Peru was first mooted by Louis Godin to Louis XV after a suggestion in 1733 by Charles-Marie de La Condamine, that it would be useful to measure a degree of the meridian elsewhere than in France. The initial suggestion of La Condamine lay dormant because of his junior status within the Academy. Godin on the other hand although younger, was a senior Academician and as such his proposal, of the same idea, in 1734 carried much more weight and began to make progress.

The deliberations of the French Academy of Sciences at the time were under the guidance of M.Bernard Le Bovier de Fontenelle, its Permanent Secretary. He was an elderly 77 but in his element with the Academy.

In the ensuing enthusiasm the proposals of Godin were latched on to by Grandjean de Fouchy and La Condamine although the former of these was unable to go on the expedition. Doubtful health on such a strenuous trip would have hampered the remainder of the party considerably so the first choice leader was destined to stay at home. The next choice was Louis Godin whose push had got the scheme under way and he became nominal leader with Pierre Bouguer as next in command, although in the event most of the organization fell to La Condamine. Bouguer initially had not had any thoughts of being at the forefront of the party but several eminent mathematicians and astronomers who had been expected to join the group withdrew for a variety of reasons. Did they have some premonition of the difficulties and duration of the venture?

Thus Bouguer and La Condamine were virtually joint leaders although quite different characters. Bouguer was always engrossed in his work and his calculations and was by far the better mathematician. He made remarkable strides in attraction of the plumb line, isostasy and related topics. La Condamine on the other hand was well versed in a wide range of scientific disciplines, indefatigable energy and intense curiosity. He even opened a trading post, or boutique, in Quito for selling silk and Holland lace.

In fact it was a rule in the French Academy that only the leader of an expedition could write up the scientific results. It is recorded that as Godin delayed his return to France it fell to Bouguer to do the reporting. However, La Condamine of his own volition, and later Juan with Ulloa, also wrote up both the scientific and other aspects of the long sojourn below the equator. Together they each complement the others in filling in the picture. Thus it would appear that the heirarchy was Godin, Bouguer and La Condamine in that order.

Instruments were obtained from the best makers of the period—George Graham in England and Claude Langlois in France, and other stores gathered together at La Rochelle with a view to departure in May 1735.

The party consisted of Louis Godin, Pierre Bouguer, Charles-Marie de La Condamine, Jean Godin des Odonais, Captain Verguin, M.De Morainville, Joseph de Jussieu, Dr. Jean Seniergues, M. Hugot, M.Mabillon and M.Couplet. Of these Verguin was a marine engineer and draftsman, de Morainville a draftsman and natural historian, de Jussieu a botanist and doctor, Seniergues a physician and surgeon, Hugot a watchmaker and instrument technician, Couplet was a newphew of the treasurer of the Academy and des Odonais was a cousin of Louis Godin.

Arrangements were also made for these to be joined in Carthagena by two Spanish naval officers. This was a political decision since Peru, at that time, was under the domination of Spain and cooperation was required from His Catholic Majesty Phillip V.

His Majesty was suitably impressed and after a favourable report from the Council of the Indies the license was granted and the patents made out on the 14th and 20th of August

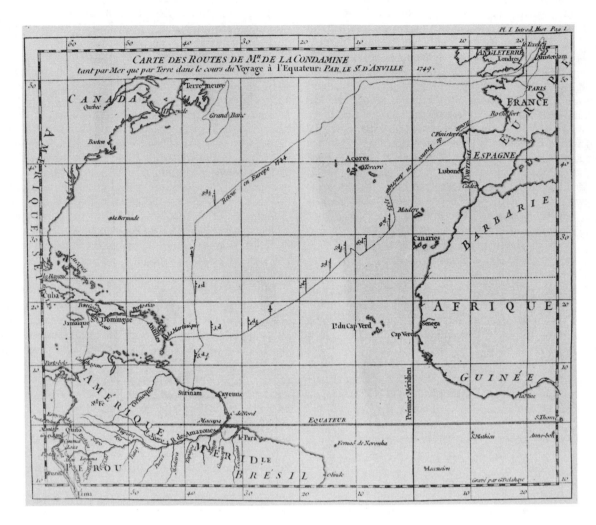

Pl. I Introd. Hist Pag 1.

CARTE DES ROUTES DE Mr DE LA CONDAMINE
tant par Mer que par Terre dans le cours du Voyage à l'Equateur: Par le Sr D'ANVILLE 1749.

14. *Map of the routes between France and Peru indicating the changing values of magnetic declination. Note that the prime meridian is through Pico Tenerife.*

1734. These contained very precise orders to the various viceroys and governors to give all required aid and assistance and to ensure that they were not overcharged for any of their needs.

In putting together the expedition the party had the assistance of M. Le Comte de Maurepas who proved adept at smoothing out difficulties. His love of the sciences and their advancement spurred him on.

La Condamine left Paris in 14 April 1735 for La Rochelle and they all embarked in one of the King's ships under Captain de Ricour and were ready to sail from La Rochelle on 16 May 1735.

Their voyage was past the Canary Islands which were visible on June 2nd and after 37 days sailing they arrived at Martinique during the night of the 22nd of June 1735. In the approach to Martinique and Dominica they came across a stretch of water of a particular white colour that distinguished it from the rest of the ocean. It was estimated to end some

100 leagues from Martinique and to cover some 40 leagues.

The few days spent in Martinique were not wasted and a variety of scientific observations were made.

Next port of call was St. Domingo where there was some delay to progress. This was waiting for permission to proceeded and stretched to some 3 months. A particularly inhospitable place to be stranded and there were various attacks of fever. Again the time was spent making numerous observations.

It was October 31, 1735 before they sailed from Petit Goave, on St. Domingo, to reach Carthagena in Colombia on the 16th of November.

The two Spanish sailors had been awaiting their arrival for some months. The chosen sailors were Don Jorge Juan, commander of Aliaga, of the Order of Malta and sub-brigadier in the Guardas Marinas, together with Don Antonio de Ulloa, a captain in the Spanish Navy.

They had both been given commissions as lieutenants of

102

men-of-war the Conquistador of 64 guns and the Incendio of 50 guns. Juan embarked on the Conquistador that was commanded by Don Francisco de Liano and Ulloa was on the Incendio, commanded by Don Augustin de Iturriaga. They initially sailed from Cadiz on the 26th of May but adverse winds entailed anchoring at Las Puercas before finally leaving on the 28th of May.

They had kept themselves occupied on the voyage with various observations among which were longitudes which they found as: Cadiz to Pico Tenerife - 10° 30′. But as Cadiz was said to be 8° 27′ from Paris Observatory and the Pico 18° 51′ from Paris so there was a variation of some 6′ against Cadiz to Pico of 10° 24′. Cadiz to Martinique was found as 59° 55′. That was some 4° at variance with the value accepted on charts at that time. Much of this difference was thought to have been due to uncertainties in the distance apart of marks on the log line. It is said that assuming the glass ran for exactly 30 seconds so the knots should have been 1/120 mile apart. The problem was to define the true length of the mile. Using Cassini's degree value of 57,060 toise a geographical mile would be 951 t or 5706 royal feet to give 1/120 as 47 ft 6½ inches. But the Paris foot varied against the London foot as 16 to 15 to make the required interval 50 ft. 8¼ inches (English).

At later dates the ratio was given variously as 864 to 811 and 144 to 135—this last equates to 16 to 15.

July the 3rd, 1735 they left Martinique for Curaçao and Uruba. By the 5th they had sight of the snow covered heights of St. Martha and were soon crossing the issue of the river de la Magdalena which stretched with prodigious rapidity several leagues from the coast. On the 8th they arrived in Carthagena Bay but could not secure until the following day. Comment is specifically made of the severity and suddenness of tornadoes and squalls and the need to rapidly prepare in readiness.

A particular point was made of recording the variation of the magnetic needle during the voyage. Both Juan and Ulloa took separate observations which all differed from the existing charts by 1° to 3½°.

On arrival at Carthagena they sought out the Governor and learnt that there was as yet no knowledge of the Academicians. Short of instruments of their own they were luckily able to find some in the city that had belonged to the local engineer Brigadier Don Juan de Herrera. They had to occupy some 4 months of waiting and spent it on determining latitudes, longitudes, needle variations and up dating plans of the city and bay.

Their impatience was rewarded on November 15th when a French man-of-war anchored under Boca Chica and they had time to repeat some of their observations with the newly arrived instruments. In summary their results were;

```
Carthagena.   Latitude    10° 25′ 48½″ N
              Longitude  282  28  36     from Paris
         or                301  19  36     from Pico Tenerife.
              Variation    8° E
```

Monte de la Popa was 84 t high. On its summit was a convent of bare-foot Augustines called Nuestra Senora de la Popa.

On 24th November the naval officers joined the Academicians together with a dozen or so maîtres and 14 domestics, on the French frigate and set sail on the 25th of November for Porto Bello where they anchored on the 29th of November at 4° 24′ longitude from Carthagena. i.e. 278° 04′ 36″ from Paris.

They observed the town of St. Philip de Porto Bello as 9° 34′ 35″ N and 277° 50′ longitude from Paris or 296° 41′ longitude from the Pico of Tenerife.

The modern equivalents to these are:

Carthagena	Latitude	10°	25′	N	longitude	282°	07′	E from Paris
Tenerife		28	20			341	00	
Porto Bello		9	35			277	58	
Panama		9	00			278	15	
Montecristi		1	00	S		277	00	

Porto Bello was the harbour discovered by Columbus on the 2nd of November, 1502 and the entrepôt for commerce for Europe.

The climate of the area was notorious for destroying the vigour of nature and of bringing life to an abrupt end.

On the 4th of December, 1735 at about 6 am the Réaumur thermometer stood at 1021, and at noon was 1023. The excessive heat was aggravated by the surrounding high mountains. The rains were sudden, tempestuous and thundrous. So inhospitable was Porto Bello that a large percentage of any ship's crew arriving there were buried before the ship was able to depart. Every possible version of fever and the like seemed to be there for the taking.

There was a choice of two routes from Carthagena to Quito - one due south up the River Magdalena for some 400 miles to the foothills of the Andes and then a further 500 miles by mule over the mountains; the other meant crossing the isthmus of Panama and sailing down the S. American coast and working inland from there. They chose the latter particularly from the point of view of carriage of the equipment which they were loathe to hump over the Andes.

All were anxious to leave such an inclement area at the earliest opportunity and on the 22nd of December 1735, embarked in two dugout vessels sent the President of Panama, Don Dionysio Martinez de la Vega, to travel the River Chagre. This river is close to where one now finds the Panama canal and even forms part of it. Formerly called the River Lagartos, its source is in the mountains near Cruces.

This form of travel was necessary because of the impracticability of travelling by land on the narrow, steep, rough tracks that then existed, with such instruments.

From Porto Bello they were able to go under sail to the mouth of the Chagre and from there it was necessary to row, or often pole, their way. The current was very fast and they measured it at velocities varying from 10 t 1 ft in 40½ seconds to 10 t in 14½ seconds. When they reached Cruces in the centre of the Isthmus it was still 10 t in 16 seconds. These are velocities up to almost 3 miles per hour. Cruces was some 43 miles from the river mouth by water and about 20 t wide at the town. The river was well known for its caymanes or alligators.

The remainder of the Isthmus to Panama was a seven hour trek on 29th December 1735. As in many other cities en route, the first call on reaching Panama was to the President.

While waiting to proceed the spare time was used for pendulum latitude observations and mapping of the locality. They found the latitude of Panama as $8° 57' 48½ ''$ N. Longitude remained doubtful; French geographers put it at East of Porto Bello while the Spanish charts put it to the West. By today's reckoning Panama is East of Porto Bello by some $17'$ while Juan and Ulloa deduced (not measured) it to be about $30'$ West. What confusion! In Panama on 5-6 January 1736 the thermometer at 6 am was at 1020½ and by 3 pm at 1025 Réaumer.

After more delays they finally embarked on the St. Christopher under Captain Don Juan Manuel Morel and sailed on 22 February 1736 towards Guayaquil. Sailing almost due south they anchored in Manta Bay on the 9th of March just south of the Equator. On going ashore and to the village of Montecristi it became obvious that the territory was of no use for survey such as required for arc measurement. There had been an original hope that the coast might be utilized rather than proceeding inland. The latitude of the anchorage was observed as $0° 56' 05½ ''$ S.

and all was ready for further local observations.

Although the season was not conducive to travelling inland the local corregidor contacted his equivalents in the neighbouring areas to obtain transport for the party to the mountains. On the 3rd of April at 3 pm, the thermometer stood at 1027. So numerous were the unpleasant insects that they would extinguish a candle in a matter of minutes. Snakes, scorpions and other such creatures abounded.

As soon as the group heard that mules sent by the corregidor of Guaranda were on their way to Caracol, they sailed on the 3rd of May up the river to that town, taking until the 11th to arrive. The mosquitos were so incessantly troublesome and the special clothes and nets so insufficient that all were swollen and covered with enormous bites.

The mules were readied for departure on the 14th of May. In the meantime Bouguer and La Condamine were lodged in the King's house (no more than a bamboo structure some 7-8 feet up on piles) in Montecristi. They set up an observatory a mile or so from the village and were able to observe the end of an eclipse of the moon on the 26th of March. They deduced from it that Porto Bello was about $54'$ West of the meridian of Panama.

Bouguer sought to observe astronomical refraction near

15. *Inscription left at the Equator in 1736.*

Messrs Bouguer and La Condamine stayed at Montecristi to extend their observations for latitude, longitude and of the pendulum. On the 13th of March the rest of the party weighed anchor again and on the 23rd anchored off the island of Puna in the bay of Guayaquil. The city was reached on the 25th by row boat. The instruments were off-loaded by the 26th

the horizon at the mouth of the River Jama. Combined with his previous observations at St. Domingo he decided that refraction reduced with height above sea level. In particular, on the 13th of April he observed two suns, very distinct, which set successively in contact with each other. The lower sun was the less bright but its edge not less determined.

104

In this very hot area the thermometer in the afternoon rose to 28 °R and was around 20 °R before sunrise. (The definitive references use Réaumur values both with and without the 1000). The lassitude of the body in the heat communicated itself to the mind, and a state of indolence that unfitted both for anything needing of application or attention.

They determined where the Equator crossed the coast and carved on a rock 'Observationibus Astronomicus Regiae Paris Scientiar Academae Hocce Promontorium Aequatori subjacere compertum est, 1736' at a place called Palmar.

An inland journey was made to Puerto-viejo and to Charapoto for further observations before Bouguer decided on the 23rd of April to travel south and join others at Guayaquil. La Condamine stayed to continue his mapping of the area before going direct to Quito, hopefully by the Rio Esmeraldas.

Bouguer's return to Guyaquil was hazardous, with his horse often up to its knees in watery morass under an almost inpenetrable forest cover.

In the meantime Juan and the others were making their way along the banks of the river Ojiva (Ojibar), crossing and recrossing nine times the dangerous water before reaching Puerto de Muschitos. The roads of the area were deep in bog but the river banks were somewhat firmer.

On the 15th of May they travelled on to Columa where the Indians quickly built an overnight shelter of branches and leaves so successfully that not even the storms penetrated them. At 6 am the temperature was 1016° R and the air cooler.

In getting to Tarigagua they considered the bridges as dangerous as the rapids and fording places, and the severe precipices frightening. The air here was a little better at 1014½° R and they were beginning to feel the cold!

From Tarigagua it was up, up, up mount San Antonia. So great were some of the slopes that the mules could hardly keep their feet or in others even have room to put their feet for fear of the precipices. The graphic descriptions by Juan suggest that the route would have deterred all but the most determined. Descent was no less hazardous than ascent. On reaching Guamac or the Cross of the Canes, a stop was necessary where next morning the temperature was down to 1010° R.

On the 18th they came to the pass at Pucara (Pacara) and then descended to Chimbo where they met the corregidor of Guaranda and other eminent persons.

After the mountains, a wide treeless plain growing cereals. They were entertained at Guaranda until the 21st of May during which time the thermometer was at 1004½° R.

On the 22nd they crossed the Chimborazo desert to Rumi Machai, a stony cave that served as overnight accommodation. During this day their sufferings were from the cold that was aggravated by the winds but these were lightened by a view of one of the ancient Inca palaces, albeit in ruins. By the next morning the thermometer was hovering at 1000, or freezing point, for their move on to Mocha. From here it was somewhat warmer as they passed Hambato, St Miguel, Latacunga to Mula-Halo by the 27th.

The morning of the 28th the temperature was up to 1007° R as they proceeded to Chi Shinche and the next day the last stretch of their journey brought them to Quito, by the 29th.

Here they were royally entertained for 3 days by the President of the Province—Don Dionesio de Alzedo y Herrera. It had taken just a year to get to Quito from France but their task was yet to begin.

What of Bouguer and La Condamine?

Before separating to go their own ways to Quito they had a visitor—Pedro Vincente Maldonado y Sotomayor, Governor of Esmeraldas, who was enthusiastic to join the party.

La Condamine and Maldonado then took the northern route to Quito up the Esmeraldas river but not until most of May had been passed in exploring Esmeraldas and improving their map of the area. During his forays into the jungle here La Condamine was brought some stretchy material, or caoutchouc, rubber. Seeing great possibilities he is even said to have fashioned a waterproof case for his quadrant from rubber. He was not the discoverer of rubber but more the rediscoverer of it as far as South America was concerned.

Bouguer on the other hand had taken the heavier items to Guayaquil. For La Condamine and his group it took 5 days of poling to get up the river to Puerto de Quito in the country of the Colorado Indians. The distance was about 100 miles. On the way La Condamine discovered a metal, which he called platino, and which was later identified as platinum.

From Puerto de Quito they engaged Indians to carry their loads through the jungle towards Quito City. By the third day they reached Tamba de la Virgen at some 500 t elevation; then to Mindo and Nona which the barometer indicated to be at 2000 t — far higher than La Condamine had climbed before. They finally reached Quito City on 4th June 1736.

Bouguer started up the River Guayaquil and on May the 19th reached Caracol, some 3 days after the other party. Here Godin had been obliged to leave some 20% of his stores because of the treacherous travelling conditions. Bouguer remained a while at Caracol, not the least because his health had suffered during the previous months. When he did move he took seven days to climb the mountains in a distance of only 9 or 10 leagues. The mules had to rest every 7 or 8 steps and the heavy rains hampered even the slow progress. By the time he came to the foot of Chimborazo he reckoned himself to be at some 1500 t height.

He was very surprised to find the well cultivated fields and many small towns and villages after suffering first the heat then the cold to reach this height.

On the 10th of June he reached Quito and the whole party was united again. All were overwhelmed with their reception at such a remote outpost.

They had required 70 mules to transport all their equipment from the coast to Quito, where the barometer stood at 20 inch 1 line. On the recommendation of Father de Tournemine and R.P. le Vantier of the mission in Saint Dominique they lodged with the Jesuit fathers in Quito.

"All were at first considerably incommoded by the rarefaction of the air, particularly those who had delicate lungs felt the alteration most, and were subject to little hemorrhages, this no doubt arises from the lightness of the atmosphere, no longer aiding by its compression on the vessels, to the retention of the blood, which on its side maintains always the power of action." [137]

Some days were spent both acclimatizing and assessing

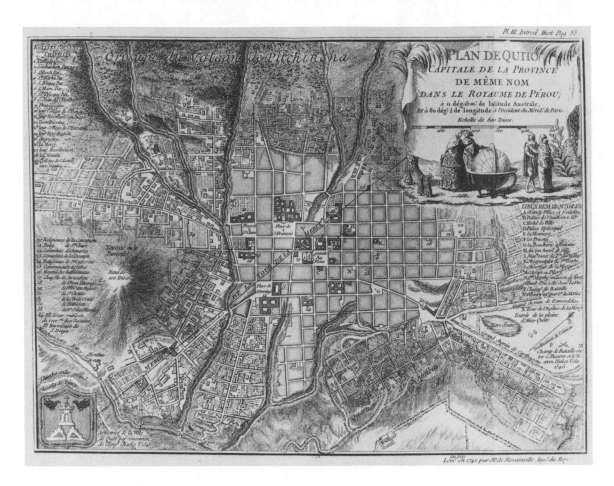

Pl. III. Introd. Hist. Pag 33

16. *Plan of Quito as drawn by members of the expedition. The diagonal line is the meridian of the Tower of Mercy. The position of Quito is quoted as 0° 15' S and 80 $1/2^o$ longitude from Paris.*

the magnitude of the task they had embarked upon among such formidable peaks and ravines.

Quito was some 800 to 900 t by 500 to 600 t, the seat of a bishop, and residence of the President of the Audience. The population was about 30,000 to 40,000, a third of whom were Spanish. The city contained two colleges, one run by the Jesuits and the other by the Dominicans, and there were many religious establishments there.

The city is at the foot of the Pitchincha mountains, part of the western chain of the cordillera. The barometer which at Quito was 20 inch 1 line was, at the top of Pitchincha, 1 line below 16 inches while at sea level it had been 28 inch 1 line.

Bouguer, from his barometric observations noticed a relation between the variations of the readings and those of a table of logarithms. Thus "look into the ordinary table of logarithms for the heights of the barometer, expressed in lines, and take 1/30th part from the difference of these logarithms, in taking with the characteristic the four first figures only which follow it, then you have the relative heights of the places in toise." [137] For Carabourou and Pitchincha the readings were 3057 twelfths of lines and 2292 twelfths of lines. Difference in logarithms = 1251. Subtract 1/30th gives a result of 1209 toise which corresponded well to the geometric solution.

And so to work.

The first task was to find a site for the base line. Initially they were disposed to using the plain of Cayambe some 12 leagues north of Quito but this proved unsatisfactory. It was while on this investigation that the party lost one of their number. M.Couplet died on the 17th September 1736 after only two days illness. He should have stayed in Quito but his keenness to be with the group would not let it appear as if he were deserting the others. To quote from [118]... "But, being of strong constitution, his zeal for the service would not permit him to be absent at our first essay. On his arrival, however, his distemper rose to such a height, that he had only two days to prepare for his passage to eternity; but we had the satisfaction to see he performed his part with exemplary devotion. This almost subitaneous death of a person in the flower of his age, was the most alarming, as none of us could discover the nature of his disease."—this was despite the attentions of Dr. Senièrgues.

The base was finally chosen on the 13th and the 14th of September in the plain of Yarouqui (Yaroqui), some 4 leagues NE of Quito and 249 toise lower. The reconnaisance prior to the choice had taken over two months but was coupled with the mapping of parts of the region. Verguin and Jean Godin des Odonais were busied near Quito while Bouguer and Morainville mapped further north.

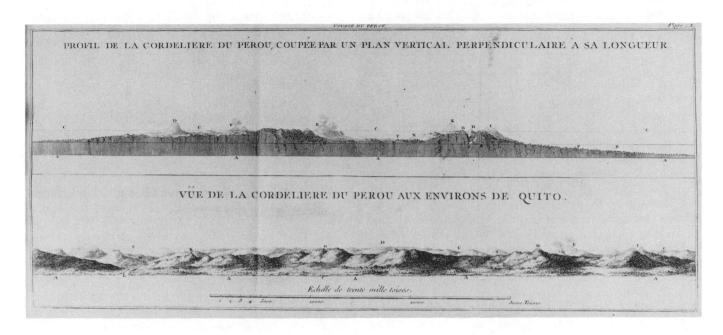

17.   *Panorama of the mountains as seen from Quito.*

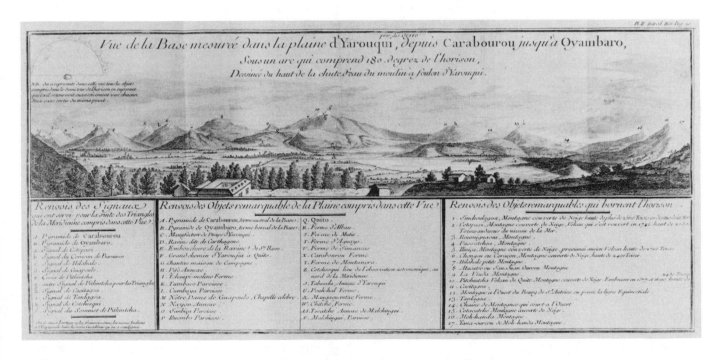

18.   *Panorama of the site of the Yarouqui baseline.*

107

The location below Quito made it slightly warmer to operate in but on the other hand it was exposed to all manner of rough weather, including whirlwinds of sand, that severely hampered progress. It was very hot by day yet freezing by night at an elevation around 1300 t.

Bouguer, Juan and La Condamine fixed the terminal points and marked the alignment with long poles. While the line was being prepared Bouguer, Ulloa and La Condamine climbed the Pic de Pitchincha that was visible from both ends of the proposed base and fixed a signal there which later proved to be one of the highest marks of the scheme at 2200 t above sea level. (over 14,000 ft or 4250m)

Eight days were spent clearing the line of the base from obstacles before measuring could begin. The group divided into two measuring parties and they took 26 days from the 3rd of October to the 3rd of November to complete the measurement. The division into parties for measurement was with Godin and Juan going one way and Bouguer, La Condamine, Ulloa and Verguin the other way. Agreement was achieved to 3 inches in 6273 toise (About 1 in 150,000). During this work however, one Indian died and Godin was often ill.

The mean measured horizontal value was accepted and this was then reduced to the chord length at the level of Carabourou (the lower base terminal) as 6272.6559 t. Details are given later.

The method of measurement was for each party to have three wooden measuring rods each of 20 ft in length and tipped with copper. When the three were end to end it gave 10 t.The rods were different colours so that they could always be placed in the same sequence. At one end of each rod the copper strip was horizontal and at the other end, vertical. Hence, when in contact this was only at a single point.

During the progress of the measurement daily comparisons were made against a field standard that had been compared with the one brought from France. This latter later became known as the toise du Pérou. The field standard was always kept in the shade. (Some references imply that the comparisons were directly against the standard iron toise brought from France rather than via an intermediate field standard but this is probably unlikely).

To keep the rods on the correct alignment a cord was stretched between successive poles. In obtaining contact it was always arranged that two rods remained in place while the third was moved. This one was then arranged to make contact in the same plane by the use of wedges. The inclination of the whole was then determined by use of a type of carpenter's level.

When it was not possible to achieve such contact then a plumb line was used to get the two ends in the same vertical plane. The terminals of the base were marked by two large millstones.

From this base line the series of triangles was set out over some 3° of meridian or 195 miles more or less due south.

After finishing the base, nearly a month was spent verifying the graduations on the quadrants. This was done by measuring angles whose known values were derived from accurately set-out distances.

On returning to Quito on the 5th of December 1736 they were shaken with a ¾ minute duration earthquake and a great many buildings were damaged. The solstice of December 20th was observed using a 12 ft. radius sector that had been designed specially for the task.

Shortly after completion of the base line La Condamine was obliged to go to Lima to obtain more funds. Not an easy or rapid exercise.

Leaving Quito on the 19th of January, 1737, he travelled via Loxa and he was doing well when able to average 7 leagues a day on the mountainous parts of the 400 league journey. It took until the 28th of February to reach Lima and this was about the same time that a new President, Don Joseph de Araujo y Rio (Don José de Araujoy Rio) was arriving in Quito as Governor and Captain General of the province.

Unhappily it was not a very opportune time to seek finance even with his letters of credit since silver was scarce but La Condamine did eventually obtain 4000 piastres (=20,000 livres).

While La Condamine was away from Quito (he did not get back there until mid-June), the new President had made difficulties with the two Spanish officers. Although Godin sought an audience with the President to plead in favour of the officers it was further contended that the party had contravened orders and were involved in illicit commerce. It was not until La Condamine returned from Lima with a letter from the Viceroy that the accusations were discharged. This was as a result of Juan following La Condamine to Lima after the initial difficulties so that he might confer with the Viceroy.

To escape arrest after the initial accusation against them the officers were looked after in the Jesuit College and things smoothed over by the Dean of the Cathedral. This, however, is what led to the illicit trading accusation against one of the servants to the party although it was later proved that all that was traded had indeed been correctly imported through the customs.

Angular observations were begun on Pitchincha where the cold was intense and started scorbutic infections and the Indians were subject to severe pains. They had every possible difficulty in managing a pendulum and they were continually in cloud that obliterated almost everything. The sky could change 3 or 4 times in half an hour with tempest followed by fine weather and then thunder.

At the same time Juan and Godin went to the summit of Pambamarca. This group also suffered from the severe cold and violent winds. On the 15th of August, 1737 at midday on Pitchincha the temperature was 1003° R and by evening 998½°R. By early morning on the 17th it stood at 996°R and by noon at 1012°R, and these were in a sheltered spot.

The summit was so small that tents were not possible and instead they used a hut so small that they could all only just creep in. The ascent had required leaving the mules part way and finishing with a four hour climb on foot, not helped by the thin atmosphere. Ulloa fell at one stage and had to return to the mule camp to recover before completing the climb the next day with the aid of some Indians.

Generally they were obliged to remain in the hut because of the cold, violent winds and continuous fog. Often it was not possible to see 6 or 8 paces. When the fog cleared the mountain top stood above a sea of clouds. When the clouds rose breathing became difficult, snow and hail were continuous and there was no venturing around.

Although they were all crowded close and lamps kept lit, they nevertheless, each had to have a chafing dish of coals. Feet swelled and were so tender that they could not even bear the heat, and walking was extremely painful, hands had chilblains, lips swelled and became chapped. Sustained on boiled rice and some fowl they had to keep the melted ice over the chafing dish to prevent re-freezing.

23 days were spent there until the 6th of September without completing the required rounds of angles to the base. Thus they decided to erect a new signal lower down but even this took until December to complete the observations.

surveyors whom they feared lost. Their re-appearance in the village caused much surprise and celebration.

It had at first been decided to use wooden pyramids as signals but these caused problems of visibility, often blown down, and destruction by Indians for the timber and rope. Hence the change to using tents.

Between August, 1737 and July, 1739 Juan and Godin occupied 32 deserts while Bouguer, La Condamine and Ulloa occupied 35. Deserts, or Paramos, in this context might best be described as barren, desolate and uninhabited areas. Although all the Cordilleras are dry or arid, some are much more so than others and the continual snow and frost make them absolutely uninhabitable by even beasts or plants.

19. Representation of the volcanic and astronomical phenomena at a mountain station.

After the rigours and difficulties of Pitchincha it was decided to camp in the tents rather than huts as used on Pitchincha. The tents were also used as signals and this , of course, exposed them to the impetuosities of the winds that sometimes tore out the stakes and blew them down. Luckily they had plenty of spare tents and their increasing dexterity at pitching them saved the danger of the party perishing. In fact, on Yasuay (Yassouai) three successively pitched tents were rendered useless and two stout poles broken. The Indians were not willing to bear the severity of the cold and, disgusted with the frequent labour of clearing snow from the tents, at the first ravages of the wind, deserted the party. Forsaken by the Indians, little or no provisions, scarcity of fuel and more or less destitute of shelter, the priest at Cannar, some 5 leagues from Sinasaguan, was offering prayers for the

In these conditions, the little cabins of the Indians and the stalls for cattle scattered up and down the mountains, and which the group used to lodge in were as spacious palaces, mean villages appeared like splendid cities and the conversation of a priest, or two or three of his companions, charmed them like a banquet of Xenophon, the little markets seemed to be filled with all the variety of a Seville fair.

Of the inhabitants of the area some admired the resolution, others did not know what interpretation to put on the perseverance, and even the more educated were utterly at a loss. The Indians were quizzed as to the activities of the surveyors and the answers they received often only tended to increase the doubts and astonishment. Some considered them little better than lunatics, others imputed the whole to covetousness and the search for rich minerals yet others thought they dealt in magic and the perplexity grew as it proved impossible to find any answer proportionate to the pains and hardships endured.

There were some pleasant interludes and Ulloa instanced two such. While at Vengotasin near the town of Latacunga, a league or so from the tent was a cow house where the group sheltered at night. One morning saw three or four Indians nearby who appeared to be on their knees, and indeed were; their hands were joined and they spoke in their own language with great fervour and by the look of their eyes it was obvious that the surveyors were the object of their address. Although several times signed to rise they did not until the group was some way distant. Hardly had they got to the signal and prepared their instruments than the same happened again. One of the Indian servants knew both Indian and Spanish languages and was directed to enquire of this strange activity. Apparently the eldest was father of the others, and his ass had either strayed or been stolen. As the group was considered to know everything they had come for commiseration at the loss and suggestion of how the beast might be recovered. This simplicity afforded no small entertainment and despite efforts to undeceive them they remained tenacious in their strange error of believing nothing was hid from the group. Finally they retired with extreme sorrow and the feeling that the group was ill-natured rather than ignorant.

His other story concerned one of the principal inhabitants of Cuença. When the group were on Boueran, Ulloa received a message from the priest of Cannar that two Jesuits of his acquaintance were passing that way and would like to see him. On beginning the journey Ulloa was overtaken by a gentleman from Cuença, inspecting his lands, and who had seen him leave his tent. He knew Ulloa's name, although previously unacquainted with him. However the Spanish officer was dressed in the garb of the Mestizos, or lowest class of person, for ease of work on the mountains. Hence the gentleman took him for a servant and began to question him. Ulloa was determined not to undeceive him. Among other things he was told that nobody believed that determining the figure and the magnitude of the earth could ever induce such a dismal and uncouth life. He was sure the group had discovered many rich minerals. Ulloa then tried to remove his prejudices but he left still unpersuaded.

THE MOUNTAINS AND PARAMOS, OR DESERTS, OF THE AREA

The two cordilleras come together at the paramos of Yasuay which, although exceedingly cold and arid, is less high than the general range. The most southern mountain is Sanquay which is very high, 2680t, and has its greater part snow covered. As an active volcano it was continually on fire and when the wind was right, its explosions could be heard in Quito almost 2° to the north. It was said to have been aflame since 1728.

On the same eastern range to the west of Riobamba was the two-crested mountain with peaks Collanes and Altar. Here the snows were much less and the height similarly so.

Besides Sanquay, many of the mountains of Peru were volcanic and north of Riobamba was the conical volcano of Tongouragoa of 2620t. This had been active in 1641 but was quiet during the survey operations. Nearby were the hot medicinal baths of Bannos. North and west of Riobamba was the highest mountain of them all, Chimborazo at 3220t (19320 ft or 5890m).

Another volcano that was active while the party was in the country was Cotapaxi, which was known to have erupted in 1533, and where a survey station was situated 180 t below the snow line. Further rumblings resulted in eruptions during 1743 and 1744, and when observed it sometimes increased and sometimes diminished in thickness of snow, but the boundary of the congelation lowered and fell below the survey station.

There were two inundations, on the 24th of June and the 9th of December, and the second was by far the greatest. Water fell 700-800 t and the first overthrew the survey station. The surges in the plain were from 60 to 120t and swept away innumerable cattle, 500 to 600 houses and 800 to 900 people. The waters ran some 17 leagues before they found an outlet at the foot of Tongouragoa, which distance it covered in less than three hours. It was a very short and rapid inundation, maybe only 15 seconds in some places but very damaging.

Other mountains had been reduced in height by volcanoes. Cargavirazo near Chimborazo inundated the village of Latacunga and its neighbourhood under water and liquid mud which later congealed. On the 20th of June, 1698, whole families were buried and not a house escaped grief of some form. Again in early December, 1736, houses in the area were damaged and many died.

The ejected materials were mixed with huge amounts of ice and snow that appeared to aid the rapid flow of material so that almost instantly many houses were destroyed and Indians killed. The countryside was a vast lake and carried away everything in its reach. Eruptions lasted several days and devastation was widespread.

After a lull of some months, May, 1744 saw even worse activity with ejections from several parts of the mountain that produced glorious night-time illuminations when seen from afar. Again on the 30th of November its ejections and inundations equalled the earlier occasions.

Five leagues to the west was Ilimissa, which was another bifid summit over 2700 t and constantly snow-covered.

As one went northward, there was Chinchulagua (Sinchoulagoa) that was active in 1660 and to the east of Quito was Cayamburo (Cayambe-ourcou) which, like Chimborazo, topped 3000 t.

A most uninviting countryside not improved by the difficulties of river crossings. The bridges took several forms. If there was a narrow place between high cliffs then four long beams, barely 4-5 ft. wide formed a path just wide enough for a loaded horse. Then there were bujucos where the crossing was wider. A bujucos was an oxhide thong and several twisted together formed a long rope. Six of these straddled the river, four of them supported a structure of sticks as a base and the other two formed hand rails. The mules were not allowed on such bridges and were instead made to swim. Because of the rapidity of the currents they often entered the water more than a mile above the bridge in order to reach the opposite side of the bridge.

Some rivers sported a tarabita. This was a form of machine

to allow animals and men to be winched over rather like a bosun's chair. The ropes for this were again of oxhide as the bujuco. One notable such site was at Alchipichi where the width was around 200 ft. and it was some 150 ft. above the water.

The 'roads' were also just as hair-raising as the tracks along the precipitous mountain slopes were hardly wide enough for the feet of the mules.

The main wealth of Peru at that time lay in the gold and silver mines and it appeared that only where there were rich mines was the surface well tended. The great number of such mines around Quito Province had, in earlier times, made many inhabitants extremely wealthy. Many mines, however, were abandoned for various reasons and the most profitable ones at the time of the expedition were around Popayan.

The manner of extraction was akin to panning and described by Ulloa thus. "The ore is dug out and layed in a large cocha, or reservoir. When this is filled, water is conveyed into it through a conduit, and the whole vigorously stirred into a mud. The lightest parts are carried away through another conduit or drain and this continues until only the most ponderous parts remain at the bottom. The worker then goes into the cocha with a wooden bucket to take up the sediment and washing it in a circular motion to separate the less ponderous parts. Any gold should be in the residue as grains although the whole is repeated on the water that was let out of the cocha to catch any small grains that may have escaped the first operation."

Aside from gold and silver, gems such as emeralds abounded

Among the many obstacles they were to face was the very inclement weather which on occasions could mean a month's wait for about a quarter of an hour's observations; destruction of signals by weather and Indians; difficult communications; scarce provisions; and always mule transport. To cap it all there was not always good accord between the various members of the group and the final difference of latitude took some 4 years to achieve. In particular it should be remembered that there were not even unreliable maps of the area. They mapped as they went.

While in the mountains they noticed a strange phenomena first seen on Pambamarca. [137] describes it thus. "A cloud in which we were enveloped, removing, opened to our view the scene of a very brilliant rising sun, the cloud passed from the other side; it was not 30 paces distant, being yet too short a one to give it that whiteness of which I have spoken, when each of us saw his own shadow projected above, and only saw his own, by reason the cloud did not present an even surface. Its proximity allowed us to distinguish every part of the shadow; we saw the arms, legs and head, but what astonished us the most was, that the head was decorated with a very lively colour, each with the same variety as the first rainbow, the red being the outward colour."

SURVEY STATIONS

The names quoted after each station indicate those who made observations at that point. The heights quoted are those given by La Condamine. For those stations he did not occupy then, the values of Juan are used. This is solely for consistency within the section but later there are details of the various values for each station.

Pitchincha (Pichinchas)
Bouguer, La Condamine and Ulloa.   2222t
Signal initially on the summit but later had to be set 210t lower. Observations started the 14th of August, 1737 and continued until December.

Cotchesqui   1491t
At the country house of Don Emanuel Frayre and chosen as the astronomical observatory for the north end of the chain. The point was on a tola or ancient Indian sepulchre on a small hillock a few toise SE of the house, and 25.06t north of the signal.

Oyambaro (Ayambara)
Bouguer, La Condamine and Ulloa.   1352t
The southern extremity of the Yarouqui base. Observations from the 20th of December, 1737 to the 29th of December, 1737.
Observations were also taken here from the 20th to the 27th of August, 1737 by Godin and Juan.

Carabourou (Caraburu)
Bouguer, La Condamine and Ulloa.   1226t
The northern extremity of the Yarouqui base. Occupied from the 30th of December, 1737 to the 24th of January, 1738. Very inclement weather and lack of visibility to distant signals. The barometer was at 21 inches 2¾ lines.
Observations were also taken here from the 20th to the 27th of August, 1737 by Godin and Juan.

Pambamarca
Bouguer, La Condamine and Ulloa.   2110t
This had been occupied in 1736 prior to the base measurement but subsequently a second signal was erected and observations taken from the 26th of January, 1738 to the 8th of February, 1738. Very strong winds made standing difficult as well as causing problems with the stability of the instrument.
Observations were also taken here by Godin and Juan from the 28th of August to the 1st of September, 1737. In 1744, March to May, Juan, in wishing to add extra northern stations occupied this point again.
In total this station was destroyed some seven times during the remeasure in 1900 but there is no record as to whether the original expedition had to rebuilt it as often as this.

Tanlagoa (Tanlagua)
Bouguer and La Condamine.   1743t
Ulloa observed separately. Extreme steepness made climbing difficult with hand and foot holds scarce. It took four hours of steady climbing. Observations were started on the 12th of February, 1738 and finished the next day.
On the 5th of September, 1737 Godin and Juan, without the regular Indians who did not wish to endure the rigours of this mountain, had to use other Indians who had been prevailed upon by the priest. Climbing unencumbered took them a whole day but for the loaded Indians two days were necessary. This was only to find some of the distant signals missing so a return to Quito was necessary. Finally they took their observations from the 20th of December to the 27th

of December, 1737.

Schangailli (Changalli). Bouguer and La Condamine.
Ulloa observed separately.   1405t
   This was a very pleasant site with no rough weather.
Bouguer and La Condamine observed in February, 1738 and
Ulloa from the 7th to the 20th of March. They were able
to lodge in the nearby town of Pintac. Again, however, some
signals had been blown down. This prompted the idea of us-
ing the tents as signals.

Goapoulo (Guapulo) Godin and Juan   1542t
   This point was conveniently near Quito so that no field
accommodation was necessary and the equipment was left
overnight in a field tent. Faulty reconnaissance necessitated
the re-siting of this station. Observing was during the first
three weeks of January, 1738, finishing on the 24th.
   In 1744 this point was occupied by Juan and Ulloa when
adding extra northern stations.
   It was used in the connection to the observatory at
Cotchesqui.

Goamani (Guamani) Godin and Juan   2080t
   Two visits were necessary here because the first signal that
was erected could not be seen from El Coraçon. These visits
were on the 28th of January and then on the 7th of February
when all was completed by the 8th.

Mira. Godin and Juan. Ulloa   1334t
   As this was a village near Quito it presented no difficulties.
Observations were in the middle of May, 1744 with com-
pletion on the 23rd. Near the observatory of Pueblo Viejo.

Cosin. Godin and Juan.
   Difficult of access. Completed the 23rd of May, 1744. One
of the stations of the northern extension.

Campanario. Godin and Juan. Ulloa.   1901t
   Difficult of access. Completed the 23rd of May, 1744.
Formed part of the northward extension.

Cuicocha. Godin and Juan.   2128t
   One of the stations in the northern extension. Observations
in 1744.

El Coraçon (Corazon)
Bouguer, La Condamine and Ulloa.   2212t
   This was occupied from the 12th of July to the 9th of
August, 1738. The signal was erected 250t below the lof-
tiest pinnacle.
   The point was also visited twice by Godin and Juan on
the 20th of January and the 12th of March, 1738.
   Another alternative name was Le Caraçon de Barionuero.
On the summit of 2476t the barometer was found to be 15
inch 9 lines.

Kota-pacsi (Cotopaxi, Limpie-pongo, Pouca-ouaicou,
Pucaguaico)
Bouguer, La Condamine and Ulloa.   2264t
   Climbed on the 21st of March, 1738, it was descended,

unfinished on the 4th of April as the frost, snow and winds
made observing impossible. A new signal was erected and
observations made from the 16th to the 22nd of August.
   Godin and Juan ascended the part called Limpie-pongo.
They spent from the 16th to the 31st of March, 1738, there,
but found that Guamani was not visible and that a new signal
was necessary. This was observed from the 9th to the 13th
of August. By moving the signal, some previously measured
angles had to be repeated.
   From barometric experiments on the peak called
Pucaguaico the mercury was at 16 inch 5 1/8 line at a height
above sea level of 2291 t. The summit of Cotapaxi was at
3126 t.
   On the return journey to Quito, Juan fell into a 25 ft. ravine
with his mule.

Chinchulagua. Godin and Juan.
   First observed on the 8th of August, 1738 but revisited
after some doubt about one angle. A snow covered peak.

Papa-Ourcou (Papa-urco)
Bouguer, La Condamine and Ulloa.   1828t
   Ascended on the 11th of August, 1738, the observations
were completed on the 16th. A relatively easy and restful
station.
   Godin and Juan similarly made their observations here in
August 1738. While there Godin had to return to Quito on
important business.

Milan. (Milin)
Bouguer, La Condamine and Ulloa.   1794t
   Of similar dimensions to Papa-ourcou, the observations
took from the 23rd to the 29th of August, 1738.
   Godin and Juan observed here after the return of Godin
from Quito. Their observations were from the 1st to the 7th
of September.

Ouangotassin (Vengotasin)
Bouguer, La Condamine and Ulloa.   2086t
   They stayed here longer than expected because of difficulties with one signal. They arrived on the 4th of
September and departed on the 18th. During this time they
were able to avail themselves of facilities in the nearby village
of Latacungo.

Tchoulapou (Chalapu, Chulapu)
Bouguer, La Condamine and Ulloa.   1953t
   Occupied from the 20th to the 23rd of September, 1738.
It was in the neighbourhood of the town of Hambato and
among numerous farms.
   Godin and Juan occupied this after they had been to Milan.
They finished here on the 18th September. This was the last
station where both groups measured all angles in each
triangle. From here onward each group measured only two
angles in each triangle.

Chitchitchoco (Chichichoco)
Bouguer, La Condamine and Ulloa.   1824t
   Completed during the period of the 24th to the 29th of
September, 1738. The signal was on the side of the moun-

tain which is a branch of the mountain called Carguairaso; it was windswept and while there they felt an earth tremor that rocked the tent from side to side.

Hivicatsou (Tivicatsu, Ivicatsu)
Godin and Juan.   1575t

A very pleasant site near the town of Pilaro which was able to provide useful facilities. Occupied from the 18th to the 26th of September, 1738.

Moulmoul (Mulmul)
Bouguer, La Condamine and Ulloa.   2006t

Signals in this neighbourhood were visited several times and there were several positional changes to ensure unambiguous bisections. On the 8th of November, Ulloa was taken critically ill and moved to a cowshed and later to Riobamba. He had first become ill at Chitchitchoco but worsened at Moulmoul. The observations were during October and November and completed on the 8th of November, 1738.

Godin and Juan had occupied the same cowshed as Ulloa for their stay which began on the 30th of September, 1738. From this position they were also able to get to Guayama and observe there at the same time. This station was completed on the 20th of October.

Amoula (Amula) Godin and Juan.   1790t
Completed on the 7th of November, 1738. From here the group returned to Quito for additional supplies.

Limal (Ilmal) Bouguer and La Condamine.   1942t
This was observed while Ulloa was ill. Occupied October to early November, 1738.

Nabouco (Nabuso) Bouguer and La Condamine   1716t
This was observed while Ulloa was ill. Occupied October to early November, 1738.

Ygoalata (Guayama, Goayama)
Bouguer and La Condamine.   2244t

This was observed while Ulloa was ill. Occupied October to early November, 1738.

Godin and Juan observed here at the same time as at Moulmoul. This was during October, 1738.

Dolomboc (Sisa-pongo)
Bouguer, La Condamine and Ulloa.   2099t

Occupied from the 9th to the 30th of November, 1738.

From the 7th of November, Godin and Juan had to return to Quito and while there Godin contracted a fever that was a long while going. The pair returned from Quito on the 2nd of February 1739 and were occupied on Dolomboc until the 19th of February.

Zagroum (Sesgum) Bouguer and La Condamine.   1814t
Occupied from the 17th to the 21st of January, 1739.

Godin and Juan were there from the 20th to the 23rd of February.

This station was in a dip in the mountain and this shelter was aided by fine weather.

Lalangouco (Lalanguso)
Bouguer, La Condamine and Ulloa.   2237t

Occupied from the 24th to the 31st of January, 1739.

Senegoalap (Senegualap) Bouguer and Ulloa.   2172t
This was occupied in early February, but Ulloa left before the end of the observations.

Godin and Juan had an extended stay there from the 23rd of February to the 13th of March, 1739.

Choujai (Chusay) La Condamine and Ulloa.   1958t
Ulloa returned during the observations. This mountain was particularly difficult for locating a signal so that it could be readily distinguished from subsequent points. The weather was harsh, the winds strong and the site was far from convenient shelter or source of provisions. They were delayed here from the 3rd of February to the 24th of March, 1739.

Godin and Juan occupied this station from the 14th of March to the 23rd of April.

Sacha tian Loma (Tialoma) Bouguer, La Condamine and Ulloa   2206t
Occupied from the 21st of March to the 25th of April.

Sinacaouan (Sinasaguan, Sinazahuan)
Bouguer, La Condamine and Ulloa.   2337t

A most inhospitable site. It formed one of the summits of Yasouay on the divide between Riobamba and Cuença jurisdictions. Even at such height the Incas had built a road. Such were the rigours of the weather here that public prayers were said in the village of Atun-Cannar, 3-4 leagues away, for the safety of the party. During the period from the 27th of April to the 9th of May there was only one clear day.

Godin and Juan occupied this station during the same period.

Quinoa Loma. Godin and Juan.   2037t
One of the most disagreeable of the stations. Occupied from the 9th to the 31st of May, 1739.

Boueran (Bueran)
Bouguer, La Condamine and Ulloa   1978t

This was near the town of Cannar where supplies were available. The station was occupied from the 10th of May to the 1st of June, 1739. Although the temperature was mild, there were three severe thunderstorms during their stay.

Yassouai (Yasuay)
Bouguer, La Condamine and Ulloa.   1882t

Completion here was not until the 16th of July, mainly because of the time spent reconnoitering an area for the south baseline. There were two possibilities — Tarqui and Los Bannos, the first of which was finally chosen. It was then possible to locate the neighboring signals. Observing on Yassouai was from the 7th to the 16th of July. This was a very steep mountain for foot slog only.

Godin and Juan occupied it from the 15th of June to the 11th of July, 1739. This group preferred the proposed baseline site in Los Bannos and arranged forward signals accordingly.

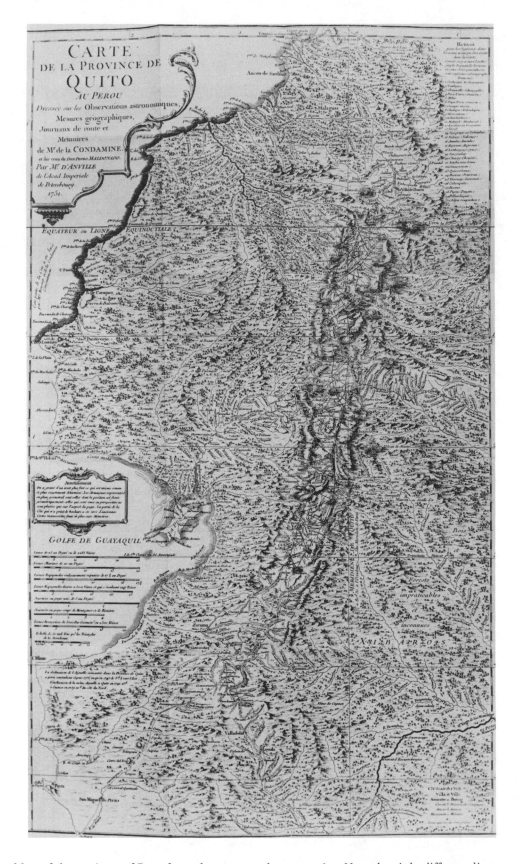

20.   *Map of the territory of Peru from the coast to the mountains. Note the eight different linear scales.*

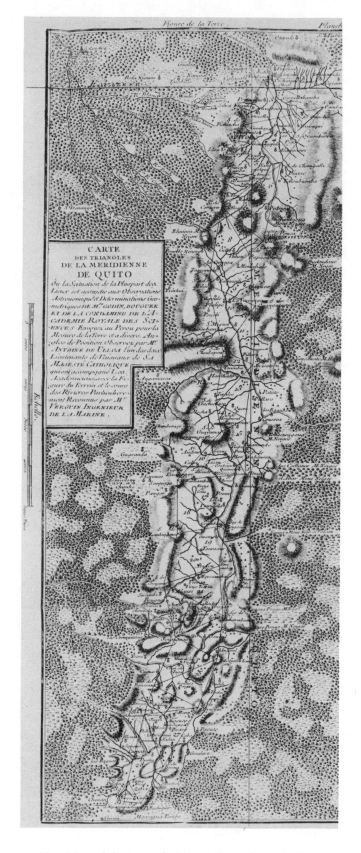

21.   Map of the area of triangulation scheme in Peru.

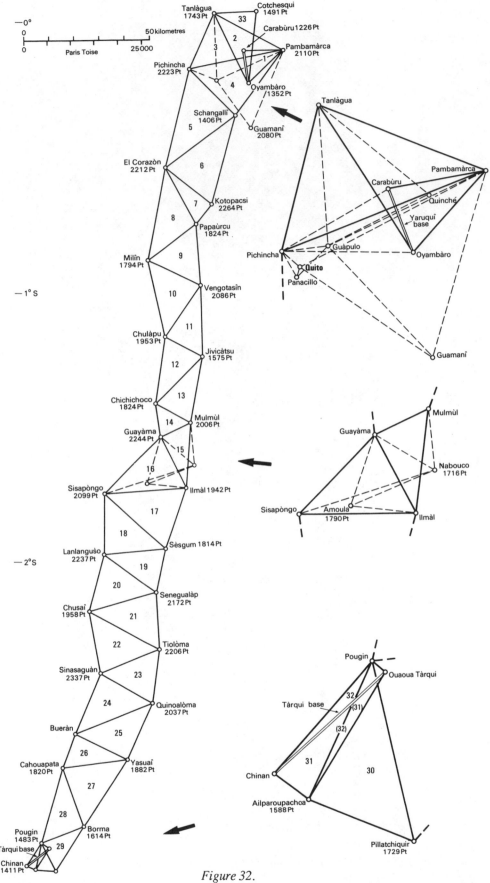

Figure 32.
Diagram of the scheme of Bouguer and La Condamine.

116

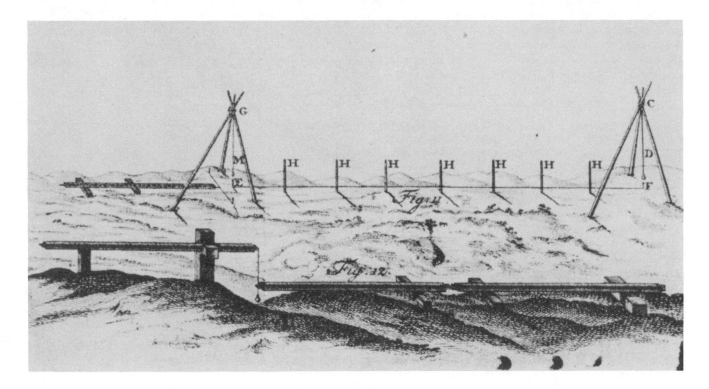

22. *Base line measurement by wooden bars placed end to end.*

Cahouapata (Surampalte) Bouguer, La Condamine and Ulloa. 1820t

    Occupied from the 24th to the 28th of July, 1739.

Borma. Bouguer, La Condamine and Ulloa. 1614t

    This station was below the general cloud line and there were clear sights on the 19th of July, to allow ready completion.

Pougin (Pugin)

Bouguer, La Condamine and Ulloa 1483t

    One of the base extension stations with no difficulties. Observed during early August, 1739.

Pillatchiquir (Pillachiquir)

Bouguer, La Condamine and Ulloa. 1729t

    One of the base extension stations with no difficulties. Observed during early August, 1739.

Ailparoupachca (Alparupasca)

Bouguer, La Condamine and Ulloa. 1588t

    One of the base extension stations with no difficulties. Observed during early August, 1739.

Chinan. Bouguer, La Condamine and Ulloa. 1411t

    South end of the Tarqui baseline. Angular observations during August, 1739.

Ouaoua Tarqui (Huahua Tarqui)

Bouguer and Ulloa. 1365t

    North end of the Tarqui baseline. Angular observations August, 1739.

Guanacauri Ulloa, Godin and Juan. 1334t

    East end of the baseline proposed by Godin and Juan. Observations at the end of 1739.

Cuença. Ulloa, Godin and Juan. 1350t

    Tower of the great church. The city lies at 2° 53′ 49″ S and 29′ 25″ W of Quito. It stands in a spacious plain and to the north of it runs the river Machangara and to the south of the city the river Matadero. The plain is some 6 leagues north-south and its great variety of trees has a delightful appearance. Occupied at the end of 1739.

Namurelte Godin and Juan

    Occupied in late 1739.

Los Bannos. Godin and Juan.

    The west end of the baseline proposed by Godin and Juan. The village of Bannos is so called because of its hot medicinal baths.

    Occupied in late 1739.

THE BASELINE AT YAROUQUI

This line, between Carabourou and Oyambaro was measured over 26 days from the 3rd of October to the 3rd of November, 1736. It was divided into 7 sections with the heights of the intermediate points from A to G referred to the height of Carabourou through the measurement of vertical angles. See Figure 31

0 = Oyambaro and K = Carabourou.

Section	Length	Observed V.A.	Arc length.
O - A	500t	−2° 13′ 45″	0′ 32″
O - B	948	−2 01 10	1′ 00
O - C	1998	−1 47 04	2′ 07
O - D	2787	−1 32 21	2′ 57
O - E	4081	−1 22 31	4′ 19
K - F	2093	+0 47 22.5	2′ 23
K - G	1400	+0 35 47 5	1′ 29
O - K	6272	−1 11 53	6′ 38
K - O	6272	+1 06 19	6′ 38

The reciprocal observations between O and K would give a coefficient of refraction of 0.08 but some other reduction was used in this case. The effects of refraction were improperly understood at this time and would certainly not have been allowed for in the same way as today.

Heights of the intermediate points related to Carbourou were given as:

O	126.08t
A	106.72
B	92.96
C	65.03
D	53.45
E	32.87
F	30.29
G	15.13
K	0.00

Each successive section was then reduced to its equivalent horizontal distance at the level of Carabourou.

The required reductions were obtained by La Condamine from the relation:

$$\text{Correction} = 2 \cdot dh \cdot \cos\left(90° - \frac{c}{2}\right)$$

Where
- dh = mean height of line to be reduced
- $\frac{c}{2}$ = half angle subtended by the arc at the centre of the earth.

Section	Mean height	Length	Req. reduction.
O - A	116.4t	500t	0.0177t
A - B	99.8	448	.0137
B - C	79.0	1050	.0254
C - D	59.2	789	.0143
D - E	43.2	1294	.0170
E - F	31.6	98	.0009
F - G	22.7	693	.0052
G - K	7.6	1400	.0032
		6272	.0974

Thus

Measured length of base[1]	=	6272t 4 pied 7 pouce 4½ lines
	=	6272. 769lt
Correction of ht. of Carbourou[2]	=	− .0974
		6272. 6717
Reduction to chord length	=	− .0158
Chord length at Carbourou	=	6272. 6559t[4]

When the inclined length OK can be found from the cosine formula to be:

$$6274.045t[5]$$
$$= 6274.t\ 0\ \text{pied}\ 3\ \text{pouce}\ 2.9\ \text{lines}[3]$$

This inclined value was required as initially all sides were computed from the angles as derived by the quadrant i.e. slanting angles. The temperature during the two measures of this baseline was given as an average of 10.5°R although Juan quotes 1016°R.

Notes.

1. There are several variations quoted for this [116] and [26] give:

   Bouguer   6272t 4 ft. 5 inch 'after adjusting by 11 inch'
   Godin   6272t 4 ft. 2⅙ inch 'after adjustment'.

   Mean   6272t 4 ft. 3⁷/₁₂

   Corrected to 7¼ inch by La Condamine.
   [135] gives:

6272t 4 ft. 7 inch 4½ lines	
-7 inch 1½ lines	
6272t 4 ft. 0 inch 3 lines	

   This correction item is really for the reduction to the height of Carabourou.

2. Carabourou was taken as 1226t above sea level and the radius of the earth for the reduction as 3,268,219t.

3. [116] gives   6274t 0 ft. 2 inch 1 line
   Godin and Juan   6274t 0 ft. 3 inch 8 line

4. In the calculations this was rounded to 6272.66t

5. In the calculations this was rounded to 6274.05

An alternative reduction by Bouguer went as follows:
The profile from Carabourou (A) to Oyambaro (B) was divided into sections. (Note that he used different letters to those of La Condamine). See Figure 32

A - C	817t
C - D	583
D - E	697
E - F	1390
F - G	798
G - H	1050
H - B	948
	6283

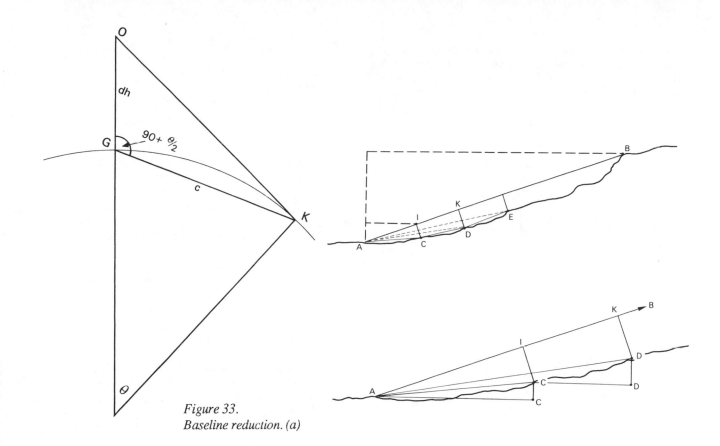

Figure 33.
Baseline reduction. (a)

Figure 34.
Baseline reduction. (b)

From a straight line AB the positions of the intermediate points were noted together with the perpendicular distances from AB to each point.

Section	Angle	Offset
From AB to		
C	−0° 52′ 36″	12.42t
D	−0 30 32	12.28
E	−0 18 57	11.38
From BA to		
F	−0 20 28	16.37
G	−0 35 11	20.24
H	−0 49 17	13.47

Then the slant lengths AC etc are greater than their projections AI etc on to the line AB by:

A - C	0.0944t
C - D	.0000
D - E	.0006
E - F	.0090
F - G	.0094
G - H	.0218
H - B	.0957
	0.2309t

The slope angle A - B was 1° 09′ 15″ and dh for AB was 126.08t
Bouguer then states that the slant sections are longer than the horizontal equivalents by a total of 1.521t and subtracts the 0.231t from it to give 1.290t. i.e.

AC′, CD′ etc = measured horizontal distances
AC , CD etc = ground distances
AI , IK   etc = required distances.

AC etc > AI etc by 0.231;
   hence AC etc = AI etc + 0.231
AC′ etc < AC etc by 1.521;
   hence AC′ etc = AI etc + 0.231 - 1.521
AI etc < AC′ etc by 1.521 - 0.231
   or AI etc = AC′ etc + 1.521 - 0.231

His summary was:

Measured length	6272t	4 pied	7 pouce	3	lines
Total corrections	1	1	8	10½	
Adjusted length	6274	0	4	1½	
	= 6274.057t				
As compared to					
La Condamine	6274.045				

119

For reducing the sections of the base line Bouguer also suggested an alternative method but did not put it into practice. The theory was: Let EK be one of the series of sections measured horizontally although successive sections were at different heights.

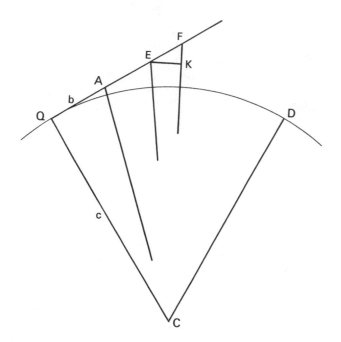

Figure 35.
Bouguer's method of baseline reduction.

Using an integral approach
Let EF = dx; EK = dz and CQ = c

$$\text{Then } z = \int \frac{c.dx}{(c^2 + x^2)^{1/2}}$$

i.e. EK = EF.cos FEK
where cos FEK = $c/(c^2 + x^2)^{1/2}$
If QA = b then the integral can be written as

$$z = \int \frac{cdx}{(c^2+b^2+2bx+x^2)^{1/2}}$$
i.e. denominator = $(c^2+(b+x)^2)^{1/2}$

Then by expansion

$$\frac{cx}{(c^2+b^2)^{1/2}} + \frac{cbx^2-\tfrac{1}{3}cx^3}{(2c^2+b^2)^{3/2}} + \frac{4cb^2x^3-3cbx^4+{}^3/_5c.x^5}{(8c^2+b^2)^{5/2}}$$

Bouguer pointed out however that because of the smallness of the quantities logarithms could not be used for the calculation. He then proceeded to derive yet another, just as complicated alternative. Both gave way to the far simpler version shown in the reductions.

REDUCTION OF OBSERVATIONS.

1. For difference of elevation.
    They used:

$$dH = \frac{L.\sin \theta_m}{\sin (90 + c/2)}$$

Where θ_m = mean vertical angle
L = slope length
c = angle subtended at the earth centre by the arc between the terminal points of a line.
or $\frac{c}{2}$ = 0.031 718 62L seconds.

The figure of 0.031 718 62 was derived from the value of 1° = 56,750t or thereabouts.

2. Reduction of quadrant angles to the horizontal.

The spherical triangle was used in the following way: Select one side as between the two signals defining the angle in question; define the other two sides by the zenith distances to the signals. This was the first instance of the use of the spherical triangle in geodesy.

Example.
For triangle I and the angle at Oyambaro.
Inclined angle at O   63° 48′ 13″
Zenith distance to P   85° 39′ 31″
to C   91° 11′ 53″

Using the relation:

$$\cos A = \frac{\cos A' - \cos b \cdot \cos c}{\sin b \cdot \sin c}$$

Where A′ = the inclined value of the angle.

Then the horizontal angle at O becomes 63°36′53″

Similarly in triangle I for the angle at Carabourou.
Inclined angle at C   77° 35′ 36″
Zenith distance to P   84° 26′ 54″
to O   88° 53′ 41″

Then the horizontal angle at C becomes 77° 38′ 30″

The most notable corrections were in triangle 31 where the angles corrected initially to 180° were then adjusted by −25′ 05″; −9′ 45″ and + 34′ 50″.

In the 33 triangles the total misclosure according to La Condamine irrespective of sign was 200″ or around 6″ per triangle. Positive and negative closures were almost equal.

While at the base, Bouguer made experiments on the relatively unknown problem of refraction and arrived at 93″ over the arc of 6′ 37.5″ which he turned into a ratio of ¼.₃. This was not, however, applied in the reductions.

THE BASELINE AT TARQUI

This was measured in August 1739 by Bouguer, La Condamine, Ulloa and Verguin. Bouguer and Ulloa measured from south to north while La Condamine and Verguin measured from north to south.

The operation was carried out as carefully as at the first base and they progressed at about 500t a day. Part of this base was, however, measured across a shallow pool with the measuring rods floating on the surface. After they had been floated into contact they were tied to stakes and hence moved forward successively in this manner by untying the rear one and floating it forward. Some 600t were measured in this way and La Condamine likened the temperature of the water as he waded in it as a comfortable 20°. Working from this he decided the mean temperature for the base was 16-17°R. As they had thermometers with them it seems unusual that they did not use them to determine the relevant values during the base measurement.

In measuring the Tarqui base there were problems also in the neighbourhood of the house of Don Pedro de Sanpertigue at Mama-Tarqui to the west of the line, where they proposed to make their astronomical observations.

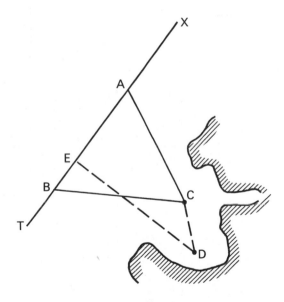

Figure 36.   The connection to Mama Tarqui observatory.

ABSTRACT OF OBSERVATIONS

<table>
<tr><td colspan="2"></td><td colspan="3" align="center">La Condamine</td><td colspan="3" align="center">Bouguer</td></tr>
<tr><td>Triangle</td><td>Stations</td><td>Observed angle corrected to 180°</td><td>Vertical angles</td><td>Reduced horizontal angles</td><td>Observed angle corrected to 180°</td><td>Vertical angles</td><td>Reduced horizontal angles</td></tr>
<tr>
<td rowspan="3">I</td>
<td>Pamba marca</td><td>38 36 11</td><td>C − 5 41 20
O − 4 30 27</td><td>38 44 42</td><td>38 36 08</td><td>C
O − 4 30 27</td><td></td></tr>
<tr><td>Carabourou</td><td>77 35 36</td><td>P + 5 33 06
O + 1 06 09</td><td>77 38 28</td><td>77 35 38</td><td>P + 5 33 05
O + 1 06 09</td><td></td></tr>
<tr><td>Oyambaro</td><td>63 48 13</td><td>P + 4 20 29
C − 1 11 53</td><td>63 36 50</td><td>63 48 14</td><td>P + 4 20 12
C − 1 12 20</td><td>63 36 57</td></tr>
<tr>
<td rowspan="3">II</td>
<td>Pamba marca</td><td>69 46 37</td><td>O − 4 30 27
T − 1 26 20</td><td>69 49 38</td><td>69 49 32</td><td>O − 4 30 27
T − 1 25 42</td><td></td></tr>
<tr><td>Oyambaro</td><td>74 10 58</td><td>P + 4 20 29
T + 1 18 39</td><td>74 14 08</td><td>74 10 57</td><td>P + 4 20 12
T + 1 18 30</td><td>74 14 03</td></tr>
<tr><td>Tanlagoa</td><td>36 02 25</td><td>P + 1 11 13
O − 1 33 48</td><td>35 56 20</td><td>36 02 25</td><td>P
O</td><td>35 56 17</td></tr>
<tr>
<td rowspan="3">III</td>
<td>Pamba Marca</td><td>38 36 32</td><td>T − 1 26 20
N + 0 09 53</td><td>38 34 46</td><td>38 36 32</td><td>T − 1 25 42
N + 0 09 53</td><td></td></tr>
<tr><td>Tanlagoa</td><td>89 14 08</td><td>P − 1 11 13
N + 2 02 56</td><td>89 16 39</td><td>89 14 04</td><td>P
N</td><td>89 16 31</td></tr>
<tr><td>Pitchincha</td><td>52 09 20</td><td>P − 0 28 36
T − 2 16 08</td><td>52 08 35</td><td>52 09 24</td><td>P − 0 28 26
T − 2 16 11</td><td>52 08 40</td></tr>
<tr>
<td rowspan="3">IV</td>
<td>Pamba marca</td><td>39 47 03</td><td>N + 0 09 53
S − 2 21 47</td><td>39 42 53</td><td>39 46 57</td><td>N + 0 09 53
S − 2 21 47</td><td></td></tr>
<tr><td>Pitchincha</td><td>61 06 24</td><td>P − 0 28 36
S − 3 38 56</td><td>61 04 34</td><td>61 06 30</td><td>P − 0 28 26
S − 3 39 11.</td><td>61 04 42</td></tr>
<tr><td>Schangailli</td><td>79 06 33</td><td>P + 2 04 55½
N + 3 25 47</td><td>79 12 33</td><td>79 06 33</td><td>P + 2 04 56
N + 3 25 47</td><td></td></tr>
<tr>
<td rowspan="3">V</td>
<td>Pitchincha</td><td>58 26 06</td><td>S − 3 38 56
C − 0 11 56</td><td>58 22 42</td><td>58 26 18</td><td>S − 3 39 11
C − 0 12 06</td><td></td></tr>
<tr><td>Schangailli</td><td>82 57 46</td><td>N + 3 25 47
C + 2 24 31½</td><td>83 05 22</td><td>82 57 38</td><td>N + 3 25 47
C + 2 24 31</td><td></td></tr>
<tr><td>El Coraçon</td><td>38 36 08</td><td>N − 0 07 59
S − 2 42 10</td><td>38 31 56</td><td>38 36 04</td><td>N − 0 07 59
S − 2 42 10</td><td>38 31 51</td></tr>
<tr>
<td rowspan="3">VI</td>
<td>Schangailli</td><td>41 15 05</td><td>C + 2 24 31½
K + 2 24 17</td><td>41 17 20</td><td>41 14 45</td><td>C + 2 24 31
K + 2 24 17</td><td></td></tr>
<tr><td>El Coraçon</td><td>74 07 48</td><td>S − 2 42 10
K + 0 06 50</td><td>74 06 23</td><td>74 08 18</td><td>S − 2 42 10
K + 0 06 50</td><td>74 06 52</td></tr>
<tr><td>Kotapacsi</td><td>64 37 07</td><td>S − 2 42 54
C − 0 19 34</td><td>64 36 17</td><td>64 36 57</td><td>S − 2 42 54
C − 0 19 34</td><td></td></tr>
<tr>
<td rowspan="3">VII</td>
<td>El Coraçon</td><td>21 22 11</td><td>K + 0 06 50
P − 1 45 19</td><td>21 17 28</td><td>21 22 14</td><td>K + 0 06 50
P − 1 45 19</td><td>21 17 31</td></tr>
<tr><td>Kotapacsi</td><td>81 46 43</td><td>C − 0 19 34
P − 5 05 50</td><td>81 46 31</td><td>81 46 54</td><td>C − 0 19 34
P</td><td></td></tr>
<tr><td>Papa-ourcou</td><td>76 51 06</td><td>C + 1 31 58
K − 5 00 47</td><td>76 56 01</td><td>76 50 52</td><td>C + 1 30 58
K + 4 59 08</td><td></td></tr>
<tr>
<td rowspan="3">VIII</td>
<td>El Coraçon</td><td>41 37 16</td><td>P − 1 45 19
M − 1 24 35</td><td>41 38 11</td><td>41 37 04</td><td>P − 1 45 19
M − 1 24 35</td><td>41 37 07</td></tr>
<tr><td>Paba-ourcou</td><td>94 06 23</td><td>C + 1 31 58
M − 0 15 32</td><td>94 06 01</td><td>94 06 41</td><td>C + 1 30 58
M − 0 16 32</td><td></td></tr>
<tr><td>Milan</td><td>44 16 21</td><td>C + 1 05 47
P + 0 03 27½</td><td>44 15 48</td><td>44 16 15</td><td>C + 1 05 50
P + 0 03 30</td><td>44 15 43</td></tr>
<tr>
<td rowspan="3">8</td>
<td>El Coraçt n</td><td>62 57 21</td><td>K + 0 06 50
M + 1 24 35</td><td>62 55 37</td><td>62 56 13</td><td>K + 0 06 50
M − 1 24 35</td><td>62 55 28</td></tr>
<tr><td>Kota-pacsi</td><td>75 17 53</td><td>C − 0 19 34
M − 1 39 14</td><td>75 18 04</td><td>75 17 45</td><td>C − 0 19 34
M − 1 39 14</td><td></td></tr>
<tr><td>Milan</td><td>41 45 46</td><td>C + 1 05 47
K + 1 23 35</td><td>41 46 19</td><td>41 46 02</td><td>C + 1 05 50
K + 1 23 35</td><td></td></tr>
</table>

Figure 37 (a). Observed angles if the Peru schemes.

Ulloa			Juan & Godin			
Observed angle corrected to 180°	Vertical angles	Reduced horizontal angles	Observed angle corrected to 180°	Vertical angles	Reduced horizontal angles	
38 36 08	C-5 43 23		38 36 46	C		
	O-4 30 27			O-4 30 27		
77 35 38	P+5 33 08½		77 35 32			
	O+1 06 30					
63 48 14	P+4 20 12		63 47 42	P+4 20 29		
	C-1 11 35			C		
69 46 38	O-4 30 27	69 49 33.5	69 46 32	O-4 03 33		
	T-1 25 42			T		
74 10 57	P+4 20 12	74 14 02.5	74 10 58	P+4 20 29	74 14 06	
	T+1 19 58			T+1 18 30		
36 02 25	P+1 11 45	35 56 24	36 02 30			
	O-1 33 48					
38 36 32	T-1 25 42	38 34 47.5				
	N+0 09 53					
89 14 04	P+1 11 45	89 16 32				
	N+2 02 52					
52 09 24	P-0 28 26	52 08 40.5				
	T-2 16 10					
39 46 57	N-0 09 53	39 42 45				
	S-2 21 47					
61 06 30	P-0 28 26	61 04 45				
	S-3 39 11					
79 06 33	P+2 04 56	79 12 30				
	N+3 25 47					
58 26 18	S-3 39 11	58 22 59				
	C-0 13 36½					
82 57 38	N+3 25 47	83 05 07				
	C+2 24 31					
38 36 04	N-0 07 59½	38 31 54				
	S-2 42 10					
41 14 45	C+2 24 31	41 16 58⅔				
	K+2 24 17					
74 08 18	S-2 42 10	74 06 52				
	K+0 06 50					
64 36 57	S-2 42 54	64 36 09⅓				
	C-0 19 34					
41 37 04	P-1 45 20	41 38 27⅓	41 36 45	P		
	M-1 24 35			M-1 24 35		
94 06 41	C+30 58	94 05 19⅓	94 06 28			
	M-0 16 32					
44 16 15	C+1 05 50	44 16 13⅓	44 16 47	C+1 50 42½	44 16 14	
	P+0 03 32			P+0 03 23		
62 56 13	K+0 06 50	62 55 03		K		
	M-1 24 35			M-1 24 35		
75 17 45	C-0 19 34	75 18 39				
	M-1 49 14					
41 46 02	C+1 05 50	41 46 18		C+1 05 42½		
	K+1 23 35			K		

Figure 37 (a). Observed angles if the Peru schemes. (continued)

ABSTRACT OF OBSERVATIONS

		La Condamine			Bouguer		
Triangle	Stations	Observed angle corrected to 180°	Vertical angles	Reduced horizontal angles	Observed angle corrected to 180°	Vertical angles	Reduced horizontal angles
IX	Papa-ourcon	60 31 30	M−0 15 31 / O+1 01 38	60 30 50	60 31 36	M−0 16 32 / 0+1 00 48	60 31 16
	Milan	60 31 54	P+0 03 28 / O+1 11 25	60 31 34	60 31 36	P+0 03 30 / O+1 11 20	
	Ouangotassin	58 56 36	P−1 14 45 / M−1 23 45	58 57 36	58 56 48	P−1 14 45 / M−1 23 45	
X	Milan	52 18 26	0+1 11 25 / T+0 24 32	52 18 26	52 18 35	0+1 11 20 / T+0 24 16	52 18 35
	Ouangotassin	78 23 32	M−1 23 45 / T−0 40 45	78 24 16	78 23 27	M−1 23 45 / T−0 40 45	
	Tchoulapon	49 18 02	M−0 40 40 / O+0 27 15	49 17 18	49 17 58	M−0 40 40 / O+0 27 15	49 17 14
XI	Ouangotassin	34 47 55	T−0 40 45 / H−2 15 05	34 46 33	34 47 55	T−0 40 45 / H−2 14 52	
	Tchoulapou	73 54 13	O+0 27 15 / H−2 42 50	73 51 44	73 54 24	O+0 27 15 / H−2 42 50	73 51 57
	Hivicatsou	71 17 52	O+2 01 00 / T+2 34 50	71 21 43	71 17 41		
XII	Tchoulapou	75 56 26	H−2 42 50 / C−0 39 53	75 57 21	75 56 22	H−2 42 50 / C−0 39 55	75 57 19
	Hivicatsou	68 53 20	T−2 34 50 / C+0 55 30	68 54 30	68 53 22		
	Chitchitchoco	35 10 14	T+0 27 05 / H−1 09 19	35 08 09	35 10 16	T+0 27 05 / H−1 09 19	35 08 11
XIII	Hivicatsou	34 29 34	C+0 55 30 / M+1 42 30	34 29 34	34 29 05		
	Chitchitchoco	72 06 03	H−1 09 19 / M+1 13 05	72 04 03	72 06 20	H−1 09 19 / M+1 13 10	72 04 17
	Moulmoul	73 24 26	H−1 54 50 / C−1 20 30	73 26 23	73 24 35	H−1 55 00 / C−1 20 30	
XIV	Chitchitchoco	48 51 21	M+1 13 05 / Y+3 29 35	48 51 00	48 51 41	M+1 13 10 / Y+3 29 35	48 51 05
	Moulmoul	54 19 21	C−1 20 30 / Y+2 07 35	54 13 18	54 19 11	C−1 20 30 / Y+2 07 35	
	Ygoalata	76 49 18	C−3 36 00 / M−2 12 58	76 55 42	76 49 08		76 55 47
XV	Moulmoul	60 49 38	Y+2 07 35 / I−0 22 25	60 47 19	60 49 30	Y+2 07 35 / I−0 22 25	
	Ygoalata	91 22 14	M−2 12 58 / I−1 33 56	91 25 57	91 22 26		91 26 11
	Ilmal	27 48 08	M+0 10 09 / Y+1 22 59	27 46 04	27 48 04	M−0 10 09 / Y+1 22 59	
1.	Moulmoul	69 54 40	Y+2 07 35 / N−1 55 50	69 48 30	69 54 45	Y+2 07 35 / N	
	Ygoalata	68 39 35	M−2 12 58 / N−3 27 05	68 44 45	68 39 41		
	Nabouço	41 25 45	M+1 47 40 / Y+3 18 20	41 26 45	41 25 34		
2.	Ygoalata	77 52 49	N−3 27 05 / A−2 21 16	77 59 34	77 53 01		
	Nabouço	59 55 41	Y+3 18 20 / A+0 12 40	59 53 12	59 55 32		
	Amoula	42 11 30	Y+2 10 20 / N−0 25 20	43 07 14	42 11 27		

Figure 37 (b). Observed angles if the Peru schemes.

Ulloa   Juan & Godin

Observed angle corrected to 180°	Vertical angles	Reduced horizontal angles	Observed angle corrected to 180°	Vertical angles	Reduced horizontal angles	
60 31 36 60 31 36 58 56 48	M − 0 16 32 O + 1 00 48 P + 0 03 32 O + 1 11 20 P − 1 14 45 M − 1 23 45	60 30 56½ 60 31 14½ 58 57 49	60 31 34 60 31 59 58 56 27	P + 0 03 23 O + 1 11 20	60 31 39	
52 18 35 78 23 27 49 17 58	O + 1 11 20 T + 0 24 29 M − 1 23 45 T − 0 40 45 M − 0 40 40 O + 0 27 15	52 18 08 78 24 33 49 17 19	52 18 06½ 78 23 42 49 18 11½	O + 1 1 20 T + 0 24 35 M T M − 0 40 40 O + 0 27 15	52 18 06½ 49 17 27	
34 47 55 73 54 24 71 17 41	T − 0 40 45 H − 2 15 08 O + 0 27 15 H − 2 42 50 O + 2 01 00 T + 2 33 29	34 46 35 73 51 54 71 21 31	34 48 21 73 54 03 71 17 36	O + 0 27 15 H − 2 42 50	73 51 34	Not measured by La Condamine. Accepted Ulloa value.
75 56 22 68 53 22 35 10 16	H − 2 42 50 C − 0 39 05 T + 2 33 29 C + 0 55 30 T + 0 27 05 H − 1 09 19	75 57 17 68 54 31 35 08 12	75 57 22 68 53 18 35 10 20	H − 2 42 50 C − 0 39 55 T + 0 27 05 H − 1 09 19	75 57 18 35 08 02	Not measured by La Condamine. Accepted Ulloa value.
34 29 05 72 06 20 73 24 35	C + 0 55 30 M + 1 42 30 H − 1 09 19 M + 1 13 05 H − 1 56 32 C − 1 20 30	34 29 09 72 04 15 73 26 36	34 29 33 72 06 00 73 24 27	H − 1 09 19 M + 1 13 05	72 03 28	Not measured by La Condamine. Accepted Ulloa value.
48 51 41 54 19 11 76 49 08	M + 1 13 05 Y + 3 29 35 C − 1 20 30 Y + 2 07 35 C − 3 35 29 M − 2 12 58	48 51 18 54 13 10 76 55 32	48 51 40 54 19 15 76 49 05	M + 1 13 05 Y + 3 29 35 C − 3 36 38½ M − 2 17 57½	48 51 04 76 56 02	
60 49 30 91 22 26 27 48 04	Y + 2 07 35 I − 0 22 25 M − 2 12 58 I − 1 33 48 M + 0 10 09 Y + 1 22 59	60 47 09⅓ 91 26 10⅓ 27 46 40⅓	60 49 38 91 22 25 27 47 57	M − 2 17 57½ I − 1 34 07½	91 26 16	

Figure 37 (b). Observed angles if the Peru schemes. (continued)

125

ABSTRACT OF OBSERVATIONS

La Condamine   Bouguer

Triangle	Stations	Observed angle corrected to 180°	Vertical angles	Reduced horizontal angles	Observed angle corrected to 180°	Vertical angles	Reduced horizontal angles
3.	Ygoalata	55 16 28	A − 2 21 16	55 18 15	55 16 46		
	Amoula	63 38 22	I − 1 33 56 Y + 2 10 20	63 38 50	63 37 58		
	Ilmal	61 05 10	I + 0 43 10 Y + 1 22 59 A − 0 53 11	61 02 55	61 05 16	Y + 1 22 59 A − 0 53 16	
XVI	Ygoalata	71 36 09	I − 1 33 56 D − 0 38 03½	71 36 44	71 35 51		71 36 27
	Ilmal	67 20 23	Y + 1 22 59 D + 0 23 39	67 20 33	67 20 40	Y + 1 22 59 D + 0 23 39	
	Dolomboc	41 03 28	Y + 0 22 43 I − 0 39 54	41 02 43	41 03 29	Y + 0 22 40 I − 0 40 15	41 02 43
16.	Ygoalata	94 15 02	N − 3 27 05 D − 0 38 03½	94 17 48	94 15 04		
	Nabouço	58 23 06	Y + 3 18 20 D + 0 59 10	58 23 16	58 23 15		
	Dolomboc	27 21 52	Y + 0 22 43½ N − 1 17 31½	27 18 56	27 21 41	Y + 0 22 40 N − 1 17 30	
XVII	Ilmal	63 39 34	D + 0 23 39 Z − 0 38 46	63 39 07	63 39 49	D + 0 23 39 Z − 0 38 46	
	Dolomboc	48 31 30	I − 0 39 54 Z − 1 07 45	48 31 45	48 31 50	I − 0 40 15 Z − 1 07 45	48 32 07
	Zagroum	67 48 56	I + 0 26 27 D + 0 51 23	67 49 08	67 48 21		
XVIII	Dolomboc	47 28 15	Z − 1 07 45 L + 0 29 45	47 26 43	47 28 29	Z − 1 07 45 L + 0 29 45	47 26 56
	Zagroum	52 01 08	D + 0 51 23 L + 1 52 36	52 01 32	52 01 15		
	Lalangouço	80 30 37	D − 0 42 35 Z − 2 04 20	80 31 45	80 30 16	D − 0 42 35 Z − 2 04 20	80 31 23
XIX	Zagroum	71 00 58	L + 1 52 36 S + 1 53 19	71 03 38	71 00 57		
	Lalangouço	47 46 33	Z − 2 04 20 S − 0 22 35	47 45 33	47 46 32	Z − 2 04 20 S − 0 22 35	47 45 52
	Senegoalap	61 12 29	Z − 2 03 51 L + 0 10 39	61 10 49	61 12 31	Z − 2 03 51 L + 0 10 39	
XX	Lalangouço	66 28 36	S − 0 22 35 C − 1 20 05	66 28 44	66 28 39	S − 0 22 35 C − 1 20 05	66 28 47
	Senegoalap	55 40 53	L + 0 10 39 C − 0 58 31	55 40 19	55 40 51	L + 0 10 39 C − 0 58 31	
	Choujai	57 50 31	L + 1 07 50 S + 0 44 07	57 50 57	57 50 30		57 50 56
XXI	Senegoalap	78 05 57	C − 0 58 31 E + 0 03 49	78 05 46	78 05 56	C − 0 58 31 E + 0 03 49	
	Choujai	45 21 27	S + 0 44 07 E + 0 42 35	45 21 40	45 21 35		45 21 49
	Sacha-tian-loma	56 32 36	S − 0 15 36 C − 0 58 30	56 32 34	56 32 29	S − 0 15 39 C	
XXII	Choujai	50 53 04	E + 0 42 35 S + 1 29 12	50 53 19	50 53 01		50 53 18
	Sacha-tian-loma	51 55 28	C − 0 58 30 S + 0 26 31	51 54 24	51 55 26	C	
	Sinaçaouan	77 11 28	C − 1 42 30 E − 0 40 20	77 12 17	77 11 33	S + 0 20 31 C − 1 42 24 E − 0 40 14	77 12 22

Figure 37 (c). Observed angles if the Peru schemes.

Ulloa   Juan & Godin

Observed angle corrected to 180°	Vertical angles	Reduced horizontal angles	Observed angle corrected to 180°	Vertical angles	Reduced horizontal angles	
71 35 57	I − 1 33 48 D − 0 38 04	71 36 32⅔	71 35 55	I − 1 34 07½ D − 0 38 52	71 36 33	
67 20 37	Y + 1 22 59 D + 0 23 39	67 20 45¼	67 20 35			
41 03 26	Y + 0 22 40 I − 0 40 15	41 02 42	41 03 30	Y + 0 22 47 I − 0 39 33	41 02 45	
63 39 49	D + 0 23 39 Z − 0 31 46	63 39 20	63 39 55			
48 31 50	I − 0 40 15 Z − 1 07 45	48 32 08½	48 31 40	I − 0 39 33 Z − 1 06 33	48 31 55	
67 48 21	I + 0 26 28 D + 0 57 35	67 48 31½	67 48 25			
47 28 29	Z − 1 07 45 L + 0 29 45	47 26 58	47 28 35	Z − 1 06 33 L + 0 29 45	47 27 06	
52 01 15	D + 0 57 35 L + 1 55 12	52 01 38	52 00 56			
80 30 16	D − 0 42 35 Z − 2 04 20	80 31 24	80 30 29	D − 0 42 35 Z − 2 04 20	80 32 16	
71 00 58	L + 1 55 12 S + 1 55 28	71 03 34⅔	71 00 57	L + 1 52 20 S	47 45 25	
47 46 32	Z − 2 04 20 S − 0 22 35	47 45 33⅓	47 46 34	Z − 2 04 20 S − 0 22 35		
61 12 30	Z − 2 03 51 L + 0 10 39	61 10 52	61 12 29			Not observed by La Condamine. Accepted Ulloa & Bouguer values.
66 28 39	S − 0 22 35 C − 1 20 05	66 28 48	66 28 27	S − 0 22 35 C − 1 20 05	66 28 35	
55 40 51	L + 0 10 39 C − 0 58 31	55 40 17	55 40 46			Not observed by La Condamine. Accepted Bouguer & Ulloa.
57 50 30	I + 1 10 03 S + 0 45 05	57 50 55	57 50 47	L + 1 07 50½ S + 0 45 05	57 51 14	
78 05 56	C − 0 58 31 E + 0 03 49	78 05 53	78 05 57½			Not observed by La Condamine. Accepted Bouguer & Ulloa.
45 21 35	S + 0 45 05 E + 0 42 35	45 21 49½	45 21 56	Se + 0 45 05 Sa + 0 42 35	45 22 10	
56 32 29	S − 0 15 39 C − 0 59 14	56 32 17½	56 32 06½			
50 53 00	E + 0 42 35 S + 1 29 02	50 53 15½	50 53 07	E + 0 42 35 S + 1 20 02	50 53 32	Not observed by La Condamine. Accepted Godin value.
51 55 27	C − 0 59 14 S + 0 22 31	51 54 23½	51 55 22			
77 11 33	C − 1 42 24 E − 0 40 14	77 12 21	77 11 31	C − 1 42 24 E − 0 40 14	77 12 21	

Figure 37 (c). Observed angles if the Peru schemes. (continued)

127

ABSTRACT OF OBSERVATIONS

La Condamine   Bouguer

Triangle	Stations	Observed angle corrected to 180°	Vertical angles	Reduced horizontal angles	Observed angle corrected to 180°	Vertical angles	Reduced horizontal angles
XXIII	Sacha-tian-loma	56 59 54	S+0 26 31	56 58 57	56 59 53	S+0 26 31	
			Q−0 58 59			Q−0 58 59	
	Sinaçaouan	50 38 45	E−0 40 20	50 38 56	50 38 45	E−0 40 14	50 38 57
			Q−1 33 12			Q−1 33 06	
	Quinoa-loma	72 21 21	E+0 48 32½	72 22 07	72 21 22	E+0 48 32	
			S+1 21 04½			S+1 21 04	
XXIV	Sinaçaouan	86 39 22	Q−1 33 12	86 42 00	86 39 20	Q−1 33 06	86 41 59
			B−1 43 10			B−1 43 04	
	Quinoa-loma	48 53 41	S+1 21 04½	48 52 08	48 53 36	S+1 21 04	
			B−0 20 57½			B−0 20 57	
	Boueran	44 26 57	S+1 30 53½	44 25 52	44 27 04	S+1 30 42	44 26 01
			Q+0 04 00½			Q+0 03 52	
XXV	Quinoa-loma	47 25 00	B−0 20 57½	47 25 01	47 24 49	B−0 20 57	
			Y−0 48 27½			Y−0 48 27	
	Boueran	47 12 04	Q+0 04 00½	47 11 52	47 11 48	Q+0 03 52	47 11 36
			Y−0 32 27½			Y−0 32 28	
	Yassouai	85 22 56	P+0 37 24	85 23 07	85 23 23	Q+0 37 23	
			B+0 21 09			B+0 21 08	
XXVI	Boueran	85 06 45	Y−0 32 27½	85 07 21	85 07 13	Y−0 32 28	85 07 51
			C−1 14 38			C−1 14 00	
	Yassouai	32 55 42	B+0 21 09	32 55 16	32 55 33	B+0 21 08	
			C−0 21 13			C−0 21 14	
	Cahouapata	61 57 33	B+1 06 55	61 57 23	61 57 14	B+1 06 55	61 57 05
			Y+0 09 08			Y+0 08 58	
XXVII	Yassouai	49 20 49	C−0 21 13	49 20 46	49 20 56	C−0 21 14	
			B−1 01 06			B−1 01 07	
	Cahouapata	77 42 19	Y+0 09 08	77 42 03	77 42 05	Y+0 08 58	77 41 48
			B−0 59 40			B−0 59 40	
	Borma	52 56 52	Y+0 45 30	52 57 11	52 56 59		
			C+0 47 27				
XXVIII	Cahouapata	34 08 45	B−0 58 40	34 09 05	34 08 37	B−0 59 40	34 08 57
			P−1 18 54			P−1 18 55	
	Borma	91 44 57	C+0 47 27	91 44 15	91 44 49	C	
			P−0 53 03			P−0 53 13	
	Pougin	54 06 18	C+1 02 30	54 06 40	54 06 34	C+1 02 30	54 06 56
			B+0 44 15			B+0 44 15	
XXIX	Borma	37 47 34	P−0 53 03	37 46 03	37 47 38		
			N+0 31 47				
	Pougin	83 53 45	B−, 44 15	83 55 05	83 53 50	B+0 44 15	83 55 09
			N+2 03 47			N+2 04 47	
	Pillatchiquir	58 18 41	B−0 41 50	58 18 52	58 18 32		
			P−2 10 20				
XXX	Pougin	38 04 38	N+2 03 47	38 04 42	38 04 42	N+2 04 47	38 04 25
			A+1 03 48			A+1 01 00	
	Pillatchiquir	54 30 07	P−2 10 20	54 32 33	54 29 49		
			A−2 00 04				
	Ailpa-roupachca	87 25 15	P−1 09 10	87 22 45	87 25 47		
			N+1 56 00				
XXXI	Pougin	16 31 31	A+1 03 48	16 25 01	16 31 07	A+1 01 00	16 24 57
			C−0 50 00				
	Ailpa-roupachca	72 49 55	P−1 09 10	72 51 00	72 50 22		
			C−6 34 40				
	Chinan	90 38 34	P+0 44 56	90 43 59	90 38 31	P+0 45 11	
			A+6 33 11			A+6 23 26	

Figure 37 (d). Observed angles if the Peru schemes.

128

Ulloa Juan & Godin

Observed angle corrected to 180°	Vertical angles	Reduced horizontal angles	Observed angle corrected to 180°	Vertical angles	Reduced horizontal angles	
56 59 53	S+0 22 31 Q−0 58 59	56 59 29	56 59 44	E−0 40 14 Q−1 33 06½	˙50 39 04	
50 38 45	E−0 40 14 Q−1 33 06	50 38 46½	50 38 52½			
72 21 22	E+0 49 19 S+1 21 26	72 21 44½	72 21 23½			Not observed by La Condamine. Accepted Juan & Ulloa
86 39 20	Q−1 33 06 B−1 43 04	86 41 28	86 39 09	Q−1 33 06½ B−1 43 04	86 41 48	
48 53 36	S+1 21 26 B−0 20 32	48 52 05	48 53 44			Not observed by La Condamine. Accepted Juan & Ulloa.
44 27 04	S+1 30 42 Q+0 03 52	44 26 27	44 27 07	S+1 30 42 Q+0 03 52	44 26 02	
47 24 49	B−0 20 32 Y−0 48 33	47 24 52½	47 24 46			Not observed by La Condamine. Accepted Juan & Ulloa.
47 11 48	Q+0 03 52 Y−0 32 28	47 11 38½	47 11 44	Q+0 03 52 Y−0 32 28	47 11 32	
85 23 23	Q+0 37 23 B+0 21 08	85 23 29	85 23 30			
85 07 13	Y−0 32 28 C−1 13 37	85 07 50	85 07 21	Y−0 32 28 C−1 14 38	85 07 59	
32 55 33	B+0 21 08 C−0 21 14	32 55 07½	32 55 17	B+0 21 08 C		
61 57 14	B+1 06 55 Y+0 08 58	61 57 02½	61 57 22	B+1 07 07½ Y+0 09 02½	61 57 13	

Figure 37 (d). Observed angles if the Peru schemes. (continued)

129

ABSTRACT OF OBSERVATIONS

La Condamine   Bouguer

Triangle	Stations	Observed angle corrected to 180°	Vertical angles	Reduced horizontal angles	Observed angle corrected to 180°	Vertical angles	Reduced horizontal angles
XXXII	Pougin	94 58 37	O − 11 36 50 C − 0 50 00	95 15 11	94 58 10		
	Ouaoua-Tarqui	78 40 54	P + 11 36 15 C + 0 27 40	78 32 18	78 41 27	P + 11 36 19 C + 0 27 44	
	Chinan	6 20 29	P + 0 44 56 O − 0 32 30	6 12 31	6 20 23	P + 0 45 11 O − 0 32 29	6 12 45
31	Pougin	75 16 06	A + 1 03 48 O − 11 36 50	78 51 01		A + 1 01 00 O	
	Ailpa-roupachca	6 09 39	P − 1 09 10 O − 2 25 44	5 59 54			
	Ouaoua-Tarqui	94 34 15	P + 11 36 15 A + 2 20 30	95 09 05		P + 11 36 19 A	
32	Ailpa-roupachca	78 41 10	O − 2 25 44 C − 6 34 40	78 50 54			
	Ouaoua-Tarqui	16 42 01	A + 2 20 30 C + 0 27 40	16 37 38		A C + 0 27 44	
	Chinan	84 36 49	A + 6 33 11 O − 0 32 30	84 31 28		A + 6 23 26 O	
XXXIII	Cotchesqui	83 25 12	T + 1 34 21½ O − 0 40 33½	83 23 53	83 25 12		
	Tanlagoa	62 37 35	C − 1 42 50 O − 1 33 48	62 39 16	62 37 33		
	Oyambaro	33 57 13	C + 0 27 34 T + 1 18 39	33 56 51	33 57 15		
(1)	Pamba-marco	47 02 44	T − 1 26 20 G − 2 10 17	47 03 54	47 02 44½	T − 1 25 42 G − 2 05 52	
	Tanlagoa	65 39 42	P + 1 11 13 G − 1 00 30	65 37 45	65 39 42	P + 1 11 45 G − 1 00 26	
	Gaapoula	67 17 34	P + 1 55 42½ T + 0 48 20½	67 18 21	67 17 33½	P + 1 51 03 T + 0 48 29	
(2)	Pamba-marca	47 57 21	G − 2 10 17 N − 0 14 12	47 55 49			
	Goapoulo	72 08 54	P + 1 55 42½ N + 2 09 22	72 12 03			
	Goamani	59 53 45	P − 0 02 40 G − 2 22 36	59 52 08			
(3)	Pamba-marco	46 41 01	N − 0 14 12 Q − 2 00 48	46 39 41			
	Goamani	72 40 04	P − 0 02 40 Q − 2 27 37	72 39 11			
	Quito	60 38 55	P + 1 41 52 N + 2 13 47	60 41 08			
27	Yassouai						
	Surampalte						
	Quanacauri						
28	Sorampalte						
	La Torre Cuenca						
	Quannacauri						

Figure 37 (e). Observed angles if the Peru schemes.

Observed angle corrected to 180°	Vertical angles	Reduced horizontal angles	Observed angle corrected to 180°	Vertical angles	Reduced horizontal angles	
47 02 44½	T – 1 25 42	47 03 25½				Not observed by La Condamine. Accepted Bouguer & Ulloa
65 39 42	G – 2 05 52 P + 1 11 45	65 38 12				Not observed by La Condamine. Accepted Godin value.
67 17 33½	G – 1 00 26 P + 1 51 03 T + 0 48 29	67 18 22½				Not observed by La Condamine. Accepted Godin value.
			47 57 18			Not observed by La Condamine. Accepted Bouguer value?
			72 08 52	P + 1 56 15 G + 2 08 30	72 12 02	Not observed by La Condamine. Accepted Godin value.
			59 53 50			Not observed by La Condamine. Accepted Juan & Ulloa value?
						Not observed by La Condamine. Accepted Juan & Ulloa value? Not observed by La Condamine. Accepted Juan & Ulloa value?
33 40 21	S – 0 21 14 G – 2 05 47	33 38 24	33 40 21			
87 14 17	Y + 0 08 58 G – 3 09 02½	87 13 24	87 14 17	Y + 0 09 02½ G – 3 09 02½	87 13 32	
59 05 22	Y + 1 48 17 S + 3 01 02	59 08 12	59 05 22			
20 33 16	C – 2 55 27½ G – 3 09 02½	20 34 58	20 33 16	C – 2 55 27½ G – 3 09 02½	20 34 57	
66 06 35	S + 2 46 08 G + 0 03 02	66 04 59	66 06 35	S G		
93 20 09	S + 3 01 02 C – 0 06 10	93 20 03	93 20 09			

Figure 37 (e). Observed angles if the Peru schemes. (continued)

ABSTRACT OF OBSERVATIONS

La Condamine   Bouguer

Triangle	Stations	Observed angle corrected to 180°	Vertical angles	Reduced horizontal angles	Observed angle corrected to 180°	Vertical angles	Reduced horizontal angles
29	Tanlagoa Guapula Pambamarca	See (1)					
30	Guapulo Pambamarca Campanario						
(31)	Pambamarca Campanario Cosin						
(32)	Campanaria Casin Cuicoćha						
32	Cosin Cuicoćha Mira						
(3)	Tanlagoa Guapulo Pambamarca						
(4)	Guapulo Guamani Pambamarca	See (2)					
(5)	Guapulo Guamani El Coraçon						
(6)	Guapulo El Coraçon Chinchulagua						

Figure 37 (f). Observed angles if the Peru schemes.

Observed angle corrected to 180°	Vertical angles	Reduced horizontal angles	Observed angle corrected to 180°	Vertical angles	Reduced horizontal angles	
65 39 42	G − 1 00 26	65 38 12				
	P + 1 11 45					
67 17 33½	T + 0 48 29	67 18 22½				
	P + 1 51 03					
47 02 44½	T − 1 25 42	47 03 25½				
	G − 2 05 52					
72 54 09	P + 1 51 03	72 56 27	72 54 10	P + 1 56 15	72 56 50	
	C + 1 46 35			C + 1 46 35		
32 02 10	G − 2 05 52	32 02 02	32 01 30			
	C − 1 10 34					
75 03 41	G − 1 55 00	75 01 31	75 04 20	G − 1 56 10	75 01 44	
	P + 0 55 50			P + 0 55 50		
96 21 15	C − 1 10 34	96 21 53	96 21 12	C		
	Ca − 0 27 03			Ca −		
38 07 35	P − 0 55 50	38 07 29	38 07 38	P + 0 55 50	38 07 34	
	Ca + 0 22 55			E + 0 22 55		
45 31 10	P + 0 12 48	45 30 38	45 31 10			
	C − 0 43 58					
38 02 09	Co + 0 22 55	38 02 39	38 02 27	E + 0 22 55	38 02 30	
	Cu + 0 21 39			Cu + 0 21 39		
75 42 02	Ca − 0 43 58	75 41 44	75 42 01½			
	Cu + 0 03 18					
66 15 49	Ca − 0 43 26	66 15 37	66 15 31½	C − 0 45 45	66 15 32	
	Co − 0 10 41			E − 0 09 58		
59 48 04	Cu + 0 03 18	59 46 49½	59 48 04			
	M − 2 03 08					
82 21 03	Co − 0 10 41	82 20 43	82 21 03	E − 0 09 58	82 21 04	
	M − 2 20 36			M − 2 22 40		
37 50 53	Co + 1 40 45	37 52 27½	37 50 53	E		
	Cu + 2 01 05			C + 2 01 05		
			65 39 42			
			67 17 33½			
			47 02 44½			
			72 08 52	Gi + 2 08 30	72 12 02	
				P + 1 56 15		
			59 53 50			
			47 57 18			
			69 25 54	Gi + 2 08 30	69 28 17	
				C + 1 34 15½		
			74 00 12	Go		
				C		
			36 33 54	Go − 1 57 08½		
				Gi		
			38 05 10	C + 1 34 15½		
				Ch		
			58 53 26	G − 1 57 08½	58 53 58	
				Ch − 0 48 39		
			83 01 24			

Figure 37 (f). Observed angles if the Peru schemes. (continued)

ABSTRACT OF OBSERVATIONS

La Condamine   Bouguer

Triangle	Stations	Observed angle corrected to 180°	Vertical angles	Reduced horizontal angles	Observed angle corrected to 180°	Vertical angles	Reduced horizontal angles
(7)	El Coraçon Chinchulagua Limpie-Pongo						
(8)	El Coraçon Limpie-Pongo Milan						

Figure 37 (g). Observed angles if the Peru schemes.

134

Observed angle corrected to 180°	Vertical angles	Reduced horizontal angles	Observed angle corrected to 180°	Vertical angles	Reduced horizontal angles	
			36 14 53	Ch – 0 48 39 L – 0 08 39	36 14 36	
			66 29 34½	C L		
			77 15 32½	C Ch		
			66 43 25½	L – 0 08 39 M – 1 24 35	66 43 12	
			73 23 35	C M		
			39 52 59½	C + 1 05 42½ L		

Figure 37 (g). Observed angles if the Peru schemes. (continued)

On one portion of the baseline AB of 439t 4 ft. a triangle was formed as ACB where TX was the southern part of the baseline. (Figure 34)
C was an Indian cabin.
The angle at C was 48° 16', and those at the baseline: A=52°08' and B = 79°36'
B was 1322t 1 ft. from Chinan (T). A was 1761t 5 ft. from T. D represents the place where the instrument was to be in the house of Mama-Tarqui.
DC was 89t 2 ft. in a direction N 22° W while the baseline TX was N 32° 26' E. La Condamine used a variation on this arrangement of Bouguer's but both were said to have found the same result, namely that D was 530½t from the base at a right angle from E, where TE was 1353t.

This can be verified by the use of coordinates:

Let B	=1000.000E		1000.000 N
Then A	=1235.801E		1371.086 N
By sine rule BC =	465.333t and bearing	112°02'	
Hence C	=1431.144E	and	825.515 N
As CD was			
given as	= 89.333t at 158°		742.687 N
Hence D	=1464.609E	and	742.687 N

From this the length BD is 531.104 and bearing 118° 59'

Then in triangle BED,
	BE	=	31.961
and	TE	=	1322t 1 ft + 32t
		=	1354t 1 ft.
	DE	=	530.141t

This will be seen to be nearly the same as the quoted values. The small variations could be due to unquoted seconds used in the angles or unmentioned corrections. Whence for the correction from T to D in the direction of the arc i.e., north,

TE' =.	1353t at 32° 26'	=	1141.954 t
TD' =	530.5t at 22° 26'	=	−284.517t
Thus difference northings		=	857.44 for TD

(La Condamine quoted 856.71 and Bouguer 857.48)

The two measurements of the total length agreed to 1ft 1 inch over 5259t. When reduced to the level of Carabourou the value became 5258.949t.

[198] quotes slightly different values of
Bouguer   5258.71t reducing to 5258.90 at ht. of Carabourou
La Condamine   5259.20t reducing to 5259.41 at ht. of Carabourou
Ulloa   5259.65t reducing to 5259.86 at ht. of Carabourou

For this measurement the rods were compared daily at a mean temperature of 16-17°R.

The vertical angles between the terminals were:

Ouaoua to Chinan	+0° 27' 40"
Chinan to Ouaoua	−0° 32' 30"
Difference	0° 04' 50"

| Arc length | 0° 05' 33" |
| Difference | 0° 00' 43" |

Depression angles were observed along the baseline in the same manner as at the northern baseline and were here initially related to Chinan (X).

Point	Height	Line	Length	Section	Length
X	0.00t	X - L	558t	X - L	558t
L	26.50t	X - K	658t	L - K	100t
K	33.33t	X - I	1138t	K - I	480t
I	34.67t	X - H	1539t	I - H	401t
H	39.50t	X - G	1759t	H - G	220t
G	37.50t	X - A	5059t	G - A	3300t
A	48.33t	X - U	5259t	A - U	200t
U	44.83t =	Ouaoua..			

Using the same formula as for the Yarouqui base, the correction to the level of Chinan was + 0.062t (Chinan was the higher end). As Carabourou was 185t lower than Chinan, the correction to its level was − 0.300t.

The correction from arc to chord was	− 0.013t
Reduction to inclined length XU	+ 0.165t
Whence reduced value for the base	5258.949t

When computed through the net from the yarouqui base the check was given as 5260.03t or an agreement to 1.08t. other calculations for the same check were however somewhat smaller and quoted in [198] as

Bouguer	0.31t difference from Yarouqui to Tarqui.
La Condamine	0.92t difference from Yarouqui to Tarqui
Ulloa	0.21t difference from Yarouqui to Cuença base
Juan	0.31t difference from Yarouqui to Cuença base

The temperature of the northern base was taken as 10.5°R and the southern 16-17°R. This gave a mean of around 13°R or the same as the temperature at which the iron toise they were using had been standardized. Because of this, no temperature corrections were applied.. How wise this was rather depends on the reliability placed on these two estimates of temperature, either, or both of which could have been considerably in error. As they had thermometers with them it is surprising that they do not appear to have been much used for the bases but more for registering the extremes of temperatures on the mountains.

With reference to the misclosure of the first base to the second it will be seen later than La Condamine decided to spread half of this as a scale factor to the overall arc length. i.e. 1.08t in 5259t is approximately 1/5000 and La Condamine adjusted his measured arc by 1/10,000.

Orientation

The general direction of the chain was about 14° west of the meridian going from north to south. What a pity they had not at that time started to use whole circle bearings! It would appear that Bouguer considered the meridians through the various stations to be parallel because of their proximity to the Equator. La Condamine on the other hand made allowance for convergence of the meridians in his final reduction.

The azimuths were found by sun observations on some 20 lines as checks against accumulation of errors. The starting value for Carabourou to Oyambaro was 19° 25′ 04″ E of South or a whole circle bearing of 160° 34′ 56″

Line	Date	Azimuth	Observer
Carabourou-Oyambaro		160°34′56″	La Condamine
	25/11/36	160 33 52	Godin (Comp. by Bouguer)
Oyambaro-Pambamarca	26/11/36	44 10 49	Godin (Comp. by Ulloa)
	25/11/36	44 10 44	Godin (Comp. by Juan)
	25/11/36	44 10 49	Godin (Comp. by Bouguer)
	26/11/36	44 11 30	Godin (Comp. by Juan)
	25/11/36	44 09 59	Godin (Comp. by Ulloa)
Oyambaro-Tanlagoa	26/11/36	329 56 59	Godin (Comp. by Juan)
	26/11/36	329 56 55′	Godin (Comp. by Ulloa)
Schangailli-El Coraçon		233 25 33	La Condamine
Pitchincha-El Coraçon		194 52 44	La Condamine
Milan-El Coraçon		190 26 49	La Condamine
	28/8/38	190 26 45	Bouguer
Chitchitchoco-Ygoalata	29/9/38	162 42 01	La Condamine
	29/9/38	162 42 28	Bouguer
	29/9/38	162 42 42	Ulloa
Moulmoul - Ilmal		178 50 33	La Condamine
	20/10/38	178 51 29	Ulloa
	21/10/38	178 52 00	Ulloa
Dolomboc - Lalangouco	20/11/38	179 42 41	La Condamine
Sicapongo-Lalangouco	20/11/38	179 43 26	Bouguer
Zagroum - Lalangouco		260 14 35	La Condamine
Choujai-Sacha-tian-Loma		117 43 20	La Condamine
Boueran-Cahouapata		199 52 17	La Condamine
Boueran - Yassouai		114 46 16	La Condamine
	8/7/39	114 45 24	Godin (Comp. by Juan)
Zagroum-Lalangouco	21/2/39	260 14 31	Godin (Comp. by Juan)
Ouaoua Tarqui-Chinan	18/10/39	212 23 47	La Condamine
	18/10/39	212 25 47	Bouguer
Campanario-Cosin	20/2/44	60 50 16	Godin (Comp. by Juan)

The results as given by La Condamine.

Station	Distance (t)	Bearing	E (t)	N (t)
Quito			0.00	0.00
	19 101.42	65°41′16″		
Pamba marca			17 407.42	7 864.22
	9 790.27	224 11 46		
Oyambaro			10 582.46	845.01
	14 000.26	3 54 29		
Cotchesqui			11 536.67	14 812.91
	8 801.91	267 18 22		
Tanlagoa			2 744.48	14 399.22
	12 679.21	203 17 57		
Pitchincha			- 2 270.66	2 753.98
	21 074.38	194 53 48		
El Coraçon			- 7 688.39	-17 612.12
	19 171.87	190 27 46		
Milan			-11 169.94	-36 465.23
	16 767.00	167 33 34		
Tchoulapou			- 7 557.96	-52 838.54
	13 216.85	186 39 57		
Chitchitchoco			- 9 092.15	-65 966.05

Station	Distance (t)	Bearing	E (t)	N (t)
Goayama	6 762.18	162°43′09″	- 7 083.41	-72 422.98
Dolomboc	16 518.42	222 41′32	-18 283.87	-84 564.12
Lalangouco	13 136.75	179 42 43	-18 217.83	-97 700.71
Choujai	12 926.82	194 28 45	-21 449.89	-110 216.96
Sinacaouan	13 585.32	168 34 41	-18 759.55	-123 533.22
Boueran	12 679.26	203 07 54	-23 740.54	-135 193.13
Cahouapata	7 644.52	199 52 59	-26 340.45	-142 381.95
Pougin	16 435.63	193 41 30	-30 230.71	-158 350.55
Ouaoua to Chinan	568.41	110 56 56	-29 697.26	-158 554.93
Chinan	5 259.94	212 23 53	-32 517.68	-162 994.96
Pamba marca			17 407.42	7 864.22
Schangailli	18 115.02	215 43 39	6 829.34	-6 841.62
Pouca ouaicou	19 245.40	192 08 24	2 782.01	-25 656.67
Papa ourcou	4 922.82	225 45 36′	- 744.14	-29 090.49
Ouangotassin	12 975.77	174 12 46	564.39	-42 000.18
Hivicatsou	13 730.55	182 04 19	67.98	-55 721.75
Moulmoul	13 640.65	187 18 32	- 1 667.36	-69 251.56
Ilmal	13 462.44	178 51 32	- 1 399.26	-82 711.26
Zagroum	13 744.61	200 05 08	- 6 119.47	-95 619.93
Senegoalap	10 372.98	189 10 50	- 7 774.44	-105 860.04
Sacha-tian-Loma	12 241.01	174 13 56	- 6 544.26	-118 039.08
Quinoa-Loma	10 867.72	188 48 01	-8 206.92	-128 778.86
Yassouai	12 370.59	200 08 45	-12 467.48	-140 392.60
Borma	17 156.32	212 29 36	-21 683.90	-154 863.16
Pillatchiquir	10 786.78	210 02 07	-27 083.04	-164 201.47
Ailpa-roupachca	4 101.81	277 10 42	-31 152.70	-163 688.92
Chinan	1 531.22	296 56 57	-32 517.62	-162 994.97

Length of arc = 162 994.96 + 14 812.91   = 177 807.87
$\frac{1}{10\ 000}$ correction to whole scheme   - 17.78
Cotchesqui signal to observatory   + 25.06
Chinan signal to observatory   - 856.71
Eccentricity of arc to meridian   - 7.97
Total length of arc at level of Carabourou   176 950.47

Note: The correction of $\frac{1}{10\ 000}$ was from the check between the two bases. The agreement was 1.08 t on 5259 or about $\frac{1}{5000}$. Thus La Condamine decided to apply a correction of half of this, or $\frac{1}{10\ 000}$, to the arc length.

The results as given by Bouguer.

Station	Distance (t)	Bearing	E (t)	N (t)
Cotchesqui			11 536.67	14 812.91
	8 802.01	267°17′31″		
Tanlagoa			2 744.43	14 397.06
	12 678.78	203 17 00		
Pitchincha			- 2 267.20	2 750.83
	21 072.93	194 53 19		
El Coraçon			- 7 681.70	-17 614.63
	19 169.57	190 27 32		
Milan			-11 161.55	-36 465.68
	16 763.84	167 33 03		
Tchoulapou			- 7 547.72	-52 835.32
	13 214.92	186 39 36		
Chichichoco			- 9 080.35	-65 961.04
	6 758.35	162 43 06		
Goayama			- 7 072.65	-72 414.19
	16 514.31	222 41 50		
Siça Pongo			-18 271.38	-84 550.79
	13 134.59	179 43 38		
Lalangouco			-18 208.87	-97 685.25
	12 926.92	194 29 25		
Chousay			-21 443.39	-110 200.98
	13 583.94	168 35 22		
Sinacaouan			-18 755.99	-123 516.36
	12 677.31	203 08 51		
Boueran			-23 739.46	-135 172.90
	7 642.82	199 54 26		
Cahouapata			-26 345.21	-142 358.97
	16 432.83	193 42 32″		
Pougin			-30 239.52	-158 322.97
	5 175.71	206 14 03″		
Chinan			-32 527.36	-162 965.50
	5 258.95	32 26 28″		
Ouaoua Tarqui			-29 706.29	-158 527.24

Length of arc = 14 812.91 + 162 965.50 = 177 778.41
Adjustment for discrepancy between baselines − 5.68
Correction to first Tarqui observatory − 857.48
Correction to La Condamine's observatory + 10.66
Correction from first to second
Tarqui observatory [26] + 14.50
176 940.41

Note: that the coordinates quoted for Bouguer are based on those of La Condamine for Cotchesqui to give ready comparison. The coordinate discrepancies can, of course, be largely accounted for by small changes in the bearings.

Connection of heights to sea level.

Considerable problems with almost continuous fog between the mountains and the sea made connection of the heights to sea level very difficult. It was not until June-August, 1740 that Bouguer managed a connection.

This was achieved trigonometrically through several points that were not part of the main chain but connected to it at Pitchincha and Papaourcou.

From the village of Niguas in the middle of the forest towards the coast it was possible to see both Pitchincha and the twin peaks of Ilimissa. This was a perpetually snow covered mountain just to the south of Coraçon. Whence:

	Bearing	Vertical angle
Niguas to Pitchincha	120°07′07″	4°30′30″
Niguas to Ilimissa	164°04′22″	2°40′10″

His measure of the barometer here was 24 inch 11¼ lines which was only ½ line higher than he had observed at Goave some years before. Since Goave had a height of 550t this put Niguas as about 542 t above sea level.

Travelling further towards the sea; at the confluence of two rivers was the I.de l'Inca. At this point Bouguer was able to determine the latitude from sun observations as 14′ 33″ N and also see the southern summit of Ilimissa. Its elevation was 1° 53′ 43″. The barometer here stood at 27 inch 9⅓ line.

From the rate of fall of the river bed and the position in relation to the sea he estimated that this point must be about 30 t above sea level. Later, however, he modified this to 42 t.

The two peaks of Ilimissa were connected to Pitchincha, Papaourcou and Quinché thus:

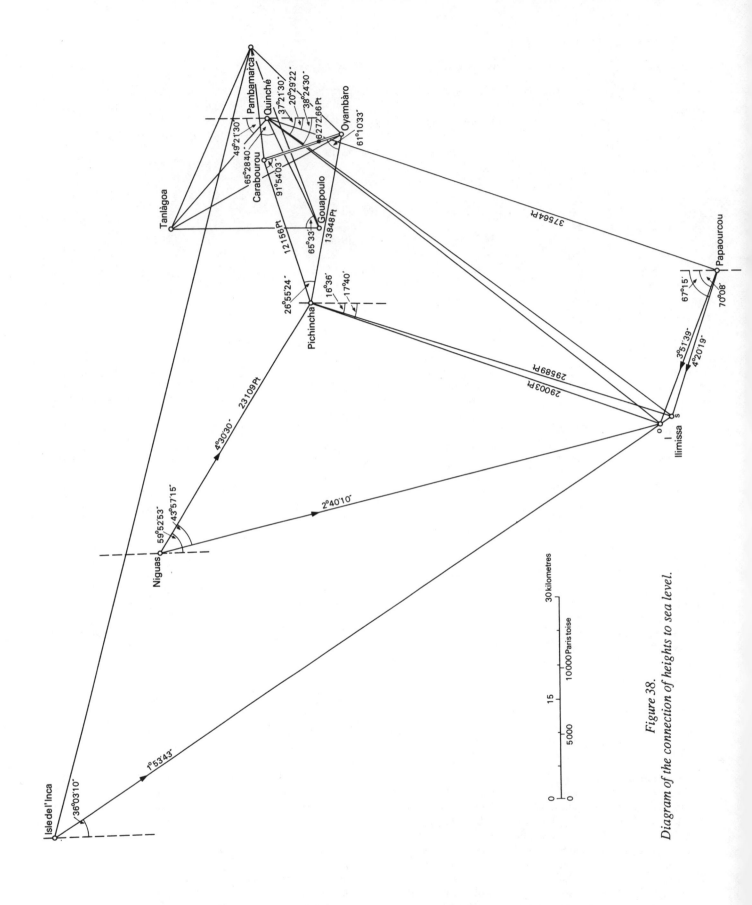

Figure 38.
Diagram of the connection of heights to sea level.

140

Northern peak from	Bearing		
Pitchincha	197° 40'		
Quinché	128 24	30"	
Papaourcou	292 45		V.A 3° 51' 39"

Southern peak from	Bearing		
Pitchincha	196 36		
Quinché	127 24	30"	
Papaourcou	289 52		V.A 4° 20' 19"

Two triangles near the base line were observed thus:

1. Carabourou 91° 54' 03"
   Oyambaro 61 10 33   Length P - O 13 848 t
   Pitchincha 26 55 24   Length P - C 12 156 t

2. Tanlagoa   Bearing from Quinché
                    310° 38' 30"
   Gouapoulo 65 33 00
   Quinché 65 28 40

Quinché to Papaourcou	200° 29' 22"	37 564t
Pitchincha- S. Ilimissa	196 36	29 589
Pitchincha- N. Ilimissa	197 40	

From these observations the following absolute heights were derived.

I. de L'Inca	42t
Niguas	538
Ilimissa	2 812
Pitchincha	2 434

Note: This is the summit whereas for the triangulation a point 211t lower was used.

The effect of these heights on the length of a degree to reduce it to sea level was found to be - 21.4t per degree or 66.4t for the whole arc.

Heights of stations above sea level.

There are three versions of some of these as will be seen from the following table.

Station	La Condamine	Bouguer	Juan	Ulloa*
Carabourou	1226.00t	1226t		1226t
Oyambaro	1352.11	1352		1356
Cotchesqui	1490.84	1487		
Pamba-marca	2109.80	2109		2112
Goapoulo (Godin mark)	1541.85	1544	1620	1566
Goamani (Godin mark)	2080.42			
Tanlagoa	1743.37	1743		1754
Pitchincha signal	2222.53	2225		2228
Schangailli	1405.53	1408		1409
El Coraçon signal	2212.18	2163	2298	2213
Pouca-ouaicou signal	2264.47	2263		2265
Milan	1793.82	1789	1879	1794
Papa-ourcou	1828.39	1827		1833
Ouango-tassin	2086.29	2080		2090
Tchoulapou	1952.58	1945	2038	1954
Hivicatsou	1579.09	1568		1579
Chitchitchoco	1824.22	1817	1909	1828
Moulmoul	2006.09	1999		2011
Ygoalata (Goayama)	2243.75	2235	2329	2248
Ilmal	1941.77	1935		1948
Dolomboc	2098.52	2094	2181	2104
Nabouco	1715.98	1708		
Amoula	1790.31	1784		
Zagroum	1813.82	1808		1818
Lalangouco	2236.74	2234	2319	2244
Senegoalap	2172.22	2168		2179
Choujai	1958.27	1956	2041	1962
Satcha-tian Loma	2206.28	2205		2212
Sinacaouan	2336.84	2334	2419	2336
Quinoa Loma	2037.18	2034		2041
Boueran	1977.51	1972	2061	1980
Yassouai	1881.70	1878		1886
Cahouapata	1819.97	1818	1904	1824
Borma	1614.06	1612		
Pougin	1482.69	1482		
Pillatchiquir	1728.94	1730		
Ailpa-roupachca	1587.75	1583		
Ouaoua Tarqui	1365.34	1366		
Chinan	1411.37	1411		
Quito Grand Place	1462	1466		

Station	La Condamine	Bouguer	Juan	Ulloa*
Riobamba	1695			
Cuença	1350		1414	1332
Guanacauri				1334
Campanario			1901	1835
Cosin				2040
Cuicocha			2128	2055
Mira			1334	1267
Chinchulagua				

*Ulloa quotes differences of height. These have been applied to a value of 1226t for Carabourou and meaned where necessary. Note that on any one mountain the different observers could well have occupied noticeably different points.

Astronomical observations

Many sets of observations were taken at both Cotchesqui and at Tarqui for the determination of the amplitude of the meridian arc. At each location an observatory was constructed to house the zenith sector and peripheral equipment. Great care was taken to adjust the sector into the plane of the meridian and this direction was retained by 18 ft long wires at the sides of the observatory that were protected against accidental disturbance. Every day the parallelism of the limb to the meridian was calculated by use of a 25 inch, closely divided scale. For each observation the instrument was turned throught 180° to help mean out some of the sources of error; very much akin to changing face on a theodolite.

In all cases the results were reduced to the 1st of January, 1743 as a datum for direct comparison and generally observations were to ϵ Orion. Although θ Antinoüs and α Verseau were also observed the results were not used.

At Tarqui.

12 - 27 November 1739	1° 40′ 43.3″
8 - 13 December 1739	52.2
30 December - 13 January 1740	52.1

The readings on each occasion were tabulated thus:—

Limb	Date	Micrometer[1]	Precession	Aberration	Nutation	1.1.43		Mean	Result
E	12.11.39	+7′ 42.5″	+9.9″	−6.9″	+8.7″	+7′ 54.2	}	+7′ 54.0″	
	13.11.39	7 42.1	9.9	−6.9	8.7	7 53.9			
									+10′ 25.5″
W	15.11.39	+2 19.4	9.9	−6.6	8.7	2 31.4	}	+2 31.5	
	19.11.39	2 17.7	9.9	−6.1	8.7	2 30.2			
	27.11.39	2 19.4	9.8	−5.1	8.7	2 32.8			

Arc of sector[2]	3° 11 01.2	
Double zenith distance	3 21 26.7	
Zenith distance	1 40 43.3″	
	by ϵ Orion at Tarqui 1.1.1743	

Note (1) on the actual micrometer there were only parts marked, not minutes and seconds.

Thus:—

E	12.11.1739	+ 1055 parts	
	13.11.1739	+ 1054	+ 1054.5
W	15.11.1739	+ 318	
	19.11.1739	+ 314	+ 316.6
	27.11.1739	+ 318	
			+ 1371 parts

Since 1000 parts had been calibrated as 7′ 18″ 24‴
Then 1371 parts = 10′ 01″

Note: (2) for the 2nd and 3rd groups of observations a limb of 3° 22′ 15″ was used.

After completion of the complete group of observations the results were approved between the observers by a certification thus:

"I certify the truth of ....." signed by Bouguer, then by La Condamine and supported by Verguin. In all their work they held a suspicion of each other and some of their certifications were before a notary.

At Tarqui.

Using ϵ Orion and a limb of 3° 22′ 15″

5 and 17 March 1741	1° 41′ 26.6″
28 July - 19 August 1741	15.0
12, 13 September 1741	10.1
9 and 28 October 1741	11.6
18 and 21 November 1741	09.2
2 to 4 December 1741	13.0

Bouguer meaned the last five of these sets after they had been corrected as 1° 41′ 08″

At Cotchesqui.

Using ϵ Orion.

19 February to 9 March 1740	1° 25′ 45.6″
11 March to 20 April 1740	44.4

Neither Bouguer nor La Condamine were very happy with the results up to this stage so finally in 1742-3 they managed to obtain simultaneous observations at both observatories onto the same star - ϵ Orion.

At Tarqui.
With a limb of 3° 22′ 15″

29 November - 3 December 1742	1° 41′ 11.0″
8 December - 21 February 1743	10.25

At Cotchesqui
With a limb of 2° 51′ 50″

9 - 16 August 1742	1° 25′ 52.4″
20 August - 8 October 1742	53.4
22 October - 2 January 1743	50.5

From these simultaneous results they extracted:

8 - 9 December 1742	Cotchesqui	1° 26′ 54.0″
	Tarqui	1 41 46.1
		3 08 40.1
17 December 1742	Cotchesqui	1 24 39.0
	Tarqui	1 40 38.8
		3 05 17.8
Mean result, allowing for refraction		3 07 00.95

Bouguer accepted a value of 3° 07′ 01″ for the amplitude to combine with his value for the arc length of 176 940t at the level of Carabourou

Whence	1°	=56 767t	
or		=56 746	at sea level.
Adjustment		+ 7	for temperature effect on the iron toise as compared to the wooden rods.
Accepted value	1°	=56 753t	
or		945.9t	per minute of arc and 3 251 707t for the radius of curvature at the equator.

Note however that La Condamine computed the arc as 176 950t i.e. 10t greater than Bouguer.

> or 1° = 56 770t at level of Carabourou
> = 56 749 at sea level.

He did not see fit to apply the same temperature correction as Bouguer.

The rounded mean value for Bouguer and La Condamine was taken as 56 750t. Godin and Juan, as shown later, from an arc of nearly 3° 27′ found a value of 56 768t.

Two reccomputations of the arc are worthy of note:
Von Zach.
[55] found the arc to be 3° 07′ 03.79″ for 176 874t or
1° = 56 732t
Delambre,
[6] found the arc to be 3° 07′ 03.2″ for 176 881t or
1° = 56 737t
Later, when Bessel [38] analysed 10 different arcs he used 3° 07′ 03.455″ and 176 875.5t as the mean between the values of Von Zach and Delambre. These values can be compared with Hayford (1909) ellipsoid of about 1° = 56 734t. It is interesting to note that on the 22nd of March 1742

Bouguer and La Condamine wrote to Godin that 945t was the whole number of the toise in 1 minute of arc near the equator. Godin preferred 946t but surprisingly said he would not communicate his full results to Bouguer and La Condamine until his return to Europe.

From the other stars that had been observed by Bouguer and La Condamine the arc length would have been:

From θ Aquillae (Antinoüs)	3° 06′ 59″
From α Aquarius (Verseau)	58

Among their other astronomical observations they determined the distance between the tropics as 46° 57′ 04″ or the obliquity of the ecliptic as 23° 28′ 32″

The latitude of their stations at Cotchesqui and Tarqui are given by Delambre [20] as:

	From La Condamine	From Bouguer
Cotchesqui	0° 02′ 32.5″ N	0° 02′ 29.95″ N
Tarqui	3 04 33.2 S	3 04 30.8 S
Amplitude	3 07 05.7	3 07 00.75
Mean	3 07 03.2	

Bessel, in his analysis of 10 arcs, gave different values again:

Cotchesqui	0° 02′ 31.377″ N	
Tarqui	3 04 32.068 S	
Amplitude	3 07 03.455 =	mean arc length from Von Zach and Delambre.

Several efforts were made over a number of years to combine the results of this arc with others to derive the earth parameters. Among these were:

(a) Combining Lapland and Peru

e²	= 0.006 435 06	e =	0.080 219
a	= 3 271 652t		
f	= 0.003 223		= 1/310.3
Quadrant	= 5 130 817t		= 10 000 150m

(b) Combining Peru with France of 1740. i.e. 1° = 57 075t
f = 1/303.6

(c) Combining central USA arc of parallel 39° with Peru [229]

a	= 3 272 400t	= 6 378 027m
b	= 3 261 518t	= 6 356 819m
f	= 1/300.7	

(d) Combining Lake Erie arc of parallel 42° with Peru [229]

a	= 3 273 321t	= 6 379 822m
b	= 3 261 466	= 6 357 716m
f	= 1/288.6	

(e) Combining Lake Superior meridian arc with Peru. [229]

a	= 3 272 169t	= 6 377 577m
b	= 3 261 394	= 6 356 577m
f	= 1/303.7	

(f) Combining the Atlantic meridian arc with Peru [229]

143

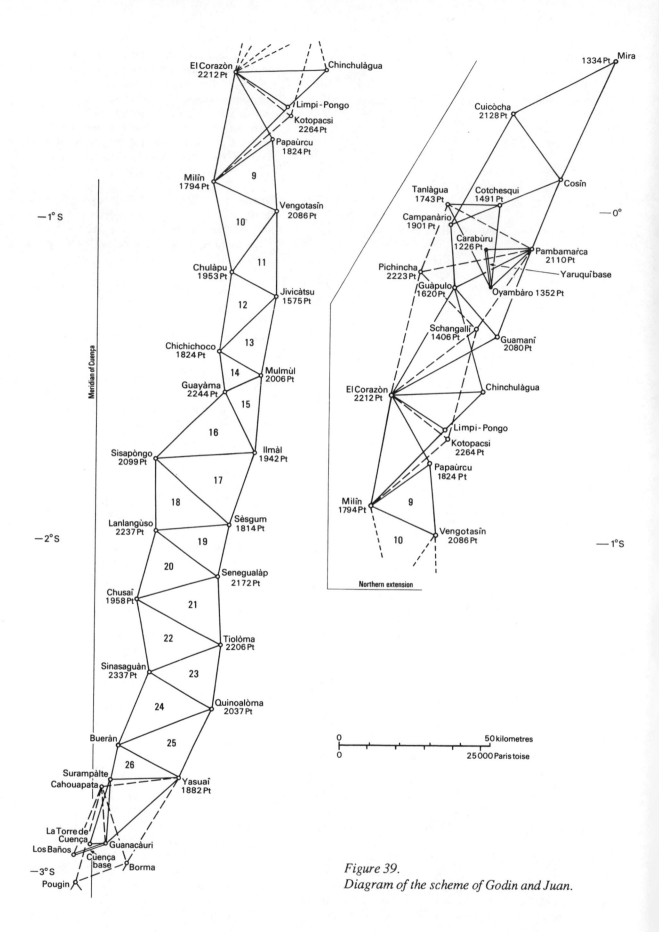

Figure 39.
Diagram of the scheme of Godin and Juan.

144

a = 3 272 414t   = 6 378 054m
b = 3 261 701   = 6 357 175
f = 1/305.5

(g) Combining Peru with eight other arcs including that of Lapland. [182]

e² = 0.007 202   f = 1/278

The extra work of Godin and Juan.

Godin and Juan made extra observations to give an alternative solution. Several triangles were added to the north; a base line was measured at Los Bannos and different astronomical observations used.

The base from Guanacauri to Los Bannos, called the Cuença base, was measured in the same way as the other baselines. Unfortunately it crossed several rivers and other obstacles that required two small triangles to be observed with quadrant. The result was:

6 197t 3 ft 8 inch at 1023°R

It took 21 days to measure. The rivers were between ¾ and 1 ell deep - or up to one's middle (1 Ell = 45 inches). The check length calculated through from Yarouqui was 6 196 t 3 ft 7 inches

For their arc from Pueblo Viejo near Mira, to Cuença, reduced to sea level, Juan found 195 725.397t and Ulloa 195 743.697t from which the mean of 195 734.547t was accepted.

The figure of 195 743.697 had been derived as 195 817.081t at the height of Carabourou and 195 747.596t at sea level. This was then reduced in the ratio 6 274 to 6274t 9 inches to achieve 195 743.697t For the amplitude 3 stars were used - ϵ Orion, θ Antinoüs and α Verseau. Results were:

3° 26′ 53″   (mean on 3 stars)
3 26 52½   (mean on 3 stars)
Mean 3 26 52¾

Whence 1° at sea level = 56 767.788t
Two measures of the latitude of Cuença gave:

− 2°54′10″ on 25.09.1740
−2 54 07¼ on 27.09.1740

They were using a 20 ft. radius sector where the limb was of copper and the ends of it were nailed and riveted to an iron bar mounted on a 20 ft. length of timber. The telescope

had a micrometer. The plumb bob crossing the limb for reading the arc was just over 1 pound weight and its string about 0.03 line diameter.

It should be noted that whereas Bouguer and La Condamine adjusted their inclined angles to the horizontal prior to computations that Godin and Juan did not. They computed using the slope angles and slope sides and subsequently reduced the distances to the horizontal.

For the reduction they used the same method as Bouguer and La Condamine.

$$\text{Horz. distance AB} = \text{slope dist. AB} \times \frac{\sin (\text{complement of mean VAs})}{\sin(90 + \tfrac{1}{2} \text{ angle at earth center})}$$

For example:

Side GK Slope distance	19 179.609t	
Vertical angle G to K	1° 05′ 42½″	
Angle at center	0 19 59	
Sum	1 25 41½	
Vertical angle from K to G	1 24 35	
Mean	1 25 08	
Complement	88 34 52	
90 + ½ center angle	90 09 59½	

Thus:

$$\text{Horz. distance} = 19179.609 \times \frac{\sin 88° 34′ 52″}{\sin 90 \ 09 \ 59½}$$

$$= 19173.809t$$

Where the angle at the center was found as (length in toise/16) seconds In this reduction however each line was reduced to the horizontal at the elevation of the lowest of the two points defining that line; not to a common datum. Those later used to compute the arc length were so reduced.

Over 29 triangles Juan quoted a total misclosure, irrespective of sign, of 351.7″ or a mean of about 12.5″.

From the published results of Juan and Ulloa it is possible to arrive at two values for the arc length from Pueblo Viejo to Cuença. The values listed are as given in [116] although recomputation of the data as listed would modify the results. In particular there appears to be a mistake with the value for Mira to Cuichocha in Ulloa's results. Even the total will be found to be incorrect by 5t. With other small discrepancies the mean value is incorrect by several toisc.
Results of Ulloa:

	Horizontal distance at level of Carabourou	Bearing	dN t
Pueblo Viejo			+ 170.620
Mira	20 707.010	234°12′17″	-12 128.372
Cuicocha	23 132.417	202 48 37	-21 323.270
Campanario	8 703.055	174 00 16	- 8 655.453
Guapulo	21 951.316	208 35 15	-19 275.239*[1]
El Coraçon	19 170.757	190 26 53½	-18 850.289
Milan	16 764.871	167 32 28½	-16 370.076
Tchoulapou	13 214.207	186 38 58	-13 125.317
Chichichoco	6 759.349	162 42 43	- 6 454.071
Goayama	16 512.612	222 40 58	-12 138.182

145

	Horizontal distance at level of Carabourou	Bearing	dN t
Sisapongo			
Lalangouco	13 134.552	179 42 46	-13 134.390
Chujai	12 925.073	194 28 31	-12 514.538
Sinacaouan	13 584.365	168 34 30	-13 315.348
Boueran	12 677.560	203 07 05	-11 659.234
Surampelte	7 643.068	199 53 00	- 7 187.278
Cuença	9 876.352	189 38 25	- 9 736.791
			- 114.853
			195 817.081
Reduction to sea level*₂			69.485
			195 747.596
Scale reduction			- 3.899
			195 743.697

*₁ The distance for Guapulo to El Coraçon was derived in three parts as:

Tanlagoa - Pitchincha	11 646.749
Pitchincha - El Coraçon	20 365.638
Tanlagoa - El Coraçon	32 012.387
Tanlagoa - Guapulo	-12 737.148
Guapulo - El Coraçon	19 275.239

*₂ Correction quoted is 76.485 which is about right. However figures given for result indicate a deduction of 69.485 - indicating a +7 adjustment that is unidentified. Alternatively, working backwards there is a printing error equivalent to 12t.

Results of Juan and Godin:

Pueblo Viejo	Horizontal distance at various levels	Bearing	dN t
Mira			+ 170.620
Cuicocha	20 703.536	234°11′07″	-12 115.006
Campanario	23 130.299	202 47 43	-21 323.709
Guapulo	8 698.453	173 59 31	- 8 650.321
El Coraçon	21 953.245	208 36 40	-19 272.536
Milan	19 173.809	190 28 26	-18 854.333
Tchoulapou	16 765.992	167 34 25½	-16 373.266
Chichichoco	13 217.175	186 40 44½	-13 127.474
Goayama	6 762.335	162 43 18½	- 6 455.956
Sisapongo	16 523.658	222 42 09½	-12 142.961
Lalangouco	13 141.311	179 43 55½	-13 141.167
Chujai	12 931.589	194 28 31	-12 521.083
Sinacaouan	13 591.351	168 35 18	-13 322.659
Boueran	12 684.594	203 08 31	-11 663.917
Surampelte	7 645.400	199 54 04	- 7 188.828
Cuença	9 879.214	189 39 46	- 9 739.055
			- 114.845
			195 836.460
Reduction to sea level*			111.063
			195 725.397

*Each line was reduced from a different elevation. Mean result from Ulloa and Juan 195 734.547t

Recomputation

This section is based on work that formed part of an MSc thesis at University College London, 1984.

In any investigation based on published observations only there are numerous difficulties. If one had been able to refer to the original manuscripts then missing information could have been located and printing errors more easily highlighted.

This step had now materialized for the future.

Such questions come to mind as how many times was each of the angles measured? Each the same number or of sufficient variation to warrant weighting? Data was such that most had to be treated equally. But what of angles greater than 90° where the observation had to be made in two parts? There were 7 such in the scheme of La Condamine, 6 for Bouguer, and 4 for each of Juan and Ulloa. They could legitimately be given a weighting of √2 for that reason alone.

What corrections should be considered that were only

uncertainly known, or not known at all, in the 18th century? It could be that many corrections that would be considered today are of such a magnitude that they would have been swamped by observational errors at that time. This would indeed be the case for such corrections as that for skew normals.

Refraction in particular was a problem. During the expedition Bouguer experimented at length to determine some of its properties. He appreciated that it might vary with height but could not quantify the effect. He derived a table of values as a function of angles of elevation but the corrections he applied did not always seem to correspond with that table.

No account was made for solar parallax which could amount to several seconds.

Until Bouguer managed to connect the scheme to sea level the Academicians considered a range of heights for Carabourou (their initial station). Juan used 1402t and Ulloa

1268t but later adopted Bouguer's value of 1226t.

The effect now called deviation of the vertical was then unknown and in its neglect many tens of seconds of error were possible in the amplitude of the arc. Unfortunately, without knowledge of the deviation at the stations the relevant corrections still cannot be applied.

For some aspects of the work insufficient information is published to allow more than a simple check of the quoted calculations. Lack of temperature values for instance foil any attempt to verify the base lengths. The recomputation of the Yarouqui base agreed with La Condamine to the second decimal place as 6274.05t for the inclined length. That at Tarqui came to 5259.20t against 5258.95t by La Condamine. By calculating from the first base to the second the value for Tarqui became 5260.00 against 5260.03 by La Condamine.

Recomputed results:

	Academicians	Recomputation
Yarouqui base	6274.05	6274.05
Tarqui base	5258.95	5259.20
Tarqui base computed from Yarouqui	5260.03	5260.00
Godin observations computed by Juan		
Bearing Oyambaro - Pambamarca	44°10'44"	44°10'42.8"
Oyambaro - Pambamarca	44 11 30	44 11 34.3
Oyambaro - Tanlagoa	329 56 59	329 56 51.2
Observations by Juan		
Zagroum - Lalangouco	260 14 31	260 00 02.8
Yassouai - Boueran	294 45 24	295 14 04.4
Campanario - Cosin	60 50 16	60 50 03.3
Godin observations computed by Ulloa		
Oyambaro - Pambamarca	44 09 59	44 11 07.9
Oyambaro - Pambamarca	44 10 49	44 11 58.2
Oyambaro - Tanlagoa	329 56 55	329 56 52.4
Observations by Ulloa		
Chitchitchoco - Goayama	162 42 42	162 44 43.9
Ilmal - Mulmul	358 51 29	358 49 24.3
Ilmal - Mulmul	358 52 00	358 49 56.1
Amplitude of arc.		
Juan	3°26'38"00'"	
Ulloa	3 26 37 27	
But they ignored these results and used		
	3 26 52 45	3°26'49"43.5'"
La Condamine and Bouguer	3 07 00.95	3 07 00.95

For the meridian arc there are four possible comparisons from available data: the computations of La Condamine, Bouguer, Juan and Ulloa.

These are given in terms of the northing coordinates given by La Condamine using Cotchesqui as datum for the results of La Condamine and Bouguer and El Coraçon for Juan and Ulloa.

1. La Condamine.

Cumulative meridian length

	Original	Recomputation
Quito	0.00	0.62
Pambamarca	7 864.22	7 864.29

	Original	Recomputation
Oyambaro	845.01	845.05
Cotchesqui	14 812.91	14 812.91
Tanlagoa	14 399.22	14 399.24
Pitchincha	2 753.98	2 754.23
El Coraçon	-17 612.12	-17 610.94
Milan	-36 465.23	-36 463.57
Tchoulapou	-52 838.54	-52 836.40
Chitchitchoco	-65 966.05	-65 963.50
Goayama	-72 422.98	-72 419.58
Dolomboc	-84 564.12	-84 563.50
Lalangouco	-97 700.71	-97 699.56

	Original	Recomputation
Choujai	-110 216.96	-110 216.33
Sinacaouan	-123 533.22	-123 531.41
Boueran	-135 193.13	-135 192.50
Cahouapata	-142 381.95	-142 381.82
Pougin	-158 350.55	-158 352.65
Chinan	-162 994.96	-162 997.61
	177 807.87	177 810.52
	- 857.40	- 857.40
	176 950.47	176 953.12 ˙

2. Bouguer

Cumulative meridian length

	Original	Recomputation
Cotchesqui	14 812.91	14 812.91
Tanlagoa	14 397.06	14 399.24
Pitchincha	2 750.83	2 753.14
El Coraçon	- 17 614.63	-17 612.30
Milan	- 36 465.68	- 36 464.68
Tchoulapou	- 52 835.32	- 52 835.61
Chichichoco	- 65 961.04	- 65 962.30
Goayama	- 72 414.19	- 72 418.64
Sisapongo	- 84 550.79	- 84 550.55
Lalangouco	- 97 685.25	- 97 706.19
Choujai	-110 200.98	-110 219.54
Sinacaouan	−123 516.36	-123 537.56
Boueran	-135 172.90	-135 192.64
Cahouapata	-142 358.97	-142 378.17
Pougin	-158 322.97	-158 342.62
Chinan	-162 965.50	-162 984.84
Oua-Oua Tarqui	-158 527.24	-158 547.29
	177 778.41	177 797.75
	- 838.00	- 838.00
	176 940.41	176 959.75

3. Ulloa

Cumulative meridian length

	Original	Recomputation
Pueblo Viejo	43 599.59	43 562.71
Mira	43 770.21	43 733.33
Cuicocha	31 641.84	31 639.47
Campanario	10 318.57	10 317.06
Guapulo	1 663.12	1 661.15
El Coraçon	- 17 612.12	- 17 610.94
Milan	- 36 462.41	- 36 461.50
Tchoulapou	- 52 832.49	- 52 832.80
Chichichoco	- 65 957.81	- 65 959.43
Goayama	- 72 411.88	- 72 415.11
Sisapongo	- 84 550.06	- 84 552.53
Lalangouco	- 97 684.45	- 97 688.72
Choujai	-110 198.99	-110 200.32
Sinacaouan	-123 514.34	-123 520.69
Boueran	-135 173.57	-135 173.75
Surampelte	-142 360.85	-142 358.67
Cuença	-152 097.64	-152 094.60
	195 697.23	195 657.31

	Original	Recomputation
	114.85	114.85
	195 812.08	195 772.16

4. Juan

Cumulative meridian length

	Original	Recomputation
Pueblo Viejo	43 578.84	43 577.70
Mira	43 749.46	43 748.32
Cuicocha	31 634.45	31 634.20
Campanario	10 310.74	10 314.24
Guapulo	1 660.42	1 660.41
El Coraçon	- 17 612.12	- 17 610.94
Milan	- 36 466.45	- 36 462.79
Tchoulapou	- 52 839.72	- 52 833.82
Chichichoco	- 65 967.19	- 65 957.47
Goayama	- 72 423.15	- 72 413.42
Sisapongo	- 84 566.11	- 84 545.34
Lalangouco	- 97 707.28	- 97 681.68
Choujai	-110 228.36	-110 194.18
Sinacaouan	-123 551.02	-123 516.07
Boueran	-135 214.94	-135 170.93
Surampelte	-142 403.77	-142 333.63
Cuença	-152 142.83	-152 055.78
	195 721.66	195 633.48
	114.84	114.84
	195 836.50	195 748.32

Taking the four values for the arcs and reducing each to the length of 1° at sea level gives:

	Original	Recomputation
La Condamine	56 749.44	56 750.29
Bouguer	56 746.21	56 752.42
Ulloa	56 770.12	56 772.38
Juan	56 765.15	56 765.44
Mean of La Condamine and Bouguer	56 747.83	56 751.36
Mean of Ulloa and Juan	56 767.64	56 768.91
Overall mean	56 757.73	56 760.13

Equipment

Quadrant (Quart de Cercle).

Several quadrants of varying sizes were used to measure the 'horizontal' angles. That used by Bouguer was of 2½ ft. radius; La Condamine had one of 3 ft. radius that had previously been in the possession of le Chevalier de Louville and was the first use of a micrometer on the telescope; Godin had one of 21 inch radius and the Spanish Officers had one by Langlois that was a little larger than that of Godin at 2 ft. radius. Considering the rough and windy terrain it was important for them to be as rigid as possible and this was achieved both by use of heavy iron bars in the construction and in a very sturdy frame.

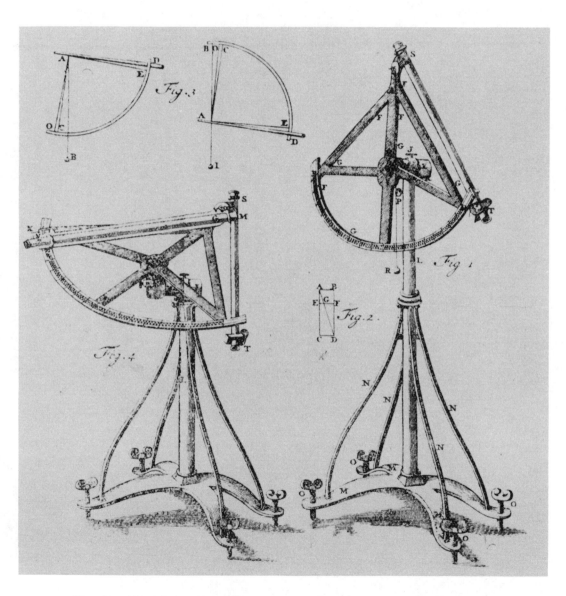

23.   *Details of the quadrant illustrating how it can be used in various attitudes.*

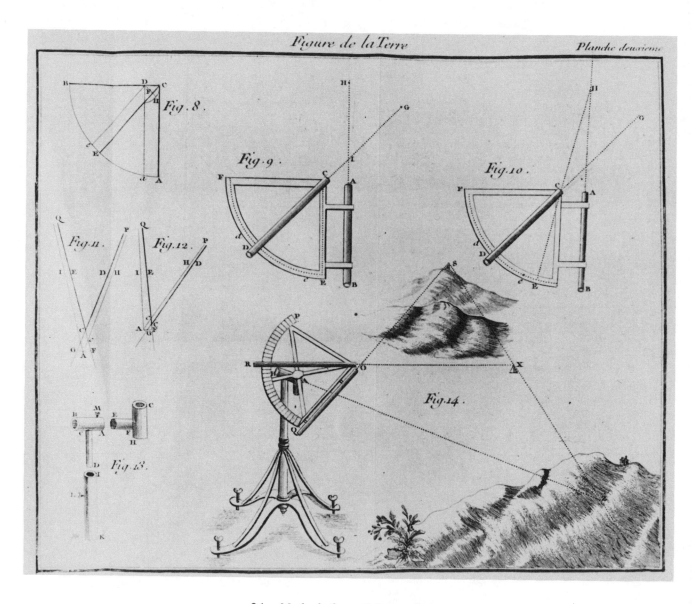

24. Method of use of the quadrant.

Each instrument was meticulously checked for adjustment and accuracy as will be seen from the following procedures.

The quadrants were among the earliest to have telescopes attached and with such attachments it was essential for the optical axis to be parallel to the radius. This obviously applied to both fixed and movable telescopes where the various axes should pass through the centre of the graduated arc. That this was so was checked before and after each angle and if there was any small variation the mean was accepted.

Of just as much importance was verification of the divisions of the arc. With the manual methods of graduation of that period there could be variations that were random, systematic, isolated or even gross.

Several methods were used, two accepted ones and others that were developed specifically in Peru. Remembering that, unlike a theodolite, the maximum angle possible with the instrument in one position was 90°, then simply to measure a series of angles that completed a circle was a check on gross errors of either graduation or/and bisection of targets. Similarly the angles of a closed figure, particularly the triangle, performed a similar check. In such terrain the points would, of necessity, be at different elevations and before closing a round of 6 or 7 angles to 360° much reduction was required. Such observations could be considered good to about 20-40" per angle.

La Condamine's new method took two forms. He set out a long, straight line of, say, 1500t and treated this as the diameter of a circle. The quadrant, correctly centered, was put at the center of the circle. The semi-circle was subdivided into different numbers of equal parts i.e., 3, 4, etc., so that successive checks could be made on angles of 60°, 45° etc. For the smaller angles he devised a method of checking each successive degree mark.

For this he set out a distance of exactly 500t from the quadrant position and a cord was stretched at right angles at the far end. On this cord were marks to indicate the tangent value for each successive degree.

e.g. for	1°	at	8.73t
	....		
	5°		43.74
	....		
	20°		181.99

Two particular sources of error were possible here — the markings on the cord and the setting out of the right angle. For the marks on the cord one might assume that they could be placed to within 5mm which is about 0.0025t. For a 20° angle at the instrument an error of this magnitude would affect the angle by only about 1".

For the setting out of the right angle, if carefully done this could be good to 1'. On the 20° angle this would be equivalent to about 7" in the result. Thus it would seem that the method was not necessarily as crude as it might appear.

The fixed telescope would be set in the zero direction and then as each degree point on the cord was sighted any error could be quantified by the micrometer. In this way a complete table of graduation errors for each instrument was compiled. It was also feasible to intercompare their various instruments in this way. Up to a month is quoted as the time necessary to complete such checks.

In one instance an alidade was found to be 1/8 line too long (at 12 to the inch this is around 1/100 inch) and this affected its correct rotation about the center of the arc. The magnitude of the effect was variable but could be more than a minute of arc.

Considering that the graduations of the arc were by dividers and that setting accuracies better than about 1/25 line could not be expected, then an error in the total arc by this means would reflect proportionately in the subdivisions.

With the form that the instrument took it was very difficult to center correctly over a station and thus essential to indulge in reduction to center at each station. This could be achieved by recording two measurements and applying them as corrections at the computation stage. This problem was aggravated because at most stations two instruments were used simultaneously and so it was impossible to get more than one of them anywhere near the correct position.

Since the angles could be recorded into the seconds there were several comments on the problem of the log sines of parts of minutes. Although not specifically stated it is obvious that the tables used were only graduated to 1 minute and the variations of differences with size of angle caused concern.

The sector

Although the party had a 12 ft. radius sector with them it was known to be faulty and their instrument man, Hugot, was charged with constructing a new one. The most obvious change was to be the replacement of the 30° arc with one of only about 5°-10°. An additional problem was the stability of the whole since such long telescopes and arms were very liable to flexure.

A sector had three main parts; a graduated scale or limb of copper; an arm that was joined to the limb and of a length equivalent to the radius; and a suspension system. The whole was supported from the roof of a purpose built observatory. Full details of the construction are given in [26] and [135].

What is of particular note is that La Condamine and Bouguer spent a month checking the values of all parts of the micrometer that was attached to the telescope of the sector. This they did rather along the same lines as for the quadrant.

The sector was placed horizontally at one end of the Tarqui base and at the other end, 5260t distant, 2 targets, 80 ft. apart were set at right angles to the baseline. One might liken this to a subtense bar measurement. The ratio of the distances gave an angle of about 8' 43" which was calculated to correspond to 1196 parts (or 1193) of the micrometer. When 1000 parts were equal to about 7' 18" and a table could be compiled for the equivalent of each division. There were approximately three divisions to a second.

The most difficult part of constructing a sector was the graduation of the arc since to subdivide a given length into a large number of equal parts was fraught with sources of error. The Academicians found a way around this. For the star they intended to use at Tarqui — ε Orion — they calculated that an arc of 3° 22' would suffice. Now an angle only 15" less than this was equivalent to a chord of 1/17th the radius. Working backwards from this, they found a radius which was near.. e.g. 8 inch 6 line x 17 = 12 ft. 6 line. Thus from the arc of length 8 inch 6 line it was possible to step off this amount 17 times to determine the center. The small variations of the observed angle from 3° 21′ 45″ were accommodated by the micrometer.

A shorter arc of 1/18th of the radius was also used. See page 51. For the Cotchesqui base the required arc was 2° 52′. Hence Bouguer decided to use 1/20th x radius which gave 2° 51′ 54″ and the remainder by micrometer.

To overcome any eccentricities of the optics it was possible to reverse the unit and take the mean result.

Among the possible causes of variation in the results obtained were:

1. The effect of expansion of the metal. By experimentation it was found that a 10° R rise in temperature on the iron component lengthened it by 0.012 line. However the attached copper arc would lengthen rather more. It was this difference that mattered, not the absolute amount.
2. If there were 1/6 line flexure on the telescope this would affect the angle by 20 seconds.
3. Stability of the whole structure since earthquake vibrations were frequent.
4. Parallelism of the telescope to the plane of the sector.
5. Movement of the sector in the plane of the meridian.
6. Oscillation of the plumb line was dampened, but not fully eliminated, by use of a container of water.
7. The plumb line was of a thickness equivalent to 3 seconds.
8. Parallax of the cross wires varied with observer and time.
9. Effect of long and short sightedness of the observers. Juan, Ulloa and Godin later used one of 20 ft. radius at Cuença and Pueblo Viejo.

Standard toise.

The group took with them a toise bar of polished iron that was 17 line wide and 4½ line thick. (One reference quotes 8 line wide by 3½ line thick). Before departure from Paris Godin had compared this with the toise étalon that had been determined in 1668 and set in the foot of the stairs of the Grand Châtelet of Paris. A second toise was deposited at the Academy of Sciences and was similar to that taken to Lapland by Maupertuis. The toise taken by Godin to Peru came to be referred to as the toise du Peru and was of much better quality than that of the Châtelet. They also had with them:

- ½ toise of medium quality steel 6 line wide by 3 line thick
- ½ toise of beaten copper 8 line wide by 3 line thick.
- A forged and polished brass plate on which a ½ toise had been marked; this was 4 inch wide and ½ line thick.
- ½ toise of aged brass, beaten and polished, 6 line wide and 2 line thick.
- A 35 inch glass tube, 2 line exterior and 1 line interior diameter.
- A length of freestone found in the courtyard of a house.

Once in Peru a very novel method was used for determining the coefficient of expansion of the toise bar. The bar was modified at one end by La Condamine so that it could be mounted on a knife edge and allowed to vibrate in the same way as a pendulum. A pendulum was mounted similarly nearby so as to be sychronous at a particular temperature. Although within sight, the pendulum was in a separate room so that the temperatures of each could be made different.

It was then arranged for the circulating air to be warmed to 55° R so that the toise lengthened and hence vibrated more slowly than the pendulum. The time interval until they again became synchronous was noted and from this it was possible to calculate the amount of extension. For the toise bar its center of oscillation was 582.56 line from the suspension point at 13°R. It was found that the extension was 0.0115 line per °R. (Some references give 0.0117 line). The 0.0117 was the mean of 3 results by La Condamine of 0.0115, 0.0118 and 0.0119 although he was happy to accept 0.0115. La Condamine's result compared well with other experiments by Juan and Godin in Quito.

Other experiments on the change of length using the other bars they had with them gave the following accepted values for 10°R change in temperature.

toise of iron	0.265 line
½ toise of iron	0.132
½ toise of steel	0.123
½ toise of copper	0.192
Plate of brass	0.240
½ toise of brass	0.200
35 inch glass tube	0.032
Freestone	0.020

The conclusion, however, was that the extension due to heat above 13°R was more than double the contraction caused by the cold that lowered the temperature of the alcohol below 13°R.

For his final calculation of the length of a degree Bouguer applied a correction of +7t for the variation of the toise as opposed to the wooden rods. This was a correction of approximately 0.1 line per toise.

[98] discusses a test of the above pendulum experiment on the return to Paris. The effect there was 58×10^{-7} for 1°F. The bar was the toise du Nord. This would be equivalent to 0.0113 line for 1°R.

The barometer.

Much barometric work was done during the expeditions particularly for deriving the heights of the stations. The range of values was considerable since at sea level it stood at 28 inch 1 line and on the mountains it fell to below 16 inch. The summit of El Coraçon reduced it to 15 inch 10 line. The observations were reduced in the following way.

Let		inch	twelfths of lines	
Carabourou	= a	= 21	39	= 3 063
Oyambaro	= b	= 20	93	= 2 973
Pambamarca	= d	= 17	40	= 2 488
	A = 126t = height of b above a			

152

Then difference of height from a to d = $A.(L_a - L_d)/(L_a - L_b)$
Where L = log.

a = 3063	log	3.48614	69968
b = 2973		3.47319	49092
$L_a - L_b$		1295	20876
a = 3063		3.48614	69968
d = 2488		3.39585	03760
$L_a - L_d$		9029	66208
colog $(L_a - L_b)$ = 1295.2		6.88766	31643
log $(L_a - L_d)$ = 9029.66		3.95567	13972
A = 126		2.10037	05451
x		2.94370	51066 = 878.4

Thus height of Pambamarca above Carabourou =   878.4 t
Geometric measure   =   882.5

Difference   4.1

Note the use of ten figure logarithms throughout where 6 or 7 would have sufficed.

While there are no notable references in the main works there are scattered references to related aspects that were noted. For example, Godin watched the daily variation of the barometer and found it highest about 0900 hour and some 1¼ line smaller by 1500 hour.

Speed of sound.

Among the many auxiliary experiments carried out were some on the speed of sound.

Before the group left France it had been reported that tests in England by Derham using cannon, muskets and clocks at ranges from 1 to 8 miles gave a result of 1142 ft. sec. or 178½ t.sec. Experiments in France by Cassini, La Caille and Maraldi gave a similar figure of 173t.sec.

Would there be any difference at high elevations? Quito, for instance, was 1517 toise above sea level and the mercury struggled to reach 20 inch.

For their experiment they placed a 4½ ft. cannon on mount Pancecillo with observers at two other positions, one to the Augustines farm near Pambamarca, 5736 t away and the other to Saguanche at 6820 toise. They used cannon balls of 8-9 pounds. The time taken for the sound to arrive was measured in half seconds and found as
For 5736 t, 66 half secs = 173 + 9/11 t.sec = 173.8t
For 6820 t, 76½ half secs = 178 + 46/153 t.sec = 178.3t

Length of seconds pendulum.

La Condamine made pendulum observations at various locations including Quito, at Para and on Mount Pitchincha where the latter was 750t above Quito and Para was at sea level. He found that a pendulum of 28 inches length vibrated 31-32 times more often at Para than at Quito and 50-51 times more often than on Pitchincha. He concluded from this that at the equator two bodies of 1600 and 1000 pounds respectively at sea level when carried to heights of 1450t and 2200t respectively would each have lost 1 pound.

His values for the length of the seconds pendulum were:

At the equator   3 ft 0 inch 7.07 line

At Quito   3 ft 0 inch 6.761 line
At sea level   3 ft 0 inch 7.173 line

Deflection of the plumb line.
(Deviation of the vertical).

While occupied on the arc measurement Pierre Bouguer harboured ideas that large masses such as mountains could have an influence on the plumb line.

In December 1738, the opportunity arose to test out his theories. So anxious was he for the scientific world to hear of his results that they were in a Memoire at the Academy of Sciences by October 1739. Not, it should be pointed out, delivered in person, since it was to be February 1743 before Bouguer left Quito and June 1744 before he was back in France. Considering the slow travelling rate of the period it was remarkable that the results were delivered in 10 months.

He chose the volcanic Mount Chimborazo as an ideal location to carry out experiments. The height he initially gave to it was 3100-3200 toise above sea level but he later adjusted this to 3217t. The particular advantage was that the mountain was an isolated one on a plateau and, hopefully, little affected by the more distant mountains. See figure 37

The volume of the mountain was calculated to be about 2×10^{10} cubic toise or about $1/7.4 \times 10^9$ part of the volume of the globe. Note that he considered only volumes to the exclusion of any effect from different densities of different rock constituents. He equated the shape of Chimborazo to that of a cone but as is pointed out in [96], he used an incorrect formula for the attraction of a cone. He deduced that if he were 1700-1800 t from the center of the mountain, which would be 1900 times less than the earth radius; the attraction would be 1/2000 of the attraction of the earth. In terms of deflection towards the mountain this would be about 1′ 43″.

To test this he chose two stations more or less on the same parallel. One, A, was 1753t at 196° from the center of the mountain and the other, B, at 3750t and 261° 51′ 29″ from A. Station B was calculated to be 505t to the south of the parallel through A, equivalent to 32 seconds of arc.

Meridian height observations were taken on two separate days at each station to a number of stars, balanced to the north and south. The mean corrected results differed by 15″; or an attraction of 7½″.

A difference of such magnitude was considerably doubtful as such a figure was of the same order as the accuracy of the instruments. Bouguer, in fact, stated that it was for the reader to judge whether the differences were sufficiently constant for the means to be of any worth.

While nothing should be allowed to detract from his pioneering efforts in this field, as is witnessed by the retention of his name to describe certain anomalies, his results at this first attempt were not good.

Some Possible Sources of Error.

In such pioneering work there was bound to be much uncertainty and the following are but a sample of some sources of it.

25. *The zenith sector as used in Peru showing details of the suspension system and end of the arc.*

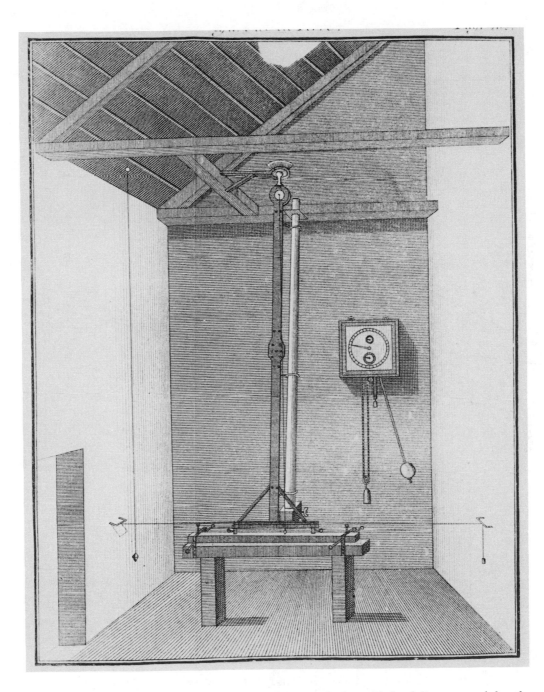

26. *The zenith sector mounted in an observatory. The horizontal wire and plumb line ensured that the arc was kept in the meridian plane.*

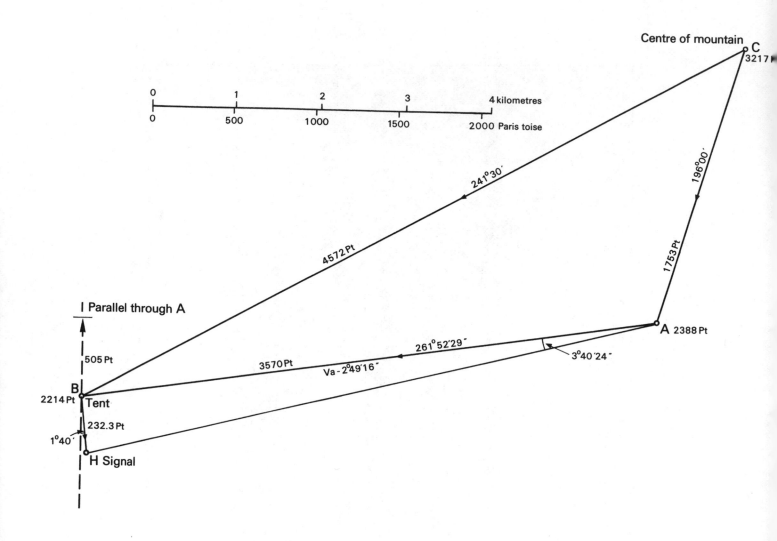

Figure 40. Scheme of points for the experiments at Chimborazo.

Base rods.

The North base was in an area of violent winds where it was more practical to lay the measuring rods on the ground than to use trestles. Two rods were always left in position as the third was carried forward. This was entrusted to unskilled Indians and almost certainly there must have been a certain amount of jarring each time they made a new contact. The effect of this would each time be to make the line appear longer than it really was—a scale error. One which, however, would occur in both measures and no doubt almost cancel out.

To achieve agreement of 3 inch or 1/150,000 must be considered to have involved a large amount of compensation. Even 150 years later such agreement would not have been easy to obtain.

Base length

For some unknown reason the party tried to ensure that the North base would be a whole number of toise but they were thwarted in this. Initially the measured value was less than 6273t and on estimation of corrections it was decided that one terminal should be moved by 3 inch 8 line to achieve a final length of 6274t. Unfortunately by the time all corrections had been applied the result was 4 inch 1½ line in excess of 6274t.

Thus, they would have been better advised not to have tinkered with the original value. In any case it would seem a somewhat pointless exercise.

Expansion.

How were the wooden rods affected by expansion? Temperature devices were not very accurate at that time but it has been suggested that the coefficient of expansion could have been anything from 0.000,015 to double that value.

Eccentricity.

For the quadrants Bouguer made a minute investigation into the possible eccentricity errors. The problem of evaluating this was different to that of a theodolite since it was impossible to compare diametrically opposite readings. The technique adopted was to set two known angles and then measure these with the quadrant. Comparison gave the actual errors but these were entered then into equations as known values with the positions (in x and y) of the center of rotation in relation to the center of graduation as unknowns. From this it was possible to determine the position of the center of rotation and to draw up a table of corrections for the whole 90°. The time spent on this exceeded a month but even so errors of 20″ and 40″ were readily possible.

Horizontal angles.

[213] took a probable error of 8″ per angle and found the uncertainty of the final side to be around 1.7t.

Sides.

At two positions in the chain, auxiliary measurements were used to modify the main chain calculations. This was at the 8th triangle where 1.1 t was involved and at the 16th where 0.3 t was applied. With variations of this magnitude only part way along the chain there would obviously be the possibility of much larger differences as the chain progressed.

The sector.

The position where a plumb line crossed the limb of the sector was important. However since 1″ on the limb was equivalent to about 0.0007 inch there was obviously good scope for a large source of error here.

Attraction.

Little was known at the time about the effects of attraction and Bouguer's experiments were inconclusive. Later investigations have suggested that there could have been at least 10″ discrepancy from this cause at each astronomical station, and that the effect could have been 30 times greater than expected.

Sea level.

The transfer of height from sea level to the cordillera was very tenuous. With a mixture of trigonometric heights and barometer heights there could have been a 50t height error.

Vertical angles.

The angles were by quadrant and while there could have been ± 30″ in each from the reading this could well be extended to ± 1′ by virtue of unaccounted attraction effects. These values were then used in the subsequent reductions.

Refraction.

Was little known and could have contributed noticeable effects.

COMMEMORATION.

A commemorative plaque was fixed to the wall of the Jesuit Church in Quito (later preserved in Quito Observatory) and two pyramids were built at the extremities of the Yarouqui base. Controversy followed.

Consider first the pyramids. These were monuments that La Condamine wished to construct at each end of the northern baseline to commemorate the expedition. A fine idea until it came to the decoration and inscription.

In size they were 2t square rising to 2½t above ground level with a 1t thick foundation set on piles because of the nature of the ground. Most impressive, but how nationalistic could they be in terms of having a French coat of arms and a pro-French inscription to the exclusion of Spain? In any case, were not structures of this size out of place in such a stark countryside?

La Condamine had a problem. Before leaving France he had been instructed to leave an inscribed monument and had even been supplied with a suggested text by the Academy of Belles Lettres, "with those alterations which circumstances of time and place required". He had himself first proposed a very simple Latin inscription of details of the length of the base, date, etc. At the most 8 or 10 lines. The problem was discussed on many occasions by that Academy.

It had taken several months and considerable effort, started on the 30th of April 1740, to construct the pyramids and he was certainly not going to sacrifice all this easily because of the inscription.

The construction was of brick and for this they had to build a special kiln near the area and for the necessary water a 2 league long canal had to be constructed.

The intention was that the monuments should be under the

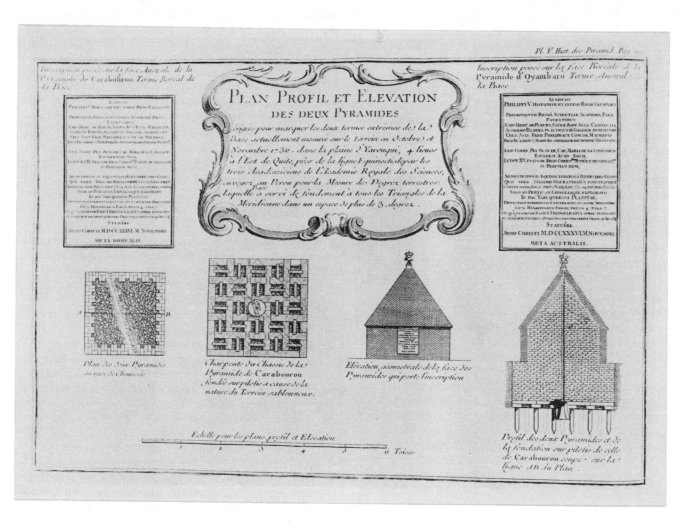

27.   *Diagram of the pyramids. Note the intricate design and overall size of the monument.*

protection of L'Audience Royale de Quito.

First to object were the Spanish Officers Juan and Ulloa who had been away on active service with the Spanish Navy off the coast of S.America while most of the work was carried out. The first inscription made no mention of them nor of the King of Spain, and what of the Fleur de Lys? One reference says that the King was prominently mentioned but that Juan and Ulloa were far down the text. There were obviously several versions in an effort to get agreement.

After numerous of these alternative suggestions had been considered under the mediation of Louis Godin, the Spanish King had pride of place at the top of the inscription and the officers were described as ... equally shared in ... This was far more to their liking than the previously suggested "assisted by" or "with cooperation of".

The Fleur de Lys still remained and aggrieved not only Juan and Ulloa but the Peruvian authorities as well. A court case with the Fiscal of Quito ensued and the withdrawal of such emblems demanded. For a long while the proceedings dragged on with La Condamine defending the Fleur de Lys as representative of both the French and Spanish royal houses.

Not until 1742 did the end of the controversy appear in sight with the decision of the Fiscal that the decoration be omitted and the Spanish Officers receive due recognition. About the same time, however, the Spanish Government was requesting the destruction of the pyramids.

On his return to Paris La Condamine raised considerable objection so that the Marquis de la Ensenada rescinded his destruction decision; but too late. They were gone.

In 1836 they were restored, albeit in different positions, on the orders of President Rocafuerte and an acceptable inscription attached. The inscription as agreed by the Marquis de la Ensenada was in Latin. The translation of which would be:

"In the reign of his Catholic Majesty, Philip V, King of Spain and the Indies; agreeable to the request of his most Christian Majesty Louis XV, King of France, and in condescension to the desire of the Royal Academy of Sciences at Paris, Louis Godin, Pierre Bouguer, Charles de la Condamine, members of that Academy, were, by the command and munificence of the most Christian King, sent into Peru, to measure the terrestrial degrees under the equinoctial in order to obtain a more accurate knowledge of the true figure of the earth. At the same time, by the command, and at the expense of his Catholic Majesty, were sent George Juan, knight of the Order of St John of Jerusalem, and Antonio de Ulloa, both lieutenants in the royal navy, and well acquainted with all the branches of the mathematics, during the whole process of this mensuration they all equally shared in the fatigues, hardships and operations and with an unanimous consent determined in the plain of Yarouqui a horizontal distance of 6272 $^{551}/_{726}$ Paris toises in a line whose direction was N 19° 25′ 36″ westerly and intercepted between the axes of this and the other obelisk as the base or side of the first triangle and a foundation for the whole work. In the month of November, 1736. In the memory of which transaction an obelisk has been erected at each extremity of the said base."

[198] reports that in 1856 a small commemorative column was erected on the hill of Pugin, dominating the plain of Tarqui but the marble plaque of La Condamine that had been placed at the astronomic station was taken in 1804 by Caldas to a museum in Bogota; returned to Ecuador in 1885 but never replaced.

A decree of the President of the Republic on the 12th of October, 1886, ordered the erection of a monument. P.Menten did some triangulation to reconstitute the exact astronomic position at Tarqui for this but it never materialized.

Maybe the most permanent reminder of the expedition was the naming of a peak Montagne des Francois, or Frances-Urca or Frances Loma in the vicinity of Pambamarca. While even this could not be found on a modern 1:50,000 map at least the name of La Condamine appears near to the Yarouqui station.

In 1880 Whymper, who was staying at a hacienda near Pifo, some 4 km south of Yarouqui, saw one of the inscriptions of La Condamine set in the stairs.

After the remeasurement of the scheme, 1899-1906, a new monument was inaugurated on the 10th of August, 1913. It was sculpted by Loiseau-Rousseau. Part of the inscription, translated from the French, read, "This monument will have, we are certain, in our epoch of progress, a happier fate than the famous pyramids erected by La Condamine on the terminals of the base of the Academicians, in the plain of Yarouqui, destroyed shortly afterwards on the order of the Royal Council of the Indies, following a lawsuit which remains famous, and reconstructed later under President Rocafuerte on some sites which are probably and unfortunately not the original sites."

It is recorded in [98] that Humboldt saw one of the original tablets in the Jesuit College in Quito.

On the 6th of July, 1742 a bronze scale, of a length equal to the seconds pendulum was set in marble and fixed by 3 crampons to the wall of the Jesuit Church in Quito. The length used was the mean of the values found by Godin, Bouguer and La Condamine. The bar terminated in 2 circles of 1 inch diameter, between the center of which, the distance was marked. The design and inscription are shown in plate 27.

Prior to all these activities, even before they initially reached Quito, Bouguer and La Condamine had left a much less controversial reminder of where the equator crossed the coast near a place called Palmar. They carved in the rock: "Observationibus Astronimicus Regiae Paris Scientiar Academae Hocce Promontorium Aequatori Subjacere Compertum est—1736".

The death of Senièrgues.

In 1739 sudden death befell Dr. Jean Senièrgues. At the end of August each year, in Cuença public square, bull fights were held over a period of four days. On the last, August 28th, trouble erupted.

There was some doubt as to how the trouble started but was thought to have arisen because the doctor had been giving free treatment to various of the local people in Cuença. Among his patients there was thought to have been the daughter of a long established local family who he befriended because of her broken engagement to a musician. Was

OBSERVATIONIBUS

LUDOVICI GODIN, PETRI BOUGUER, CAROLI-MARIÆ DE LA CONDAMINE,

è REGIÂ PARISIENSI SCIENTIARUM ACADEMIÂ,

INVENTA SUNT QUITI;

LATITUDO HUJUSCE TEMPLI, AUSTRALIS, GRAD. o, MIN. 13, SEC. 18: LONGITUDO OCCIDENTALIS AB OBSERVATORIO REGIO, GRAD. 81, MIN. 22.

DECLINATIO ACUS MAGNETICÆ, À BOREA AD ORIENTEM, EXEUNTE ANNO 1736, GRAD. 8, MIN. 45: ANNO 1742, GR. 8, MIN. 20,

INCLINATIO EJUSDEM INFRÀ HORIZONTEM, PARTE BOREALI, CONCHÆ, ANNO 1739, GRAD. 12: QUITI 1741, GRAD. 15.

ALTITUDINES SUPRÀ LIBELLAM MARIS GEOMETRICÈ COLLECTÆ, IN HEXAPEDIS PARISIENSIBUS,

SPECTABILIORUM NIVE PERENNI HUJUS PROVINCIÆ MONTIUM, QUORUM PLERIQUE FLAMMAS EVOMUERUNT,

COTA-CACHE 2567, CAYAMBUR 3028, ANTI-SANA 3016, COTO-PAXI 2952, TONGURAGUA 2623, SANGAY ETIAMNUNC ARDENTIS 2678, CHIMBORASO 3220, ILINISA 2717;

SOLI QUITENSIS IN FORO MAJORI 1462, CRUCIS IN PROXIMO PICHINCHA MONTIS VERTICE CONSPICUÆ 2042:

ACUTIORIS AC LAPIDEI CACUMINIS NIVE PLERUMQUE OPERTI 2432; UT ET NIVIS INFIMÆ PERMANENTIS IN MONTIBUS NIVOSIS.

MEDIA ELEVATIO MERCURII IN BAROMETRO SUSPENSI, IN ZONÂ TORRIDÂ, EAQUE PARUM VARIABILIS,

IN ORÂ MARITIMÂ POLLICUM 28, LINEARUM o: QUITI POLL. 20, LIN. o½: IN PICHINCHA, AD CRUCEM, POLL. 17, LIN. 7: AD NIVEM POLL. 16, LIN. o.

SPIRITÛS VINI, QUI IN THERMOMETRO REAUMURIANO, À PARTIBUS 1000, INCIPIENTE GELU, AD 1080 PARTES IN AQUÂ FERVENTE INTUMESCIT

DILATATIO, QUITI, À PARTIBUS 1008, AD PARTES 1018: JUXTÀ MARE, À 1017 AD 1029: IN FASTIGIO PICHINCHA, À 995 AD 1012.

SONI VELOCITAS, UNIUS MINUTI SECUNDI INTERVALLO, HEXAPEDARUM 175.

PENDULI SIMPLICIS ÆQUINOCTIALIS, UNIUS MINUTI SECUNDI TEMPORIS MEDII, IN ALTITUDINE SOLI QUITENSIS, ARCHETYPUS

(MENSURÆ NATURALIS EXEMPLAR, UTINAM ET UNIVERSALIS!)

ÆQUALIS HEXAPEDÆ, SEU PEDIBUS 3, POLLICIBUS o, LINEIS 6 ½: MAJOR IN PROXIMO MARIS LITTORE LIN: MINOR IN APICE PICHINCHA LIN.

REFRACTIO ASTRONOMICA HORIZONTALIS SUB ÆQUATORE MEDIA, JUXTÀ MARE 27 MIN: AD NIVEM IN CHIMBORASO 19' 51": EX QUÀ ET ALIIS OBSERVATIS, QUITI

LIMBORUM INFERIORUM SOLIS, IN TROPICIS DEC. 1736 ET JUNII 1737, DISTANTIA INSTRUMENTO DODECAPEDALI MENSURATA GRAD. 47, MIN. 28, SEC. 3

EX QUÀ, POSITIS DIAMETRIS SOLIS, MIN. 32, SEC. 37 ET 31' 33"; REFRACTIONE IN 66 GRAD. ALTITUDINIS o' 15"; PARALLAXI VERO ¼ 40",

FRUITUR OBLIQUITAS ECLIPTICÆ, CIRCA EQUINOCTIUM MARTII 1737, GRAD. 23, MIN. 28, SEC. 28.

STELLÆ TRIUM IN BALTHEO ORIONIS MEDIÆ (BAYERO 1) DECLINATIO AUSTRALIS, JULIO 1737, GRAD. 1, MIN. 23, SEC. 40.

EX ARCU GRADUUM PLUSQUAM TRIUM REIPSÂ DIMENSO, GRADUS MERIDIANI SEU LATITUDINIS PRIMUS, AD LIBELLAM MARIS REDACTUS, HEXAP.

QUORUM MEMORIAM,

AD PHYSICES, ASTRONOMIÆ, GEOGRAPHIÆ, NAUTICÆ INCREMENTA,

HOC MARMORE PARIETI TEMPLI COLLEGII MAXIMI QUITENSIS SOC. JESU AFFIXO, HUJUS ET POSTERI ÆVI UTILITATI V. D. C.

IPSISSIMI OBSERVATORES. ANNO CHRISTI M. DCCXLII.

Le mesure ci-dessus qui, pour représenter le Quart du Pendule équinoctial, devoit avoir près de 9 pouces 1 ligne ¾, est trop longue d'environ ½ de ligne de trop.

28. *Inscription on the wall of the Jesuit church in Quito. The horizontal line represents the length of the seconds pendulum.*

it that he went with another or did the girl have a liking for Senièrgues and become jilted because of him? no matter, the result was the same in such a close community.

The gossip spread like wildfire. Dr. Senièrgues was furious with the musician and challenged him to a duel. Unfortunately, the doctor was set upon by a mob from the bullfight. He was fatally wounded.

Murder in front of 3000 witnesses was no solace as he died in La Condamine's bed at the home of the Pére Recteur des Jesuits.

When he was set upon, Bouguer, La Condamine and the others had sprung to his defense but were hopelessly outnumbered. La Condamine was naturally furious and insisted upon justice being served on those responsible. Progress was painfully slow but after some 2 years of tortuous effort the murderers were named, convicted, but so casual was life there that they were not sentenced. Proceedings started in August, 1741 but it was on the 19th of May, 1742, when an order was signed for the banishment of the culprits. Not a very satisfactory outcome.

Even more outraged, as soon as he returned to Paris, La Condamine wrote a book on the murder and had it published in 1746 as "Lettre sur l'emeute populaire de Cuença". [131]

An epitaph to Senièrgues was placed on the tomb in the church at Cuença.

Return to Europe.

Before the party fully split up for the various return routes to Europe Louis Godin announced that he was going to take a post as astronomer at the University of San Marcos in Lima. His cousin Jean Godin des Odonais who had been little more than a chainman to the Academicians stated that he was in love with Isabela, the 13 year old daughter of Don Pedro Manuel de Grandmaison from Riobamba. She was a fluent linguist. They were married in the pueblo of Guzmán near Riobamba on the 27th of December, 1741.

To multiply the opportunities of making observations Bouguer, La Condamine and Godin planned different routes for their return. La Condamine had had his imagination fired by the sight of a rough map of the Amazon area and decided to return via that great river.

He left Tarqui on the 11th of May, 1743, and first went via Zaruma to Lima which route proved difficult with the swollen river Los Jubones but he learned later that assassins had lain in wait for him on the main route. At Zaruma the barometer stood at 24 inch 2 line which equated to 700t according to La Condamine's calculations.

On his route he had to cross many bridges made of cords, bark and lianas. Loxa, where he repeated his observations of 1737, was some 350t lower than Quito and noticeably warmer.

On 3rd June on one of the mountains near Loxa he spent all day finding 8 or 9 young guinquina plants which he planted in soil and had them carried to Cayenne and transported to France.

From Loxa to Jaen he passed through an area of incessant rain, by Loyola and Valladolid and the river Chinchipe where 5 leagues by raft brought him to Tomependa at the confluence of three rivers. So rocky, narrow, fast and with falls as to be unnavigable. On one of the numerous river crossings in the area the mules dashed into the water fully laden and all the instruments and papers were saturated. Each sheet of paper had to be carefully dried individually.

On the 4th of July he boarded a small two-oared canoe of balsa, and after a couple of days reached the river Marañon where it begins to be navigable. Overnight the river rose ten feet and there was a hasty requirement to raise the shelter which the Indians were skilled at achieving. The added depth necessitated modification to the boat and this took three days.

By the 8th he was passing the strait of Cumbinamba and then on to Escarrebragas. Whirlpools, rapids and almost every conceivable hazard was to be met and overcome. One night spent in the boat awaiting the receding of the waters nearly came to disaster. Sleeping on the raft with but an ancient American retainer, the river, which had already fallen 25 feet in 36 hours continued to sink and in the middle of the night the splinter of a large branch under the water penetrated the raft. If La Condamine had not been awake he could well have been suspended in the air by a branch and at least, all the results of 8 years labour would have been lost.

In this part of the river Pongo near Borja La Condamine reckoned the velocity to be two toise a second. Use of a raft with its extra flexibility compared to a canoe allowed travel over the thousands of hazards on the river bottom. Tying together of the timbers by lianas instead of nails or dovetails gave a special springiness and shock absorbing feature.

The area of Borja was different again with a maze of rivers and lakes in every direction. The latitude he found her to be 4° 28′ S. In Borja the Reverend Father Magnin, a Jesuit missionary from Fribourg was expecting La Condamine and he later accompanied him on the next leg of the journey to Laguna where they arrived on the 19th of July. During this section they had found, on the 17th, near Pastaça, a letter attached to a tree. It was dated the 1st of June and had been left by Don Pedro Maldonado.

It was at Laguna that Don Pedro Maldonado, governor of the Province of Esmeraldos had been waiting six weeks. He had similarly experienced many dangers and fatigues in getting thus far by way of the Pastaca river. As was La Condamine, so Maldonado made observations as he went, principally with a compass and portable gnomon.

Laguna was a village of more than 1000 Americans and was the chief mission of the whole province of Maynas. On elevated ground, on the edge of a great lake, 5 leagues from the mouth of the river Guallaga. The latitude was observed to be 5° 14′ S.

On the 23rd of July, Maldonado and La Condamine moved on in two 42 ft. long canoes each from the trunk of a single tree. Rowers from prow to middle, with travellers and baggage at the poop under shelter from both sun and rain. Continually they mapped the course by compass and watch, noting deflections of the route, breadths, confluences, islands, current flow, depths and the barometer at every stop.

On the 25th of July, they passed to the south of the river Tiger and the next day on the north side of the mouth of the Ucayale, one of the largest of the rivers that swell the Marañon. By the 27th they had reached the mission station of Saint Joachim. Seeds were collected of the plant Curupa which effected intoxication and extraordinary visions and was taken rather like snuff but up a tube and into the nostrils. In fact there were so many plants of vast variety along the banks of the Amazon and its tributaries that it would take years of toil to collect and describe them all. La Condamine contented himself with collecting seed at every possible opportunity. In particular he was taken by the range of lianas and their size up to the thickness of a man's arm and greater.

The gums, resins, balsams and juices of every sort as well as many oils were extracted from trees of numberless variety. In particular the resin called cahouchou (rubber) could be moulded to any shape, was impervious to rain and most remarkable in its elasticity.

The group moved from St. Joachim on the 29th of July with fresh canoes and crews, for the mouth of the river Napo where La Condamine was particularly desirous of fixing the position. This he did with the 18 ft. long telescope transported over mountains, through forests and down rivers and rapids for 150 leagues. From the emersion of the first satellite of Jupiter he found Napo as 4¾ hours from the Paris meridian on the 31st of July, 1743.

From Pevas, the last Spanish missionary settlement to St. Pablo was 3 days and 3 nights journey in which time they did not see a single dwelling. The greatest peril here was in the numerous uprooted trees concealed below the river surface. It proved safest to keep as far from the banks as possible both because of this and to avoid the actual falling trees.

St. Pablo had buildings of stone, brick and plaster instead of reed bowers. The missionary, having heard in advance of the visitations prepared two large canoes with crews and a Portuguese guide. Such canoes could be 60 feet long by 7 feet broad and 3½ feet deep.

In 5 days and nights of rowing they arrived at Coari, a settlement of Portuguese Carmelite missionaries. A number of large confluences were negotiated during the journey.

On the 20th of August they left Coari in another fresh canoe. At Coari they had been unable to find a language in which to converse other than the use of signs. This was despite Maldonado knowing Peruvian and the ability of a Portuguese guide to speak Brazilian.

The 23rd of August saw them enter the Rio Negro where the width was found to be some 1200t and the latitude 3° 09′ S, and by the 28th they passed the Jamundas and came to the Portuguese fort of Pauxis where they were received with much pleasure. The Commandant detained the party for four days at the fort and a further day at his country house before accompanying them for 6 or 7 days as far as the fortress at Curupa.

The 4th of September provided the strange sight of mountains in the distance, something not seen during the previous two months of sailing down the various rivers. Two days later they had to divert from the main Amazon stream to avoid the strong currents known to be at the mouth of the river Xingu. Here the width could be measured in many miles and the torment of mosquitos, gnats and other vicious insects abated. By the 9th they reached the Portuguese fortress of Curupa where they were royally entertained for 3 days by king's lieutenant. From this point boats only move

with the ebb and flow of the tide and a few leagues below Curupa several large rivers converge, the Para, the Tocantin and the Muju, which at 2 leagues from its mouth was 749t wide. Another 8 days sail would bring them to Para on the coast and this they did by the 27th of September, 1743. Here a comodious and richly furnished house was awaiting them and it had a garden overlooking the sea so that observations could be readily made. This was a large city trading directly with Lisbon. La Condamine found the latitude as 1° 28′ S. For the longitude an eclipse of the moon on the 1st of November, 1743 and immersions of the first satellite of Jupiter on the 6th and 29th of December suggested a difference from Paris of 3 hours 24 minutes. Paris today is given as 48° 50′ N and 2° 20′ E while Para (Belem) is 1° 20′ S and 48° 30′ W i.e. it would be 50° 50′ from Paris. Thus 3 hours 24 minutes or 51° was a very commendable agreement.

Of more importance in relation to the expedition was the length of the pendulum as compared to that at Quito. A difference of 14-1500t in elevation yet with 1½° of latitude. From 9 experiments varying by only 3 oscillations in 98,740 he found that at Para there were 31-32 more oscillations than at Quito and 50-51 more than at Pitchincha.

Maldonado was able to accept the opportunity of sailing with a Portuguese fleet which departed for Lisbon on the 3rd of December 1743. He reached Lisbon in February 1744 where he was received by the Charge des Affaires de France—de Beauchamp. Death arrived on the 17th of November, 1748 in London.

La Condamine on the other hand wished to complete his chart of the area and to achieve this he needed to see the true mouth of the Amazon and for this he travelled first to Cayenne, where a French ship waited to take him to France. To get to Cayenne he embarked on the 29th of December, 1743 in a 22 oar dug-out canoe (pirogue pontée) complete with stores, furnishings and recommendations from the Franciscan Fathers. He took 2 months to make a journey that ought to have lasted only a fortnight. Some of this time was spent on the deserted island of Isle de la Pénitence and on various mud banks.

From his journey it became evident that the Cayenne region might well have been a better site for the arc. With Cayenne at 5° 00′ N one could have proceeded possibly for 4° southwards along a meridian without leaving French territory. Cayenne is at 5° 00′ N, 52° 18′ W and Macapa at 0° 05′ N, 51° 10′ W. In fact, with Portuguese permission this could have been extended to the equator. Such a plan would have been much easier to execute than had seemed possible when La Condamine had proposed it to the Academy a year before the projected Quito trip. Additionally of course, it would have been completed far quicker.

The party reached Cayenne on the 26th of February, 1744. While La Condamine made pendulum observations here to complement those of Richer, he also made experiments on the speed of sound. From 5 experiments he had 4 results agreeing within ½ second in 110 over a distance computed trigonometrically as 20,230t from a base of 1900t measured on the beach. The result found for the speed of sound was 183½t per second whereas at Quito he had found it as 175t. For the experiment a 12 pound cannon was used.

After a 6 month wait for a French ship and advancing lowness of spirits and jaundice he accepted a Dutch offer to go to Surinam and have a passage to Holland. He was in a convalescent state when he left Cayenne on the 22nd August, 1744 and he reached Surinam on the 27th, and Paramaibo the following day. The Governor furnished him with a passport in case of rupture between France and the States General.

On the 3rd of September, 1744 he embarked on a merchant ship of 14 cannon, full of coffee and with a crew of 12, bound for Amsterdam. On the 29th they were approached by an English cruiser but luckily bad weather dispensed with any possible trouble. On the 6th of November a similar cruiser from St. Malo required assurance before allowing passage to continue. A pilot was taken on at Texel on the 16th but more bad weather delayed landing at Amsterdam until the 30th of November, 1744. It took a further two months to obtain the necessary passport to cross the Low Countries. Finally Paris was reached on the 23rd of February, 1745 where Bouguer and Verguin were met. Almost 10 years after the departure he was back home.

All was not over however as controversy erupted between Bouguer and La Condamine. Bouguer had the sympathy of La Gournerie for his complaints while La Condamine had Delambre as an ally. Some 44 detailed objections by Bouguer were explained or refuted by La Condamine in a special publication.

While Maldonado and La Condamine were making their way down the Amazon, Juan and Ulloa were occupied with the Spanish Navy off the coast of South America. It was not until the 22nd of October, 1744 that they were able to embark, Juan in the Lys and Ulloa in the Delivrance, at Callao bound towards Chile. The ships kept company until November when they separated for the Lys to call at Valpariso. The Delivrance continued to Conception Bay where it arrived on the 21st. The Lys reached there on the 6th of January, 1745 and they both sailed again on the 27th. and had to put in for repairs. Although the Delivrance was not much better the Captain wished to use the best weather to round the Cape, which they did during early March. However by the 19th of March they were shipping a great deal of water and the continuous manning of the pumps made their thoughts turn to abandoning ship which they may well have done but for their valuable cargo of 2 million Peru dollars, half of which was in gold and silver.

The Captain had the possibility of putting into Montevideo but declined on the 25th of March despite the very low state of their provisions and drinking water. On the 21st of May they did finally put in to land at Fernando de Norona, to end a voyage of 150 days of extreme fatigue and anxiety. Norona was in latitude 4° S and longitude 29° 56' E of the meridian of Conception according to Ulloa. A French map of the time placed it at 42° 32'30" E of Conception. Could Ulloa had been over 12° out in his reckoning? ·   ·

In 1744 Juan and Ulloa had found the position of Conception as 36° 43' 15" S and longitude 303° 18' 30" from Teneriff. Today's value for Teneriff is 28° 20' N and 16° 40' W which would place Conception at 286° 38' 30" from Greenwich, which agrees closely with the present day value. i.e. this would place Fernando de Norona at 333° 14' 30"

from Teneriff or 316° 34' 30" from Greenwich, as opposed to its accepted value of 4° S, 33° 10' W (or 326° 50'). i.e. Ulloa was in error by some 10½° in longitude and the French map by some 2½°.

Repairs were carried out to the ship to the extent that the amount of pumping would be reduced but not done away with. Supplies of wood, water, calves and hogs were taken on board and they sailed again on the 10th of June. On the 27th of June they were in the Sargasso seaweed area; a plant which, when pickled, they said was equal to samphire. Just before entering the Sargasso the inflow of unwanted water had increased again but decreased considerably as they passed through the weed. No doubt it helped to repair some of the punctures, if only temporarily.

On the 21st of July at 43° 57' N and 39° 44' E of Conception, (326° 22' 30" from Greenwich) north west of the Azores, two English ships came in sight and fired shots. The Delivrance had three frigates in company but this was still inferior to the attackers who later turned out to be privateers. These were the Prince Frederick and the Duke and they were able to capture two of the French ships and their cargo of millions of dollars worth of gold, silver, ingots and wrought plate as well as cacao, quinquina and vigonia wool. The Delivrance escaped and made for Loisburg in the Cape Breton Islands, Newfoundland, as a safer route than to Spain.

On the 13th of August a brigantine was seen plying the coast and the Delivrance hoisted the French ensign. This drew some gun fire but did not give immediate cause for anxiety. Later however, two men of war appeared which were mistakenly thought to be French but turned out to be English, the Sunderland and the Chester, and capture was inevitable.

Ulloa, upon capture appearing certain, threw overboard all his papers that could be of use to the enemy but retained those relating to the meridian measurement. The discard had been previously decreed to the captain and officers in case of Ulloa being killed in such action.

The Lys, which had left the group on the 5th of February, 1745 as she was shipping 6 inches of water an hour, went to Valparaiso for repairs and departed on the 1st of March. On the 8th of July they reached Cape François or Guarico on the north side of St. Domingo. By the 6th of September there was a whole fleet of vessels ready to sail to Europe, 53 including men of war, frigates, brigantines and bilanders. They arrived in Brest bay on the 31st of September, 1745 and Juan was able to pay his respects to the Royal Academy of Sciences, to which he was admitted as a corresponding member. He then set out for Madrid where he arrived in early 1746.

Meanwhile in Louisburg, Ulloa was in captivity but praised the humanity of Mr. Warren, commodore of the English squadron, who several times dined with him. Mr. Warren put Ulloa's papers in the care of the English captain who was to take him to England. Captain Brett of the Sunderland was commanded to remit them to the Admiralty on arrival.

The group left Louisburg on the 19th of October for Newfoundland where they arrived on the 24th. They sailed for England in November, 1745 and anchored at Spithead on the 29th of December.

Ulloa was taken to Fareham for captivity while others were

put in prison at Porchester Castle. He expressed his gratitude to the humanity Captain Brett and his officers had shown on the voyage. The commissary of the Spanish prisoners was Mr. William Rickman who lived at a farm house called Pesbrook, 3 miles from Fareham and a mile or two from Titchfield. Would this be the present day Posbrook? Ulloa asked Mr. Rickman to solicit of the Duke of Bedford and the Admiralty for the return of his scientific papers. This was graciously acceded to and on the 12th of April, 1746 Ulloa was granted audience with Lord Harrington and given assurances as to the return of the papers.

Then Martin Folkes, President of the Royal Society, heard of the situation and requested the papers be handed to him rather than fall into the hands of ignorant people at the Admiralty.

The papers were indeed referred to him for examination and he also introduced Ulloa to the meetings of the Society. The papers were returned to Ulloa on the 25th of May as a result of the testimony of Folkes to the Admiralty. It was then proposed to Earl Stanhope that Ulloa be admitted as a member of the Royal Society.

He embarked at Falmouth in a Lisbon packet boat and then hastened to Madrid where he arrived on the 25th of July, 1746. He became a commodore in the Spanish navy and was at one stage Governor of Louisiana.

Juan was made a post-captain on his return to Spain and later commandant of the Naval Guards, commodore and ambassador to Morocco.

Perhaps the most remarkable of all stories of the return to Europe was that of Madame Godin.

On the 28th of July, 1773, M.Godin des Odonais wrote to La Condamine from St. Amand in Berry in reply to a request for details of the journey Madame Dona Isabela Godin made down the Amazon. It was then some 30 years since Godin and La Condamine had parted in Quito.

Godin's idea at that time was to seek the best route for his wife to travel as an alternative to mules over the mountains. This he did not do immediately after La Condamine had left because his wife had several pregnancies. In 1748 however there was news of the death of Godin's father and this prompted him to make the route to Cayenne and in so doing prepare the way for his wife to follow later.

Godin left Quito in March 1749 leaving behind his wife who was yet again pregnant, and reached Cayenne in April. From here he made plans to voyage the 1500 leagues up the Amazon to collect his family and return with them.

While in Cayenne he made efforts to get the requisite passports but despite numerous letters, many of which were apparently lost or intercepted, none appeared. In 1765 things did begin to move and 10 months after contacting the Comte de Herouville, a galliot of 30 oars arrived to take him to the first Spanish settlement above Para and to wait there for his return with the family.

Cayenne was left in November 1765 but by the time of reaching Oyapac Godin was seriously ill. Despite six weeks' wait he was little improved so he arranged for one Tristan D'Oreasaval to take care of various papers and orders and to proceed to Las Lagunas and pass them on at the mission for onward transmission to Riobamba.

Tristan left Oyapac on the 24th of January, 1766 but instead of carrying out his instructions as detailed by Godin he passed the papers and letters on to a Jesuit, Father Yesquen in September, 1766. By some strange route Madame Godin got wind of the existence of the letters addressed to her and of the waiting vessel. Much effort was made to extract the letters from the Jesuits but they never did materialize.

Should Madame Godin risk such a hazardous journey solely on rumour of a vessel awaiting her? What of all the time that had elapsed since September, 1766 and the Jesuit?

She was not ready to leave Guzman near Riobamba until the 1st of October, 1769. With her went M.R., a physician who pleaded for a passage on the representation that she might well need his services. Her two brothers were aiming respectively for Rome and Spain and one had a son of 9-10 years whom he wished to get an education in France. The father-in-law of Godin went ahead to seek suitable accommodations, and make arrangements for canoes, paddlers and guides. Madame Godin was thus accompanied by her two brothers, a physician, her negro Joachim, 3 female mulattoes and 31 baggage carriers.

Even on the first leg of the journey, to Canelos, there was disaster. When her father-in-law had passed there it was a thriving village, a month later when the party arrived it was deserted. Smallpox had hit. The population had scattered to individual huts in the jungle. The bearers decided likewise and disappeared.

Madame Godin decided to brave all the perils and proceed by canoe down the Pastaza. She engaged the only two non-infected Indians to build a canoe and pilot it to Andoas. After only two days travel, out of probably 12 needed to get to Andoas, the Indians quit. Going ahead without anyone to steer, some three days later one fell overboard on trying to retrieve a hat and was drowned.

Soon after the canoe capsized. M.R. and the negro went on to Andoas to obtain aid and another canoe, promising to return in a fortnight. After 25 days they resorted to constructing a raft on which to venture farther when there was no sign of return from the others. This raft hit an obstruction, upset, and all effects perished. Luckily, no one drowned but little was salvaged. They had to resort to using the banks which were difficult in the extreme because of their undergrowth and lianas. After only a few days they were lost, feet torn to shreds, provisions exhausted, dying of thirst they could but sit and await death one by one.

Madame Godin, when the only one left, delirious and choking, summoned up enough strength to drag herself in search of some salvation. This was around late December, 1769. It took some 9 days to reach the banks of the Bobonafa. Clothes in rags and shod only by the soles cut from the shoes of the dead, living on wild fruit and some eggs she found.

On reaching the river she came across some natives and persuaded them to allow her on board their canoe and to take her down river where they reached Andoas in early January 1770.

Meanwhile the negro who had gone ahead, returned to where he had left the party; traced their route through the jungle to where they lay dead; retraced his route and was back at Andoas before Madame arrived, convinced that she also must be dead. He returned to Quito and his services were then lost to the dwindling group.

Madame with a crew from Andoas reached Laguna and were well looked after during the 6 weeks she remained there to recuperate. Upon a message reaching Omaguas that Madame was, in fact, alive, M.R. hastened to join her and took with him several utensils and various clothes but gave no account of some other items that would not have been subject to rotting or decay. Madame had had enough of the gentleman, blaming all her misfortunes on him and dismissed him. Unfortunately she was imposed upon by one M.Romero to allow M.R. to stay because of his poor children.

Messages were sent forward requesting assistance and preparedness while it was pointed out that as yet Madame had by far the largest part of a dangerous journey still to cover.

A canoe was prepared, equipped and directions given that it was to stop nowhere until the Portuguese ship was reached. The canoe was met by a pirogue from the ship at the village of Pevas. The remaining 1000 leagues passed comfortably with various delicacies supplied by the two accompanying canoes.

One suffering remained, a thumb damaged by thorns in the woods was at one stage considered for amputation but this was finally avoided although the use was lost.

The major of the Para garrison sailed to meet Madame and escorted her to Fort Oyapok. It was then that M.Godin heard of his wife's approach and sailed in a galliot of his own to meet her near Mayacare. After 20 years they were united again. They anchored at Oyapok on the 22nd of July, 1770. Then to Cayenne. Godin managed to get involved in a law suit with Tristan over a non-payment of agreed monies. Godin won but could not collect because of the insolvency of Tristan.

Madame was again unwell in Cayenne so it was 1773 before they could venture back to Europe; 38 years after the expedition had left France.

What of the rest of the original party?

Pierre Bouguer had returned to Europe in 1744. He left Quito on the 20th of February, 1743, descended the valley of the river Magdalena to Carthagena and then to St. Dominique. He reached Paris at the end of June 1744 and died in 1758.

Dr. Joseph de Jussieu travelled through several of the neighbouring territories in S.America including Bolivia, Tucuman and Paraquay, collecting botanical specimens as well as acting as a doctor and road engineer. A 5 year collection of plants was lost through the carelessness of his servants and this sent him somewhat mad. In failing health he left Lima for Paris in 1770 where he arrived in 1771. He died in 1779.

Hugot, the instrument builder chose to stay in S.America since he had married a local girl in Quito. Said to have died in 1744.

de Morainville met his death in falling from scantling round a church of which he was architect, at Cicalpa near Riobamba, in 1744.

Louis Godin took a post as mathematician and astronomer at the university of San Marcos, Lima before returning to Europe in 1751.

Capt. Verguin was struck by illness in Quito and had to delay his return until 1746.

Mabillon lost his memory for a period.

Pedro Maldonado returned to Lisbon in February 1744 and died on the 17th of November, 1748.

M. Couplet had died on the plains near Quito in 1736.

Dr. Jean Seniergues had been murdered in Cuença in 1739.

Brief Chronology

	1733	Idea first put to Louis XV
14, 20 August	1734	License granted and patents made out.
14 April	1735	La Condamine left Paris
16 May	1735	Sailed from La Rochelle
26 May	1735	Juan and Ulloa sailed from Cadiz
22 June	1735	Arrived Martinique then on to St. Domingo
26 June	1735	Juan and Ulloa arrived in Martinique
3 July	1735	Sailed from Martinique
8 July	1735	Juan and Ulloa arrived at Carthagena
31 October	1735	La Condamine and party sailed from St. Domingo
16 November	1735	Arrived Carthagena. Juan and Ulloa awaiting them
25 November	1735	All sailed for Porto Bello
29 November	1735	Arrived Porto Bello
22 December	1735	Left Porto Bello
29 December	1735	Arrived Panama
22 February	1736	Sailed from Panama
25 March	1736	Reached Guayaquil
11 May	1736	Arrived Caraçol
29 May	1736	Juan and party reach Quito
4 June	1736	La Condamine and Maldonado reach Quito

Brief Chronology - Continued

10 June	1736	Bouguer reached Quito
17 September	1736	Couplet died
3 October - 3 November	1736	Yarouqui base measured. One Indian died.
19 January	1737	La Condamine goes to Lima from Quito
28 February	1737	Reached Lima. Juan followed him there.
20 June	1737	La Condamine returned to Quito.
14 August	1737 to	
July	1739	Observation of the chain
August	1739	Tarqui base measured
28 August	1739	Senièrques murdered in Cuença
16 January	1740	Party left Tarqui for return to Quito
17 February	1740	Bouguer and La Condamine at Cotchesqui
1 May	1740	Last angle measured in extra triangle
June - August	1740	Bouguer connected scheme to sea level
3 October	1740	Godin des Odonais left for Carthagena
30 October	1740	Juan and Ulloa leave Quito for Lima
18 December	1740	Arrived in Lima
8 August	1741	Juan and Ulloa left Callao for Guayaquil
5 September	1741	Juan and Ulloa reached Quito
16 December	1741	Juan and Ulloa leave Quito again for Guayaquil
24 December	1741	Arrived at Guayaquil
27 December	1741	Godin des Odonais married
5 January	1742	Ulloa left Guayaquil for Quito
19 January	1742	Arrived at Quito
22 January	1742	Ulloa leaves again for Guayaquil
26 January	1742	Arrived Guayaquil. Joined Juan to go on to Lima.
20 February	1743	Bouguer left Quito for Carthagena and France
11 May	1743	La Condamine left Cuença for Lima
4 December	1742 to	
6 July	1743	Juan and Ulloa with Navy off S.America
19 September	1743	La Condamine and Maldonado at Para
3 December	1743	Maldonado embarked for Lisbon
30 December	1743	La Condamine embarked for Cayenne
27 January	1744	Ulloa reaches Quito again from Callao
22 March - 22 May	1744	Ulloa and Godin at Mira repeating astronomy
June	1744	Bouguer reached France
27 August	1744	La Condamine arrived in Surinam
22 October	1744	Juan and Ulloa embark in separate ships
21 November	1744	Ulloa in 'Delivrance' at Conception Bay. Juan in 'Lys' to Valparaiso
30 November	1744	La Condamine arrived in Amsterdam
6 January	1745	Ships of Juan and Ulloa at Conception
27 January	1745	Departed from Conception
23 February	1745	La Condamine reached Paris
	1745	Verguin left Quito
1 March	1745	'Lys' left Valparaiso
13 August	1745	Ulloa captured at Louisburg
31 September	1745	Juan in 'Lys' arrived in Brest
19 October	1745	Ulloa left Louisburg
29 December	1745	Ulloa arrived at Spithead
	1746	Verguin reached Paris
25 May	1746	Ulloa had his papers returned
25 July	1746	Ulloa reached Madrid
September	1749	Godin des Odonais in Para
	1751	Louis Godin returned to Paris from Lima
	1771	Jussieu returned to Paris
21 April	1773	The Godins left Cayenne
26 June	1773	Isabel and Jean Godin arrived at La Rochelle
15 August	1758	Bouguer died in Paris

Brief Chronology - Continued

11 September	1760	Louis Godin died in Cadiz
21 July	1773	Juan died in Madrid
4 February	1774	La Condamine died in Paris
11 April	1779	Jussieu died in Paris
5 July	1795	Ulloa died in Cadiz

References.
See in particular the following: 1, 2, 3, 6, 20, 22, 26, 33, 38, 40, 44, 45, 47, 55, 72, 78, 86, 97, 98, 103, 104, 106, 113, 116, 118, 122, 126, 130, 132, 133, 134, 135, 137, 160, 182, 198, 202, 203, 208, 213, 222, 228, 252, 256, 262, 263, 266 and 267.

Chapter 21

THE EXPEDITION TO LAPLAND

As with the Peru expedition, before describing the work of the group, it will be appropriate to have some details of the main cast.

Pierre Louis Moreau de Maupertuis
Born 28th September, 1698 at St. Malo, Ile-et-Vilaine
Died 27th July, 1759 at Basel.

He was said to have been a spoilt child and this resulted in a certain intransigence and unwillingness to be criticized that later led him into difficulties. In 1708 his father was enobled by Louis XIV to René Moreau Sier de Maupertuys.

From private schooling he went to Paris at 16 to study first philosophy then music before settling for mathematics. By 1718 he had entered the French army and risen to Captain of the Dragoons but tired of the life and returned to his mathematics.

In 1728 he visited London for some 4 months and reinforced his belief in Newton's ideas as to the shape of the earth. The same year he was elected to the Royal Society. He became a member of the Academy of Sciences in 1731. He was very keen on biology and his home was said to have been a veritable menagerie. He felt that one of his greatest achievements was the principle of least action which he developed and which later became a famous principle throughout scientific fields. He was among the first to introduce the laws of probability to heredity.

The expedition he led to Lapland (The Arctic Circle) was away for some 18 months, far shorter than that to Peru. The results led to his being dubbed earth flattener. Most of the Lapland group, including Maupertuis, helped in the measure of a French arc in the late 1730s from Paris to Amiens. The result was announced to the Academy in December 1739.

Both before and after the expedition he exchanged numerous letters with Emilie, the Marquise du Châtelet, a lady of scientific bent who corresponded with many but none more so than Maupertuis who was a favourite of hers. On his return from Lapland he was accompanied by Christine and Elisabeth Planström, the daughters of a Torneå merchant. They went later to Paris and while the former joined a nunnery, the later was unfortunate to marry a swindler.

In 1741 he became a member of the Berlin Academy of Sciences. In the course of his first visit to Berlin in 1740-41, he became caught up in the war between Prussia and Austria and at the battle of Mollwitz, although only a spectator, he was taken prisoner. Luckily he was later released by an officer who recognized him and sent on his way back to France via Vienna.

In 1743, he was elected to the Académie Française and two years later accepted the offer of Frederick the Great to go to the Academy of Science in Berlin. There he met Mlle. Eleonore Catherine von Borck whom he married on the 8th of October, 1745.

From 1746-1753 he was President of the Berlin Academy of Sciences.

He attracted several eminent scientists to Berlin but at the same time became involved in various controversies. Not least of these was the turning against him of his old friend Voltaire over the treatment of Samuel Koenig, a student of Maupertuis' who suggested that Leibniz had first developed the principle of least action. For this Koenig found himself out of the Berlin Academy.

During the winter of 1746-47 he had a severe attack of tuberculosis which no doubt hastened his death. By 1752 he was being attacked from all sides but particularly by Voltaire who also derided the Lapland expedition. He was then forced by ill health to return to St. Malo where he remained un-

29.   *Pierre Louis Moreau de Maupertuis leader of the northern expedition.*

til early 1754 when he went back to Berlin but by mid 1756 he was back in France. While yet again returning to Berlin in 1758 he stopped at the home in Basel of his friend Johann Bernouilli II where he stayed until his death, before his wife could reach him there following a call from Bernouilli.

For details of his publications see the references.

Alexis-Claude Clairaut

Born 7th May, 1713 in Paris.
Died 17th May, 1765 in Paris.

His father was Jean-Baptiste Clairaut, a mathematics teacher in Paris who was also corresponding member for the Berlin Academy. His mother Catherine Petit bore some twenty children of whom Alexis was the second. Of the others only one survived the father.

Alexis was a child prodigy and by the age of ten was reading calculus. At thirteen, he presented a paper on geometry to the Academy of Sciences. Soon after he combined with other young students in founding a Society of Arts.

By eighteen he had published a mathematics book and was elected to the Academy of Sciences.

He joined the Lapland expedition in 1736 and it would appear that he did most of the computational work on the expedition. At about the same time he became entangled with the Marquise du Châtelet, who also had the attentions of Maupertuis. Neither married her as she already had a husband and she died at the age of forty-three in 1749.

In 1743 he published a work on the figure of the earth and in 1752 won a prize for a paper that discussed the moon. His name is still retained for the equation he derived on differentials. In particular he is remembered for calculating the return date of Halley's Comet in 1759.

For details of his publications see the references.

Anders Celsius

Born 27th November, 1701 in Uppsala.
Died 25th April, 1744 in Uppsala.

He was the son of Nils Celsius, Professor of astronomy at the University of Uppsala 1719-24 — a post that had previously been held by the grandfather of Anders, Petrus Elvius, 1699-1718.

In 1725, he became secretary of the Uppsala Scientific Society and on the 25th of April, 1730 was made Professor of astronomy. Such an appointment obliged him to travel widely to broaden his knowledge and this he did from 1732 in Berlin, Leipsig, Nuremberg, Bologna, Rome, Paris and London among others.

He arrived in Paris in late 1734 just when there was earnest discussion in scientific circles on the shape of the earth. It was during this time but after departure of the Peru expedition, that Maupertuis raised the idea of a second expedition and no doubt as a result of discussions with Celsius decided upon northern Scandinavia. After the initial arrangements had been put under way for this expedition, Celsius went to London and while there commissioned some instruments from Graham to take to the north.

Celsius left England in April, 1736 and rejoined Maupertuis and the others in Dunkirk from where they sailed on the 3rd of May, 1736. After his return, Celsius was rewarded for his efforts with a life pension of 1000 Francs per year

and he then set about founding an observatory, which he did near Svartbäcksgatan (Blackbrook street). It was opened in December, 1742.

Among his other achievements was work on developing the thermometer. This no doubt arose from his extensive use of them on the expedition. He made his first such instrument in 1741 and marked it with a scale from 0 to 100 but with the 0 indicating the boiling point and 100 the freezing point of water.

By the mid 1740s the scale had been reversed to its present form, not by Celsius who had died at an early age, but by Linné.

It became known by various names until it was incorrrectly credited to Celsius in a textbook and the name was perpetuated.

His life ended early through consumption but his travels had formed an important link between Uppsala and other scientific organisations throughout Europe.

For details of his publications see the references.

Réginald Outhier

Born 16th August, 1694 at La Marre-Jousserans near Poligny.
Died 12th April, 1774 At Bayeux.

For many years he was canon of the cathedral of Bayeux. He was an amateur scientist who supplied the Academicians in Paris with observations. His principal interests were astronomy, meteorology and cartography. In 1727, he invented a celestial globe and presented it to the Academy of Sciences. It was rotated by clockwork.

He was a correspondent with the Cassini family and joined them on the survey of the arc from St Malo to Caen. At that time he was secretary to Paul d'Albert de Luynes, a very scientific bishop at Bayeux.

On the expedition to Lapland Outhier assisted in the astronomical observations, drew 18 maps of the lands through which they passed, and studied the religious and social customs of the Lapps.

For details of his publications see the references.

Pierre-Charles Le Monnier

Born 20th November, 1715 in Paris.
Died 3rd April, 1799 in Herils, Calvados, France.

His father was Professor of philosophy at the Collège d'Harcourt and was also a member of the Academy of Sciences. His brother Louis-Guillaume Le Monnier, also a member of the Academy, was Professor of botany at Jardin Du Roi and Premiere Médecin Ordinaire to Louis XV.

In 1763 he married Mlle. Cussy, from a prominent Norman family, and had three daughters, one of whom married Lagrange. His second daughter married his brother Louis-Guillaume.

Le Monnier was presented to Louis XV by the Duc de Noailles and remained a great favourite of the King and became his Royal Astronomer.

He began his astronomical career as assistant to his father. In 1735 he presented an elaborate lunar map to the Academy of Sciences and was admitted as an Adjoint Géometre on the 23rd of April, 1736 at the age of 20. He rose to Pensionnaire by 1746, became Professor at the Collège Royal and admitted to the Royal Society, Berlin Academy and Naval Academy.

He joined the expedition under Maupertuis and proved a skilful observer. He made observations for the calculation of refraction at various latitudes and different seasons.

In 1743 he constructed a great meridian quadrant at the Church of St. Sulpice. He observed transits of Mercury in 1753 and Venus in 1761 across the sun. He did much other detailed scientific work over some 50 years particularly with Clairaut, d'Alembert, Euler, Bradley and their generation.

An attack of paralysis on the 10th of November, 1791 ended his practical career.

Charles-Etienne-Louis Camus

Born 25th August, 1699 at Crécy-en-Brie.
Died 4th May, 1768 in Paris.

He was the son of Marguerite Maillard and Etienne Camus, a surgeon. He was educated at the Collège de Navarre.

In 1727 he entered an Academy of Science prize competition and this resulted in his election on the 13th of August, 1727. He served 40 years as an administrator at the Academy and as its Director in 1750 and 1761.

He went on the Lapland expedition and subsequently with many of the same people helped in the determination of the remeasure of the arc of Picard.

Anders Hellant

Born 1717 Pello
Died 1789

Was interpreter on the expedition. He went to school in Torneå after his parents had fled there in 1717 to escape the Russians. Studied at Uppsala he then joined the office of the Governor in Umeå. Fluent in Swedish, Finnish and French, he did the Swedish translation of *La Figure de la Terre*. He became a leading figure in the Torneå valley as Economy Director. He was a member of the Swedish Academy of Sciences.

The title *Lapland* although often used to describe this expedition is really a misnomer. While it represented the exotic landscape of the far north the survey was actually to the south of Lapland. *Polar Circle* is a more apt description but *Lapland* is retained here because of its wide acceptance in English language publications and because of its brevity.

A year after Bouguer and La Condamine left France for Peru a second expedition, under Pierre Maupertuis, left for Lapland. On the principle that if the oblate-prolate argument was to be settled once and for all, then it was necessary for an equatorial expedition to be balanced with another as far north as possible.

Although the Polar Circle area at the north of the Gulf of Bothnia was the venue it was not the only one that was contemplated. Thought had also been given to Iceland, but since that is little more than 2° extent north-south they would have had little room for manoeuvre; and to the Norwegian coast. The final choice was influenced by Celsius in the hope of utilizing the many islands that lie in the Gulf of Bothnia to the south of Torneå. Unfortunately, these proved useless to this expedition although Svanberg did use them some 60 years later.

The group was Pierre Louis Moreau de Maupertuis, Alexis-Claude Clairaut, Charles-Etienne-Louis Camus, Pierre-Charles Le Monnier, Abbé Réginald Outhier and later joined by Anders Celsius, de Sommereux went as Secretary and d'Herbelot as draughtsman. The addition of Anders Celsius was both political and scientific. With the measurements to be taken on Swedish territory it was natural for the Swedish King to require one of his countrymen to accompany the party. In Celsius he had a Professor of astronomy from Uppsala who was also an able mathematician and observer. Le Monnier also had a keen interest in astronomy and Camus was a clock maker who later spent 40 years as an administrator, and some years as Director, of the French Academy of Sciences.

The idea of this second arc was fully supported by d'Anville, geographer to the King, the Count de Maurepas and Cardinal de Fleury who pursued it with Louis XV until the necessary supplies and finance were made available. This was despite there being a major war in hand for the French.

Unlike the Peru group where there was considerable friction between the members, maybe the terrain, length of time and conditions were enough to rouse anyone, the Lapland one appears to have been far more harmonious. As Outhier said, even though he had previously been prejudiced against the climate of such regions he had no hesitation at deciding to go with the expedition. It meant leaving the favourable company of the Lord Bishop of Bayeux but he was able to return to the ecclesiastical life immediately upon the return of the group to Paris.

Some five months were spent assembling equipment and stores and carrying out the sort of experiments they hoped to repeat at the Polar Circle.

The party left Paris on the 20th of April, 1736 on the St. Omer post coach. They took five servants with them. Their route to Dunkirk included a stop to dine with two members of the Academy of Science in Louvre; the next day dining at a fine castle in Gournay; then to Myancour and on to Bapaume on the fourth day. Here they had time for sightseeing at the abbey of St. Vaast where was held a thorn of the holy crown. On through Bethune and Aire to St. Omer. From here they took a canal boat via Bourbourg to Dunkirk. Some 150 miles in seven days, a nice leisurely pace, or was it, considering the probable road conditions.

On arrival at Dunkirk they were joined by two others who were Swedes, Dr Lythenius, a 30 year old physician and Le Comte de Chronihelm a 72 year old nobleman and important author. On the 29th of April, Celsius arrived in Dunkirk from London where he had collected some specially constructed instruments. On the 2nd of May, they all embarked on the Prudent under Captain Francis Bernard. In addition they now had a cook who had been supplied by De la Haye d'Anglemont, the marine commissary at Dunkirk.

Among the equipment brought from London by Celsius was a small quadrant designed for use at sea, and this they were able to test out. On one test the quadrant gave a latitude of 54°34′ while the forestaff gave 54°36′. Good agreement for such equipment or just fortuitous?

Despite an initial calm sea almost all were seasick and the cramped conditions did not help, sleeping in barely a 3 foot height between decks. The sea journey was generally not a good one with very bad weather and contents flying everywhere. By the fourth day the squalls had started and the sea "very much swelled." Continual sickness and general desire to do nothing prevailed. At one stage they were for a while pursued by a Norwegian ship "of bad appearance" but luck was with them.

Eleven days out of Dunkirk they anchored off Helsingör (Elsineur). From here Celsius and Le Monnier travelled overland to Stockholm (sensible chaps) while the two Swedes travelled to Gotenburg, the remaining members of the group sailed on. The route took them past Hven, the site of Tycho Brahe's former observatory of which nothing remained. The weather was much improved and for several days were able to cover a league an hour so that by the 21st of May they entered Stockholm; an event they celebrated with three cannon, with others responding from vessel and castle. They were reunited with Celsius and Le Monnier and for the first time found they were able to read easily all night.

Outhier reported how Maupertuis busied himself as a father with a large family dealing with customs clearance, transport, money and numerous other problems. On the 23rd of May, they were all presented by Le Comte de Casteja, the French Ambassador in Sweden, to His Majesty the King of Sweden and his Queen and invited to dine with him.

Discussions here persuaded Maupertuis to take an overland route to Torneå with the servants and equipment continuing by sea. The King described the wild game of the territory to Maupertuis and substantiated the abundance of this by presenting him with a fowling piece. Further discussions in Stockholm at the office of geographic maps brought to light the information that many measurements could be made on ice surfaces in the winter months.

When Outhier was taken to the Lutheran church of St Claire in Stockholm he appreciated both the fine singing and the idea of a beadle during the sermon whose job it was, to hit with a cane those heads that were sleeping! In yet another church, that of the Muscovite Schismatics, he noticed how slovenly the ceremony was and that the priest there blew his nose through his fingers. How careful one has to be else the smallest indiscretion is recorded for all time.

Those going by sea left Stockholm on the 3rd of June carrying on board a small boat, given by the Ambassador, for coastal use. By Monday the 4th of June the others were able to take their leave of the King.

For the overland journey Maupertuis purchased two coaches and left Stockholm on the 5th of June in one of them with Clairaut and Celsius; the rest departed in the other coach on the 6th of June. The journey took them until the 21st of June.

The switch from sea to land travel was at the insistence of all those who had suffered sickness on the journey from Dunkirk. Was it perhaps a case of out of the frying pan into the fire? To maintain reasonable progress a servant was sent forward to arrange sufficient horses at each change.

A few days out of Uppsala near Gnarp they were joined by Mr Meldecreutz, a French speaking Swedish mathematician, and by the son of M.de Cederstrom, the Secretary of State. They travelled through Gävle (Geffle), Söderhamn (Soderham) Hudiksvall (Hudswikswald) to Gnarp.

It was on this journey that they came across the unusual Swedish mile which still exists today. It was equivalent to 4 French leagues and consisted of 18,000 ells, where each ell was 1 ft. 10 inch of French measure. It is somewhat disconcerting to read of '7 miles from Stockholm to Uppsala' when this is some 44 English miles. Travellers be warned! The old Swedish mile was in fact equivalent to 10,609m but is now taken as 10,000m.

It was also on this stretch that they had their first encounter with the troublesome insects. Fresh human flesh was a delicacy eagerly sought after by the gnats.

On several wide rivers they came to, they were advised against taking the coach across. However, to improve their chances they tied two boats together and arranged the coach to have two wheels on each boat. It was then a case of hard rowing to cross the river.

They crossed the river Indal (Lindal), through Härnösand (Hernosand) and under mount Skule to Onska where the ship carrying M. Sommereux and the luggage was anchored some 3 leagues away. Maupertuis having become tired of coach travel rejoined the ship. The others travelled on to Umeå (Uhma), Skellefteå (Sialefstadt) and Piteå (Pitea) to Luleå. By Umeå they had emerged from the fir and beech woods and were then among fields and more open country. Often the horses they required had to be caught from their roaming in the woods and were quite unaccustomed to pulling coaches.

Thence by Kalix (Calix) and Sangis to the river Torneå, a total distance they estimated from Stockholm to be $107\frac{3}{8}$ Swedish miles. Today's tourist guide puts the distance at 1135 km. Maupertuis had arrived some two days earlier by sea and during the voyage was lucky enough to see the sun through a whole night.

Maupertuis' first excursion was to go to the summit of Aavasaksa, some 15 leagues north of Torneå and one of the highest mountains in the area, to witness the midnight sun and no doubt assess the terrain at the same time.

The original idea had been to utilize the numerous islands at the northern end of the gulf of Bothnia but on coming in by sea it was apparent to Maupertuis that they were all too low lying to be used. What he did notice however on his visit to Aavasaksa, was how the river Torneå more or less flowed in the meridian direction and that there were high mountains on both sides densely coverd with trees.

After studying the neighbourhood of Torneå, Camus, Outhier and Sommereux decided to take a boat along the east coast of the Gulf of Bothnia and its islands as far as Brahestad. The name Brahestad (Brakestadt in [194]) or Raahe in Finnish, was named after the original Brahe family to which Tycho Brahe belonged. Count Per Brahe junior was General Governor of Finland 1637-41 and 1648-54. In 1649 he founded Brahestad. When Outhier and his companions were travelling from Stockholm to Helsingbord they passed the ruins of a castle called Brahuss = Bon Maison as Bra = good, bon, and Hus = house, maison. In this case however, the name derived from Count Per Brahe who had built the castle in 1647. In 1708 it was damaged by fire. To quote from Outhier, "they took seven men who were to row and steer the boat, which was a common one, in which we embarked with two servants and provisions for a fortnight; that is to say, biscuit and some bottles of wine remaining of the stock laid in at Dunkirk." Taken literally one can but

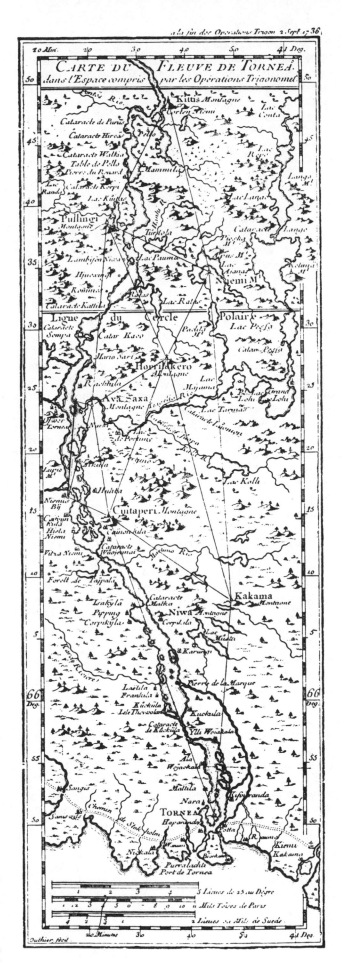

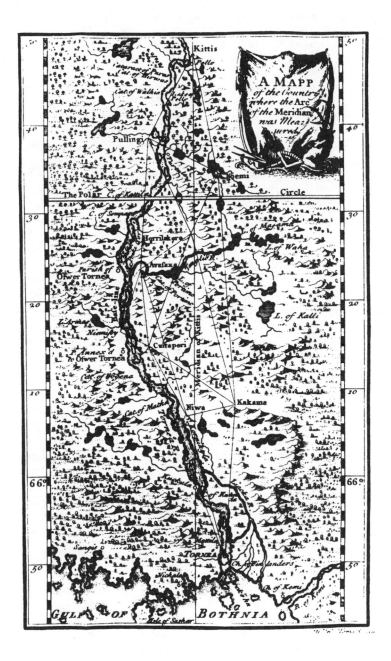

30. Maps of the area of the triangulation scheme in the
    north. Although they cover the same area note the
    number of variations between the two representations

173

wonder at their state of well-being after a fortnight! In reality they were only away a week on this occasion.

The reconnaissance began on the 25th of June and first stop was Uleåborg on the east coast of the Gulf of Bothnia. It was soon obvious that this area was in no way suitable for triangulation but they continued so as to look at the islands of Sandön (Sandhon) and Karlön (Carloohn). They returned to Torneå on th 2nd of July. They were in a quandary.

No one could tell them if the Gulf of Bothnia froze sufficiently to support survey parties although they were aware that a south wind could open up and pile up the ice. If they were to wait three months until the winter to verify the ice situation what would they do in the meantime?

Might it be possible to cut a north-south line through the forest for twenty leagues? No, this would have been quite impractical. Either they had to make use of the wooded hills along the river Torneå or move south into a suitable part of Sweden. The latter could not be contemplated so it had to be the former.

Maupertuis arranged through the French speaking Carl Magnus Du Rietz, lieutenant colonel and military chief of the Torneå region, that several soldiers would be made available to row the party to various parts of the river Torneå as they made reconnaissance. These were not regular soldiers but more a type of reservist ready to be called upon as required.

Boat was to be by far the easiest form of travel, despite frightening cataracts, because on land there was a form of thick moss that grew so tall as to make it extremely difficult to pass. On the 6th of July 1736, they set off from Torneå with their company of Finnish speaking soldiers from the Swedish Vasterbottens Regemente and seven boats. They were joined by M. Pipping and M. Hellant who both spoke Finnish and were to act as interpreters. Again Outhier only mentions biscuits and wine for provisions! For equipment there were two quadrants, a plane table, pendulum, thermometers and other items.

After passing the island of Suensaari at latitude 65°51', where the river divides, they saw few people among the deserts and mountains. As they reached cataracts at Vojakkala (Wojackala) and Kukkola (Kuckula) the sailors were left to manoeuvre the boats while the scientists walked along the banks.

The first mountains they came to were Nivavaara and Kaakamavaara, both of which had bare summits and were to the east of the river. As they progressed the conditions became harsh in the extreme — dreadful roads, snow, marsh, fog, lakes, fatigue, biting insects, cataracts, poor lodging, scarce and poor food — the difficulties were endless. An example was the eight hours it took to cover three French leagues from Karungi to Kaakamavaara.

Where the mountains had rocky summits, for example, Nivavaara, Kaakamavaara and Huitaperi, tent sites were difficult to find. While on Huitaperi the observers had both an 18 inch and 24 inch quadrant and found a discrepancy between them. The smaller one apparently closed the horizon to +4' while the larger one was correct. Since both were by Langlois it would seem surprising that they differed so, although, of course considering the methods used at that time for circle graduation it would be exceptional if they did not

differ by such an amount.

Particularly tiresome throughout the whole of the observations were the biting insects. One of the few partial deterrents proved to be the smoke from a fire but this did not prevent all from suffering severe bites. As an added precaution they covered their faces with a veil of gauze and if this touched the face the gnats were instantly there. Eating, which required lifting the veil had to be done in thick smoke for safety.

Continually during the description of the survey work the insects take pride of place. "...we were there almost devoured by gnats, which scarcely allowed us the liberty of examining the neighbourhood." "In order to eat our bread, for we had nothing else, we were obliged to be very quick in passing our hand under the veils which covered our faces; without this precaution we should...have swallowed as many of them as crumbs of bread".

Throughout the whole scheme travel proved hazardous. The marshes had only very bad tracks or paths through them and often boats had to be used. Where travel was on a main river the cataracts were severe; passengers and baggage were normally unloaded before boats negotiated the rough water. The worst such area was at Vuojennankoski (Vuojenna). The boats were unwieldly affairs of thin fir board but light and flexible to accommodate the continuous buffeting. The planks were nailed to the sides but mostly thread from reindeer sinews was used. Most had a mast for taking a sail or failing this the local sailors often raised a small fir tree complete with its branches as a substitute. There were more cataracts between Huitaperi and Korpikylä. So frightened was one passenger, M. Pipping, that he returned next day to Torneå pleading business.

For the whole two months spent in the mountains they were only clear of fog when there was a northerly wind. Imagine the problems of a survey party in such conditions where the main instrument was a 2 ft. radius quadrant of such weight that six soldiers took turns to carry it on their shoulders!

Where they found habitation the fare was usually frugal. The population were very poor people and the sudden arrival of eight or twelve extra bodies stretched their benevolence to the limit. Fish, sheep and the sparce fruits of the woods were the most that could be expected.

At almost every stop the description was of gnats even more violent and troublesome that before, immediately drawing blood! By the time they reached Pullinki the hardened soldiers were wrapping their faces and covering them with tar. "No sooner was a dish served but it was quite covered over with them, while another swarm, with all the rapaciousness of birds of prey, was fluttering around to carry off some pieces of mutton".

The northernmost station position was not easy to select from a scatter of low hills in the vicinity of Pello. Many were examined before they decided on Mount Kittis since nothing further to the north appeared to be higher or to offer any more advantage. A large part of the summit had to be cleared of trees so that lines could be seen to Pullinki and Niemisvaara.

The rough living got the better of Le Monnier and his health deteriorated. He made his way back down river on the 8th of August to Över-Torneå — 20 km south of the Polar Circle — and the house of the curate, so that he might recover. They made special friends in Över-Tornea of the curate M. Brunnius and his wife and were pleased to be able to converse with them in latin.

The first visit to Aavasaksa by the main party for observations was convenient for them to visit the Brunnius household in nearby Över-Tornea and the Provincial Governor, de Gyllengrip who was visiting the parsonage. As they were wined and dined there, it was felt prudent and courteous to reciprocate the gesture next day. This required the hasty construction on the mountain of a suitable shelter and furniture. Specific mention is made that it was here that they finished their wine!! One wonders how they managed to find time to take sufficient observations.

The house of M. Brunnius was very convenient to the site of the baseline and served excellently as lodgings while they were measuring it, despite his large family. Such hospitality was widespread through the area whether the need was for a room, food or warmth.

They had yet to measure the baseline but once that was accomplished it would be a race against time to obtain the astronomical observations before the hardest weather. Camus and Herbelot went ahead to make all necessary arrangements at Pello while last readings were observed at the base terminals.

To travel to Pello the new sector was of such bulk that three boats were required for it. In addition there were three one-second pendulums, several simple ones, quadrants, barometers, thermometers and miscellaneous items. In total they had some 18 or 19 boats, made available for the journey but were able to manage with 15. On the way Maupertuis, Celsius and Outhier digressed to go to the summit of Kaakamavaara, where it was very wet and slippery and caused Maupertuis to have a fall. Although it was thought his thigh might be broken, luck was with them and he quickly recovered.

They made their way to Pello where all was made ready and it remained for the weather to be kind. While in Pello, Outhier made particular mention of the very hot baths of the inhabitants — up to 44 °R. In essence he described a sauna. First snow fell on the 21st of September and observations proceeded slowly. Night after night the skies were obscured and impatience increased. By the 30th of September the thermometer was at −8 °R and the edges of the river were freezing. Not until the 2nd of October was it possible to make sector observations. From the 11th of October the weather deteriorated again and by the 21st the team agreed to move to Tornea while it was still possible.

Back in Tornea the sparse accommodation in this small place required all to lodge at different houses. The whole town only amounted to some 70 houses, all of wood, built along three small streets. Even so there were two churches, one of which, south of the Suensaari insel on the insel Björkön, conducted services in Finnish; the other in the north of the town used Swedish. Swedish was generally for the Burghers and Finnish for the servants and peasants.

By the 1st of November when they were preparing their observatory the river was frozen over. The sign of the start of a long stay in Tornea. The sledges appeared and ways across the ice were marked with fir sticks. Then a south wind in mid-November made the ice dangerous and flood water

entered the small observatory near sea level to a foot depth. Maupertuis, Le Monnier and Outhier used a boat to rescue the equipment.

Unlike the Peru expedition where there was suspicion and mistrust among the members, who can wonder at what happens in such inhospitable territory over such a long period, it would appear that those who lodged in Tornea were a friendly, happy group that mixed well with the local inhabitants.

Should the base be measured during the short, very cold days of winter or left until spring when there was a chance of a sudden thaw upsetting everything? Some local advice went one way, some the other. A site inspection by Clairaut, Celsius and Outhier settled for an immediate start.

Attempts at using a form of horse drawn snow plough to clear the line were not entirely satisfactory. Such was the cold at −18 °R on the first day of measurement, that Le Monnier when drinking brandy out of a silver cup had his tongue glued to it in such a manner as to tear off the skin. Maybe plastics do have their advantages.

Many fingers and toes felt the pain of near frost bite yet they all continued relentlessly. Even during a lull while the last part of the base was staked, time was not wasted. There was a tree between Niva and Hottilankero that had been included in a round of values from Aavasaka but which had not had its height recorded. Despite the fact that the height was not likely to make any sensible change to the results, Maupertuis could not contemplate the omission of its value from the records.

A rapid journey by reindeer sledge to the foot of the mountain and then a laborious climb up to the top where the sledge had to be led by a Finn on snow shoes, or boards fastened to the feet to overcome the deep snow. An alternative description is of skates or planks four or five inches wide and eight feet long- skis? The descent proved even more treacherous.

Into January and even the bottles of strong French brandy were freezing solid in unheated rooms. The spirit-of-wine thermometers became frozen around −29 °R and only the mercury ones could accommodate the depth of cold.

The long winter was spent fair copying the town plan they had made of Tornea or passing round the many books they had wisely taken with them. Others set to making pendulums from balls of various metals for a range of experiments. On the 16th of March they set up a telescope at Granvik (Granwijk) and were able to observe an eclipse of the moon. It was also thought prudent to use their enforced stay to make additional astronomical observations.

By early June preparations were under way for the return to Stockholm. So shallow was the water here that the anchorage was some two leagues offshore. Departure day was the 9th of June, 1737. Maupertuis, Le Monnier, Sommereux and Herbelot were to go by sea, while Clairaut, Camus, Celsius and Outhier were to travel overland. A last minute change resulted in Clairaut going by sea instead of Le Monnier.

The coach party left Tornea on the 10th of June. Passing through Sangis, Kalix to Old Lulea. On the way to Pitea, they noted that it took seven minutes thirty-five seconds to pass at a good rate over a wooden bridge of 102 arches. It was at Pitea that they received news of the vessel having run aground two miles from the town and they hastened to the aid of Maupertuis. A change of wind soon after they had weighed anchor became a storm and the vessel took on much water. They could scarce keep pace with the use of buckets and one poor pump. On the evening of the 11th of June the pilot was able to ground the vessel safely.

After salvaging their luggage and equipment Maupertuis continued by land on the 14th of June. Sommereux, Le Monnier and Outhier remained to await repairs to the vessel. The waiting time was usefully occupied with making a plan of the locality. By the 20th of June, the baggage was re-loaded for shipment to Stockholm and the following day the travellers set off again overland. Through Selet and Savar to Umeå, and by the 26th they reached the river Ångerman. Then on to Sundsvall and Gnarp and on to Gävle by the 29th.

On July 1st they visited the coppermine at Fahlun, some 640 feet deep and graphically described as to deter the faint-hearted from attempting. They could not have had any safety laws at that time. Special clothing for the occasion was breeches, jacket, waistcoat, wig and hat.

By the 9th of July they were at Salsberg and by the 11th of July in Stockholm. The luggage had arrived already and was in the care of the rest of the party. They were dined by the ambassador and presented to their Majesties. Time was spent cleaning the instruments that had got wet in the vessel.

Maupertuis and Sommereux took a vessel to Amsterdam on the 18th of July. Herbelot remained in Stockholm to await passage to Rouen for the instruments; Clairaut, Camus, Le Monnier and Outhier set out by coach towards Amsterdam on the same day. By Norrköping, Linköping, Gränna, to Jönköping by the 22nd. Then to Hälsingborg by the 24th. The coach was dismantled and put on board a boat for Helsingör, when they continued to Copenhagen. Hamburg was reached on the 4th of August then on to Bremen, Klappenborg, Lingen, Delden and Amersford to Amsterdam by the 13th where they again met Maupertuis. They continued by Leyden, Delft, Rotterdam, Antwerp and Brussels to Paris by the 20th of August, 1737.

The expedition concluded its work on the 21st of August, 1737 by visiting Versailles and giving an account to His Majesty, The Cardinal and Le Comte de Maurepas.

THE SURVEY STATIONS

All except two of the stations were in what is present-day Finland.

Kaakamavaara. (Kukama, Kakama)

The approach was from Karungi and was extremely difficult with dreadful tracks, recently thawed snow, and impassable marshes. Only means of crossing was on felled trees, otherwise one sank up to the knees. Two lakes had to be crossed on makeshift rafts. The insects and heat were unbearable. It took eight hours to cover three French leagues.

The hilltop was bare and craggy. The rock was of stratified white quartzite where the vertical planes collected considerable water. On the summit they put a fir pole stripped of its branches and supported by other such trees.

The point was established on the 7th of July, 1736 by Outhier, Sommereux and Hellant. It was later rebuilt in the

form used on Nivavaara. It was rebuilt yet again on the 16th of August by Camus, Le Monnier and Celsius, after being blown down. The last observations were on the 5th of September between Horrilankero and Niemisvaara by Maupertuis, Celsius and Outhier. It took the eight assistants relieving each other to carry the two ft. quadrant to the top.

Nivavaara (Nieva, Niwa)

Established by Maupertuis, Camus and Pipping on the 7th of July, 1736, this had a pyramid of stripped trees fastened at the top and such that besides forming a signal it could also be used as an observatory without any need for reduction to centre. Observations were by Maupertuis and Clairaut with the 18 inch quadrant. Such targets were readily visible at 10 or 12 leagues. They always left various reference marks in case of disturbance of the target. This was the first signal and used as a prototype for all the others. Observations were made on the 12th and 13th of July, 1736, by all members of the party, and again on the 17th by Maupertuis and Clairaut.

Horrilankero (Horrilakero)

Established by Maupertuis and Camus on the 10th of July, 1736. It was a tree covered summit and the mountain was of reddish stone interspersed with white crystal. It was difficult to approach through marshlands. Observations were made by Maupertuis, Clairaut, Camus and Celsius from the 14th to the 17th of August, 1736. On the 19th a forest fire destroyed the signal and it was rebuilt from witness marks on the 21st. 22 men were used to put out the fire and seven left on watch in case it were to break out again.

Aavasaksa (Avasaxa)

Established by Maupertuis and Camus on the 8th of July, 1736. It was a tree covered summit. Access was from the river Torneå by difficult route through the woods and up slippery rocks. It was very precipitous to the N.E. Superb views were found from the summit. It was occupied for observations from the 20th to the 30th of July, 1736, and on the 19th and 21st of August.

Huitaperi (Cuitaperi)

Established by Maupertuis and Camus on the 11th of July, 1736. It had a bare summit. The access was four leagues from Aavasaksa and when the summit was reached it gave a good view of Torneå as well as to the already established marks. The route to the summit was rough and difficult, and there was a wide covering of moss.

Clairaut, Outhier and Maupertuis took observations there with 18 and 24 inch quadrants on the 16th to the 20th of July, 1736. Sommereux and Herbelot were in attendance. The observation between Aavasaksa and the south end of the base was made on the 26th of August.

Pullinki (Pullingi)

This was reached by Outhier and Hellant on the 17th of July, 1736. It was entirely tree covered. The eastern extremity was the highest part but much clearing was necessary before a signal could be raised the next day. Observations were by Maupertuis, Le Monnier, Outhier and Celsius from the 31st of July to the 5th of August, 1736. The point is in present-day Sweden.

Ketima

A signal was established here on the 27th of July, 1736 by Outhier and Hellant but was not used in the scheme.

Kittisvaara (Pello)

This was established on the 2nd of August, 1736 by Outhier and Hellant. Unfortunately it was not possible to arrange it so that Horrilankero could be seen. To make the signal more visible against a poor background they split the trees and arranged them with the white side outwards. Observations started on the 6th of August, 1736 and were completed on the same day.

A monument has been erected in Pello to commemorate the expedition and its unveiling was witnessed by representatives from the Academies of Science in Paris and Helsinki together with the University and the Royal Society of Science in Uppsala.

Kukas

A point was established here by Outhier and Hellant on the 4th of August 1736. Avenues were cut through the trees in the directions to Horrilankero, Pullinki and Kittisvaara. It was not used in the scheme as Niemisvaara proved more useful.

Niemisvaara (Niemi)

Established by a Swedish officer and his men. Observations were by Maupertuis, Celsius and Outhier with the 2 ft. quadrant on the 11th of August, 1736. It was difficult of access from Pello where the way had to be cut through entangled growth with deep moss and fallen trees. This point was in present-day Sweden. They had arrived here on the 8th of August but bad weather persisted until the 11th.

Baseline terminals

Established by Clairaut and Camus. On the 22nd of August, 1736 the northern mark was made more substantial and at the same time the angle was observed between Aavasaksa and the south end of the base.

The following day they moved to the south end of the base and took observations there. The north end was at Poiky-Torneå and the south end at Niemisby.

Torneå

The belfry and spire. Observations began on the 29th of August when Huitaperi came into view but results were not satisfactory. Adverse weather delayed any more attempts until the 2nd of September. To avoid spectator problems, Celsius shut himself in the belfry while Maupertuis and Outhier took a walk to distract attention.

While waiting to travel to Pello for the astronomical work in early September, 1736 observations were made at the spire of the Torneå church on the island of Suensaari between the mountain targets.

With the stations they made a point of taking as many witness measurements as possible to facilitate re-positioning.

ANGULAR OBSERVATIONS.

These were taken with a quadrant of two feet radius built by Langlois and fitted with micrometers. When checked round the horizon against one of 18 inch it gave a zero misclosure and the other 4'. At each point care was taken to centre the instrument over the mark. This is not as naive a statement as it may seem because with a quadrant it is the centre of the graduated arc that has to be centred, not the support stand.

Each observer recorded his values separately. The mean was then accepted since the values always differed slightly. At the same time the vertical angles to each point were recorded for use in the reductions of the inclined angles to the horizontal.

Angle	Observed value	Reduced to the horizontal	Adjusted for closure
CTK	24° 23' 00.2"	24° 22' 58.8"	24° 22' 50"
KTn	19 38 20.9	19 38 20.1	
The above angles were		24 22 54.5	
reduced to centre as		19 38 17.8	
TnK	87 44 24.8	87 44 19.4	
HnK	73 58 06.5	73 58 05.7	
AnK	95 29 52.8	95 29 54.4	
AnH*	21 32 16.9	21 32 16.3	
CnH	31 57 05.2	31 57 03.6	
TKn	72 37 20.8	72 37 27.8	
CKn	45 50 46.2	45 50 44.2	
HKn	89 36 00.4	89 36 02.4	
HKC*(1)	43 45 46.8	43 45 47.0	
HKC (2)	43 45 41.5	43 45 41.7	
HKN	9 41 48.1	9 41 47.7	9 41 50
KCn	28 14 56.9	28 14 54.7	
TCK	37 09 15.0	37 09 12.0	37 09 07
HCK	100 09 56.4	100 09 56.8	100 09 48

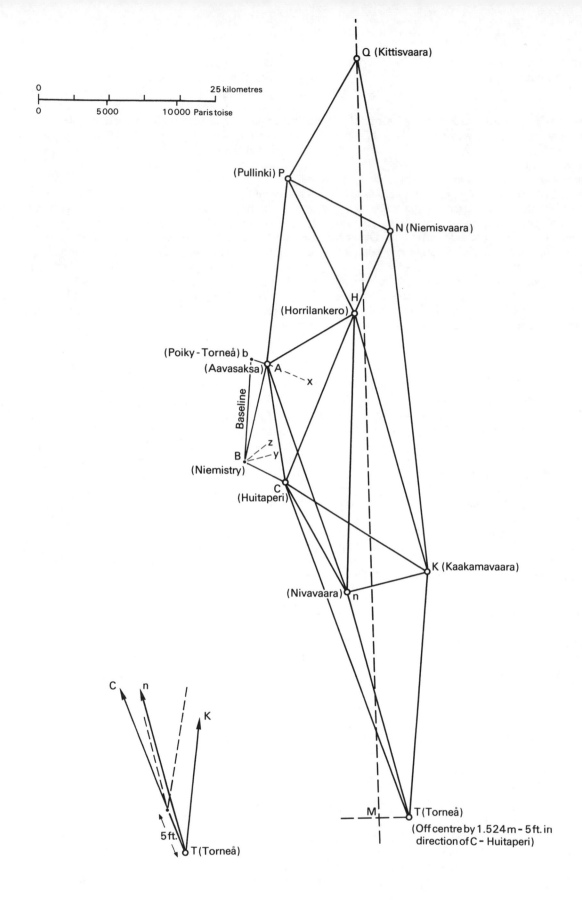

Figure 41. Diagram of the scheme of Maupertuis.

178

Angle	Observed value			Reduced to the horizontal			Adjusted for closure		
ACH	30	56	54.4	30	56	53.4	30	56	47
HAP	53	45	58.1	53	45	56.7	53	46	03
HAx	24	19	34.8	24	19	35.0			
xAn	77	47	46.7	77	47	49.5			
xAC	88	02	11.0	88	02	13.6			
CAn	10	13	54.2	10	13	52.8			
APH	31	19	53.7	31	19	55.5	31	20	01
QPN	87	52	09.7	87	52	24.3	87	52	17
NPH	37	21	58.9	37	22	02.1	37	21	56
NQP	40	14	57.3	40	14	52.7	40	14	46
PNQ	51	53	13.7	51	53	04.3	51	52	57
PNH	93	25	08.1	93	25	07.5	93	25	01
HNK	27	11	55.3	27	11	53.3	27	11	56
CHn	19	38	21.8	19	38	21.0			
CHA	36	42	04.3	36	42	03.1	36	41	56
AHP	94	53	49.7	94	53	49.7	94	53	56
PHN	49	13	11.9	49	13	09.3	49	13	03
KHn	16	26	06.7	16	26	06.3			
CHK	36	04	54.1	36	04	54.7	36	04	46

For base connection.

Angle	Observed value						Adjusted for closure		
ABb	9	21	58.0				9	22	00
AbB	77	31	48.1				77	31	50
BAb	93	06	07.2				93	06	10
ABy	61	30	05.4						
yBC	41	12	03.4						
ABz	46	07	57.5						
zBC	56	34	22.2						
ACB	54	40	28.8				54	40	28
BAC	22	37	20.6				22	37	20
ABC	102	42	13.5 (from ABy+yBC and ABz+zBC)				102	42	12

*At n mean value of AnH was 21° 32′ 02.5″

			Reduced to the horizontal			Adjusted for closure		
At K		HKC	43	45	35.6	43	45	26
		CKT	118	28	12.0	118	28	03
At A	value of HAn		102	07	24.5			
		HAC	112	21	48.6	112	21	17

Vertical angles.

At	To				At	To			
T	C	0°	00′	00″	P	H	−0°	22′	00″
	n	+0°	03	00		A	−0	18	10
	K	+0°	08	40		Q	−0	32	40
						N	−0	26	50
n	T	−0	17	40					
	K	+0	16	50	Q	P	+0	22	30
	A	+0	04	40		N	+0	01	00
	H	−0	00	30					
	C	+0	10	00	N	P	+0	18	30
						Q	−0	14	00
K	n	−0	22	50		H	−0	02	40
	C	−0	04	45		K	−0	14	00
	H	−0	05	10					
	T	−0	24	10	H	n	−0	18	15

	N	—0	08	10		A	0	00	00
						P	+0	11	50
C	K	—0	06	10		N	—0	05	00
	n	—0	19	00		K	—0	12	30
	T	—0	24	10		C	—0	10	40
	H	—0	02	40					
	A	+0	05	00	B	A	+0	40	30
						y	+1	23	30
A	P	+0	04	50		C	+1	04	05
	H	—0	08	00		z	+1	11	00
	x	—0	10	40					
	C	—0	14	15					
	n	—0	20	20					

It would appear that Celsius acted as chief observer throughout as he occupied the points with the two foot quadrant. Of those in the party he was the only one who was practiced in its use prior to the expedition. The computations were mostly carried out by Clairaut, the planning by Maupertuis and record keeping by Outhier.

BASE MEASUREMENT.

For the measurement of the baseline, 8 fir rods, each of 5 toise length, were constructed at Öfver Torneå to the design of the party members. The same person may also have constructed various supports for the rods although there is some doubt about this.

Outhier said that 'the rods for the measure of the base were made, as well as the supports" whilst [98] suggests that no allowance was made for the slope of the base since the bars, if not set level, would indicate a hypotenuse measure. [55] also assumes that the rods were simply laid upon the snow. [260] states that slope correction was ignored but this statement is probably based on the comment in [55]. [6] similarly states that the rods were placed on the snow [163] and [211] translate the same as 'that we had to place continuously upon the snow and take up again'. Records of Celsius in Uppsala [251] contain the statement 'every group had 4 rods, which were put after each other upon the snow, that could carry the rod but not us...' No mention of supports. Obviously there is some doubt here since it could have been expected that the account by Outhier was authoritative.

After a reconnaisance of the base area to ascertain the state of the ice, first moves towards assembling the equipment there were made on the 14th of December by Camus, Outhier, Hellant and Herbelot. They took a quadrant, thermometer and all necessary impedimenta. On the 18th of December they were able to stake the line in time for the arrival on the 19th of Maupertuis, Clairaut, Le Monnier, Celsius and Sommereux.

For calibrating the 5-toise rods they had with them an iron toise that had been compared with that of the Châtelet before the party left Paris. The comparison had been at 14°R. This toise came to be known as the Toise du Nord. Each end of the rods had a large round headed nail that was filed away until it exactly fitted the standard. According to Maupertuis the temperature during this adjustment procedure was 15°R or 62° of Mr. Prince.

For the eight long rods a special standard was constructed. At the apartment two large nails were hammered in the walls at a distance apart of just less than 5 toise. A series of supports were placed between these to take the 5-toise rods. (Were these supports the ones referred to by Outhier?) The nails were driven apart, and filed away until the rods just fitted between. The long rods were then adjusted between the nails as exactly as possible—in fact such that "the thickness of a leaf of the finest paper would not pass".

When all was ready to start measurement there was much snow on the ice. To try and clear this they fashioned large logs into a triangle, or snow plough shape and pulled it by horses along the line. They were not very successful and gave up after 2½ hours.

The two measuring parties each had four observers. Each had a pencil and paper or a clip board round his neck on which to mark each rod length. The rods were marked so as to be always laid in the same order. By night the cold became intense and often reached —25°R and on the last day it fell to —37°R.

The two results for the measurement were:

7,406t 5 ft. 4 inch and
7,406t 5 ft. 0 inch

During the measure which started on 21st December 1736, stakes had been left at every 100t and as a check Maupertuis Camus and Outhier used a 50t cord to determine any gross errors along the whole length. This was completed on 30th December. For the most part of the measurement the snow was two feet deep.

Maupertuis experimented in his lodgings to determine whether the toise rods of wood changed length with temperature. He kept two indoors and two outdoors and there was no sensible difference. If anything, he considered those left outside to have lengthened slightly.

The two measures of the base had agreed to 1:133,000 in appalling conditions and the mean value was accepted:

7,406t 5 ft. 2 inch = 7,406.861t.

But was this the true difference between the two measures?

In [251] it is suggested that the two teams compared their results at the end of each day and then restarted from a common point. Muapertuis wrote that the difference for each day was never more than 1 inch but it is difficult to believe that with such primitive equipment in such weather conditions that so good a daily agreement was possible. Maupertuis in fact attributes the four inches difference to the last day as they measured from the bank to the signal.

Although the whole scheme was computed in several different ways the base measure could only be incorporated as side Aavasaksa (A) to Huitaperi (C) through the two triangles AbB and ABC. Whence

$$AC = 8,659.94t$$

Meridian direction.

For the establishment of the meridian direction through the chain observations were made at Kittisvaara (Q) with a 15-inch transit telescope.

The sun was observed through the vertical planes to Pullinki (P) and Niemisvaara (N) and from this the meridian was found to be:

$$28° 51' 52'' \text{ west of the line QP}$$

i.e   208   51   52   was the azimuth of line QP

At the same point they derived the latitude as 66° 48' 20" N

A verification of the meridian was achieved in a similar way at Torneå, except that the sun was observed near the horizon .

Whence:

Torneå - Nivavaara	344° 40' 38"
and Torneå - Kaakamavaara	4 18 42½

(This compared with 4° 19' 17" found by other means). The latitude here was found to be 65° 51' 00" N

Linear length of the arc.

The length of the meridian arc was accepted as the mean of the results from the solution of two sets of triangles.
1. Using triangles ACH, CHK, CKT, AHP, HNP, and NPQ. From length AC = 8 659.94t they found:

QP	10 676.90t at bearing	208° 51' 52"
PA	14 277.43	185 26 06
AC	8 659.94	351 33 26
TC	24 302.64	339 49 08

Whence in the meridian direction:

DP	9 350.45
EA	14 213.24
AF	8 566.08
CG	22 810.62
Total	54 940.39

2. Using triangles ACH, CHK, CKT, HKN, HNP, and NPQ

QN	13 564.64t bearing	168° 37' 06"
NK	25 053.25	176 07 12
KT	16 695.84	184 11 53

Whence in the meridian direction

dN	13 297.88t
LK	24 995.83
Kg	16 651.05
Total	54 944.76

Mean	54.942.57t	
+ 73.74	Correction to centre at Torneå
+ 3.78	Correction to centre at Kittisvaara
+ 3.38	Effect of convergence
55 023.47t	Accepted value

Computations for the arc length were however carried out through ten different routes. The various results were:

54 941t	Diff. from 54 942.57 of	1½t
54 936		6½
54 942½		0
54 943½		1
54 925		17½
54 915½		27
54 912		30½
54 906½		36
54 910		32½
54 891		51½

It will be noticed that all except one of these is smaller than the accepted value. Why they selected the two routes they did is not known although the first does follow the west side of the chain and the second the east side. It is suggested that they were of the opinion that the best results would come from the routes that contained the largest angles.

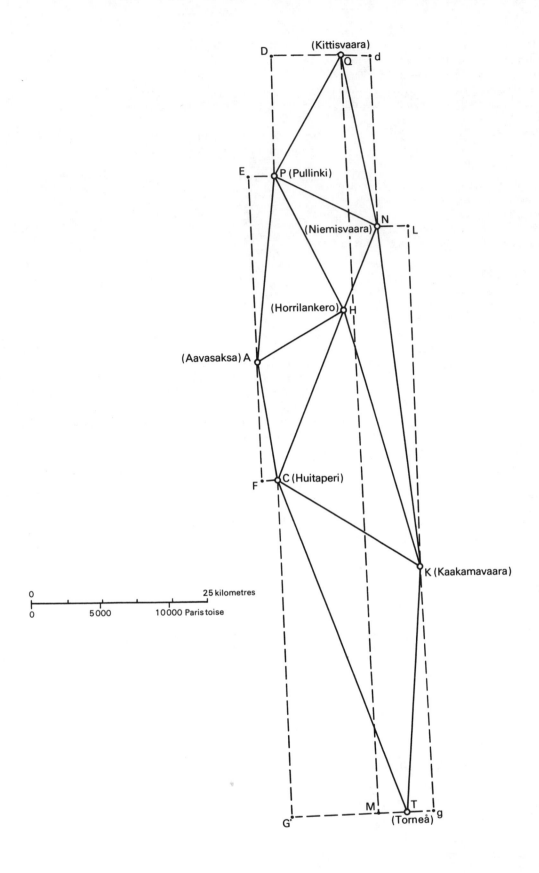

Figure 42. Diagram of the computational scheme.

Astronomy at Pello (Kittisvaara)

The group arrived at Pello on September 8th to begin the astronomical observations. Camus had previously spent two weeks preparing a room for their equipment. For this a room in the house of Korten Niemi was modified to have a stone block set in the floor to take the telescope and for the suspension of simple pendulums. The telescope was by Graham as was their clock.

For an observatory on Mount Kittis Camus purchased a cotta, or sort of hut, normally used for melting snow and warming the water for the cattle, dismantled it and re-erected it on the mountain. It there housed the sector that had been newly brought from England. Several queries are raised here.

Maupertuis describes not one but two observatories on Kittis. One housed a two ft. radius quadrant, Graham clock and transit telescope—also by Graham. This telescope was placed exactly over the station mark to allow determination of the meridian direction. The second observatory housed the zenith sector which was placed so that its limb was exactly along the meridian. Its telescope was 9 ft. long, the limb was 5° 30′ and subdivided to 7′ 30″ divisions.

Outhier persistently refers to the 'sextant' brought from England whereas it must have been a sector. (Or is this a quirk of the translator?). Outhier also said 'newly brought to Torneå from England' but there is no indication of how, when or by whom.

The pendulum by Julien Le Roi was mounted on the 13th of September. The sector was installed on the 16th together with a small Graham pendulum and transit telescope. This telescope could be pointed at the sun at noon or to a star as it crossed the meridian. Then by lowering it a point could be set on the horizon and the angles to Pullinki and Niemisvaara found.

Constant bad weather precluded use of the sector for many days. Not until 29th September did it clear sufficiently. Outhier observed the pendulum vibrations and compared them with those of the pendulum by Julien Le Roi which was regulated against the stars. The pendulum used was a bar of well polished iron, tapering towards the top and suspended on a steel knife edge.

At the same time the sector, supported in a 12 ft. high pyramid, was adjusted into the meridian line. $-8°R$ was recorded as the river edges began to freeze. Sector observations began in earnest on the 2nd of October, 1736.

On the night of the 4th of October it was ideal for Maupertuis, Camus and Outhier to observe the 'bright part of the eagle' with the fixed telescope while Clairaut, Le Monnier and Celsius observed δ of the Dragon with the sector. Similarly on the following two days.

The observing group was always of three but changed in composition, - one on the pendulum, one on the micrometer and the other manipulating the telescope and micrometer without looking at them. The final observation was made on the 10th of October. The sector results all agreed to 3 seconds but the pendulum observations were at variance. There were five of them altogether:

1, polished bar of iron. Cylindrical.
1, polished bar of iron. Lozenge shaped.
3, balls of brass filled with lead, soldered to a rod of steel and suspended from two knife edges.

One of the ball type was later changed to swing from a ring instead of from a knife edge.

On the 25th of March 1737 Clairaut, Le Monnier, Celsius and Outhier set out from Torneå to re-observe the astronomical measures at Pello. They arrived there on the 26th and the next day Sommereux and Hellant arrived with the sector. This was placed on the mountain together with a 3 ft quadrant. The Graham pendulum was set at the house of Korten Niemi as before. Maupertuis arrived on the 27th.

Observations began on the 4th of April. A variety of experiments were made with the pendulums for which great lengths were gone to in order to maintain the temperature reasonably constant. By the 11th of April all experiments were complete.

Astronomy at Torneå.

The party arrived at Torneå on the 28th of October and immediately set about constructing an observatory. They found in the house of Mr Hellander a room similar to the cotta at Pello. A firm base was dug and large stones set to take the sector. An opening was made in the ceiling and a removable cover made for it. By the 30th of October the first sector readings on δ of the Dragon were possible. Further observations were possible until the 7th of November.

A further smaller observatory was built near the river to house the meridian transit, a pendulum and a quadrant. New barometers were made and their variations noted together with those of both mercury and spirits of wine thermometers constructed on the principle of Réaumur. By the 7th of November the temperature had fallen to $-20°R$ and the river was frozen although not yet sufficiently thick to bear heavy weights. By 3rd December one reads that the spirit of wine thermometer was at $-18°R$ whilst the mercury one was at $-22°R$.

During the long winter in Torneå the idea arose of reobserving the astronomy using this time α of the Dragon. The opportunity arose for sector observations from the 17th to the 19th of March 1737. The meridian line was set out at the hut on the river bank. With a quadrant right angles were set off from the meridian to the sector observatory and to the steeple of the church. These parallels were found to be 73 t 4 ft 5 inches apart.

The arc length.

For their latitude observations at Kittisvaara and Torneå the team used a 9 foot radius zenith sector with a limb of 5½°. It had been constructed in England by Graham and the telescope had silver wires as a cross at the focus. The telescope and limb were cast in one piece and suspended from brass supports so as to be able to swing freely from a large wooden frame. The amount of movement of the telescope was recorded by a micrometer.

Suspended from the system was a long plummet that crossed the limb at the required angular reading. To steady the plummet it was immersed in a container of brandy.

The technique was to set the limb so that the plummet crossed the division expected. As the star passed so it was tracked by the micrometer to obtain small positive or negative adjustments to the reading at the plummet. The micrometer was then checked by returning it to its original position and re-noting the reading. Agreement was normally to 1 or 2 micrometer divisions.

At Kittisvaara δ Draconis
4.10.1736 Limb 2° 37′ 30″

micro before	24^r	10.7^p
micro after	24	12.5
Mean	24	11.6
Star	22	30.9
Difference	1	24.7
5.10.1736	1	26.9
6.10.1736	1	25.6
8.10.1736	1	27.3
10.10.1736	1	24.7
Mean difference	1	25.8

Note: r = revolutions of the micrometer
p = parts of the micrometer at 44 per revolution.

At Tornea δ Draconis
1.11.1736 Limb 1° 37′ 30″

Difference	1	40.3
2.11.1736	1	40.3
3.11.1736	1	41.3
4.11.1736	1	40.8
5.11.1736	1	40.3
Mean difference	1	40.6

Whence arc was   2° 37′ 30″ — 1^r 25.8^p
less   1 37 30 + 1 40.6
=   1 00 00 — 3 22.4

But on the micrometer
15′ = 20^r 23.5^p
Thus   43.83″ = 1^r
Whence   3^r 22.4^p = 2′ 33.8″
Subtracting from 1° 00′ 00″ gives
arc   = 0° 57′ 26.2″

But as the length of the chord of the sector was found to be too short by 0.002 inch a further 0.65″ was subtracted to give an arc length of   0° 57′ 25.55″

The check observations in 1737 gave:
At Kittisvaara α Draconis
4.4.1737 Limb 4° 15′ 00″

Micro difference	6^r	13.0^p
5.4.1737	6	12.3
6.4.1737	6	12.4
Mean difference	6	12.6

At Tornea α Draconis
17.3.1737 Limb 3° 15′ 00″

Micro difference	2	35.3
18.3.1737	2	35.3
19.3.1737	2	33.0
Mean difference	2	34.5

Whence arc was   4° 15′ 00″ — 6^r 12.6^p
less   3 15 00 — 2 34.5
=   1 00 00 — 3 22.1

But 3^r 22.1^p =   2 33.5
0 57 26.5
Correction for limb   — 0.65
0 57 25.85

These results however required further corrections for precession and aberration thus:

δ Draconis

Observed	0° 57′	25.55″
Precession		— 0.48
Aberration		+ 1.83
	0 57	26.90

α Draconis

Observed	0° 57	25.85″
Precession		— 0.85
Aberration		+ 5.35
	0 57	30.35

Mean result   0° 57′ 28.63

Notes.
A letter from Maupertuis to Bradley [256] quotes the amplitude, corrected for precession as 57′ 25.07″. Then an aberration correction of +1.83″ to give 57′ 27.90″. There is an error or discrepancy of 1″ here. [163] gives the correct figure of 26.90″ For α Draconis [256] gives

57′ 25.19″
+ 5.35   for aberration
57 30.54

Hence [256] gives a mean as 57′ 28.72″ which would only be correct if the δ Draconis arc were 57′ 26.90″ as suggested above. [198] gives for

δ Draconis   57′ 30.42″

α Draconis   57 26.93
Mean   57 28.67

Refraction was nowhere allowed for yet they were well aware of its existence as they experimented on its determination. Maupertuis said that if there is any at all near the zenith it must be imperceptible and cannot affect the result. The difference of 0.04″ between the two mean results above was the outcome of correspondence between Maupertuis and Bradley where the latter sent details of his latest discoveries regarding corrections that required to be applied.

With the arc length as	55 023.47 t	
the value of 1° =	57 438.60 t	using 0° 57′ 28.63″
or 1° =	57 437.94 t	using 0 57 28.67

Maupertuis quoted this latter value.
[256] quotes 1° = 57 437.1 t as 377.1 t greater than the central degree of France. This value is the one that would be obtained from taking the amplitude as 57′ 28.72″.

Heights.

Although the group observed vertical angles for the reduction of the quadrant readings nowhere is there any indication of height values. However, using the observed vertical angles and calculated distances it is possible to derive approximate heights for comparison with the values used today. It is stressed however that these are only approximate because:

a. the length of any line will vary slightly according to the route through which it is calculated and
b. the vertical angles were considered as simultaneous reciprocal values when this was clearly not the case.

Heights

Station	(a)	(b)	(c)	(d)	
T	0.0 t	0.0 m	30.0 m	-	
K	80.0	155.9	185.9	189 m	
n	47.8	93.2	123.2	124	
C	83.8	163.3	193.3	192	
H	99.5	193.9	223.9	224	
A	108.0	210.5	240.5	242	
B	20.2	39.4	69.4	-	(On river)
P	156.1	304.2	334.2	335	
N	97.3	189.6	219.6	220	
Q	69.0	134.5	164.5	165	

Notes.

(a) = Height calculated in relation to Torneå in toise
(b) = Height calculated in relation to Torneå in metres
(c) = Heights on assumption that Torneå tower was 30 m
(d) = Heights from present day maps.

Obviously very good agreement considering the accuracy of the angular observations and the fact that the targets may not have been on the highest points.

Verifying the sector.

For checking the sector the group set out a line on the ice of the river Torneå on 4 May 1737. It was 380t 1 foot 3 inches (380.21t) long.

The centre of the sector was placed at one end, supported horizontally on two trestles. At the other end a line was set out at right angles for 36t 3 feet 6 inches 6⅔ lines (36.59t).

[98] suggests rather that the test figure set out formed an isosceles triangle whereas both Maupertius and Outhier read as if it were a right angled triangle. From the centre of the instrument a fixed silver wire was arranged to just graze the limb. The angle between the two distant marks was recorded by 5 different observers to differ from 5° 30′ by

6.5 parts
8.3
7.0
7.9
6.8
Mean −7.3 parts = −7.3″

Thus observed angle was 5° 29′ 52.7″
But the sector was too
small by * 3.75
Hence observed angle was 5 29 48.95

Approximate distances.

AB	7 243 t	PN	8 768 t	CK	11 412 t
AC	8 660	HN	7 027	Cn	8 514
BC	3 415	PQ	10 677	Kn	5 616
AH	7 452	NQ	13 562	nT	15 949
CH*	13 402	NK	25 048	KT	16 698
AP	14 277	HK	19 070	CT	24 303
HP	11 559				

Calculated value was   5 29 50
         Difference   01.05

* = information supplied by Graham who constructed the sector.

[55] suggests that there is a 'curious misprint' in Maupertuis's book on this point but it is not readily seen what the author meant. Clark says it should be 5° 29′ 52.7″ which, as is shown above, is the observed value quoted by Maupertuis.

They were later criticised for not making double observations -i.e. changing face as it were, with the sector, but it would have been very difficult with this particular instrument.

In analysing the possible effects of various sources of error they investigated the effect of two angles in each triangle as too great by 20″ each and the third too small by 40″. Always in the same manner. The recalculation gave a change of only 44.05t.

Refraction.

Aside from the routine survey measurements Maupertuis had his party carry out experiments in various related fields. One of these was to determine the amount of refraction − a yet sketchily understood phenomena.

To effect his method he had to begin with the assumption of 4 quantities:

Latitude of observing station.
Obliquity of the ecliptic.
The sun's position.
The sun's parallax.

The first two were obtainable by observation, the last

two could be found in tables. For the sun's position there were tables by de Louville.

Then, for example,

At Torneå, 1200 hours, 1st December 1736

Sun's declination	21° 55′ 21″		
Colatitude	24 09 10		
Thus V.A. to sun's centre	2 13′ 49		
Sun's parallax	−10		
Thus true V.A. to sun centre	2 13 39		
Sun's semi-diameter	0 16 19		
Height of upper limb	2 29 58		
Same by observation	2 45 42		
Difference = refraction	0 15 44	at vertical angle 2° 46′	

Other observations at Torneå gave

3.12.1736	V.A. 2° 31′ 00″	refraction 0° 18′ 26″	
8.12.1736	1 56 51	0 20 03	
5.1.1737	2 09 32	0 20 03	

In April and May 1737 they also observed Venus in like manner

V.A. 3 35 00	refraction 0 13 56	
3 38 00	0 14 11	

For this it was necessary to make observations over several days to be able to calculate the vertical angle for Venus as it crossed the meridian.

At Kittis they also used Venus in a similar manner thus:

7.4.1737   V.A. 1° 25′ 20″ refraction 0° 24′ 11½″

Comparison of refraction at Torneå and Paris.

With so little known about refraction at that time it was prudent to see if it varied with latitude as did the length of the seconds pendulum. For this they observed the Pole star and Arcturus.

Zenith distance of Pole star observed November, December 1736, reduced to beginning of December 1737.

At Torneå	22° 03′ 05″
Ditto for Arcturus	45 16 09
Thus meridian arc between them at Torneå	67 19 14

Same stars at Paris, reduced to same date.

Pole star	39 02 28
Arcturus	28 16 37
Thus meridian arc between them at Paris	67 19 05

The same procedure was repeated with the Pole star at its nadir.

Pole star at Torneå	26 14 17
Arcturus at Torneå	45 16 09
Thus meridian arc between them at Torneå	71 30 26

Pole star at Paris	43 13 43
Arcturus at Paris	28 16 37
Thus meridian arc between them at Paris	71 30 20

Whence they asserted that refraction at these zenith distances was sensibly the same in Torneå and in Paris. The observations in Torneå were with 3 foot radius quadrant whilst in Paris a 2½ foot radius quadrant had been used.

Pendulum experiments.

By the early 18th century interest was being taken in gravitation not just for its relation to the figure of the earth but also for its relation to earth motion and for the gravity value at different places.

Maupertuis and his colleagues carried out various experiments at Pello. This presented considerable difficulty because of the atmospheric conditions in the area.

They used a clock by Graham with a bob fitted to a flat brass rod. The top formed two knife edges of steel. The lens was 6 inch 10¾ line diameter and 2 inch 2¾ line thick. The weight controlling the clock was 11 pounds 14½ ounces. Within the case was a mercury thermometer on which zero represented the boiling point of water. The numbers increased with decreasing temperature.

Instructions with the clock suggested that in London, when the thermometer was 138 the clock gained 24^S a day and at 127 it gained 238^S a day upon mean time. The arc of the bob at full weight was 4° 20′ or at half weight 3° 00′.

As there was no substantial construction in Pello suitable for the stable mounting of instruments a special stone pillar, 6 ft. by 3 ft. was built. A telescope was mounted and pointed towards Regulus and the times of its meridian passage recorded. With reference to these times the pendulum gained 60¾ ^{S} the first day and 54^S the next.

This variation was thought to be due to the temperature changes but the problem was one of finding a way to keep the temperature constant. To keep the room at a constant temperature in such extreme and changeable conditions meant a 24 hour vigil on the thermometers. For this they had two thermometers—one by Abbé Nolet graduated on the Réaumur scale with freezing at 0° and the other by Prince where freezing was at 32°. In both cases the numbers increased with heat.

For 5 days and 5 nights they managed to maintain the Nolet thermometer at 14-15° and that of Prince at 60-62°. Even raising or lowering it made a considerable change in the temperature. The result was that over 5 days the average gain was 53.5^S per day. The same pendulum when returned to Paris and observed at the same temperatures with the same thermometers showed a loss of 5.6^S a day.

This would suggest that there was an acceleration from Paris to Pello of 59.1^S a day. In addition, from Graham's figures it appeared that there was an acceleration from Paris to London of 7.7^S since the 127 temperature of Graham was about 14½ and 61 respectively on the thermometers used in Sweden.

Further experiments were made with globular bobs fashioned from different materials — lead, silver, iron, tin, and copper. The motions of these were compared with the Graham clock and whilst they differed among themselves, when all were later tested in the same way in Paris there was agreement to better than 2^S in each case.

They were able to compare the Paris-Pello acceleration with several other theories of the time:

Value by Maupertuis	59.1^S
From Newton's tables	52.3
From Bradley's tables	63.6
From Huygens theory	Far less

From their observations they concluded that if, as Mairan found, the seconds pendulum in Paris was 440.57 lines, at Pello it would be 441.17 lines. The principle they applied was that the increase was approximately in the ratio of the squares of the sines of the latitudes. From the table they compiled there was a close correlation between the acceleration in one revolution of the stars and the required lengthening of the pendulum. This approximated as

For 1 second acceleration = 0.01 line extension.

Svanberg and the remeasurement of the arc.

Because the results of Maupertuis, whilst verifying that the earth was oblate, were rather large in comparison to others of the same century elsewhere, the Swedish Royal Academy felt further investigation was necessary.

It was Melanderhielm who had conceived the idea for this new arc measure after giving years of thought to the earlier measures of Maupertuis and the doubts they raised. Could there be errors of 12 seconds or 200t to explain the apparent inconsistencies?

Melanderhielm (1726-1810) was Professor of astronomy in Uppsala from 1761 and in 1797 became secretary to the Academy of Science in Stockholm. Although too old himself to take part, he was spurred on when the results of the French work of Mechain and Delambre in France became available and strengthened his convictions.

It fell to the lot of Jöns Svanberg to go to the Polar Circle and carry out the further measurements. In fact Svanberg came from the Torneå area. He was born in Kalix, west of Torneå in 1771, went to school in Torneå and thence to Uppsala in 1787. In 1811 he became Professor of mathematics there. He should not be confused with Gustav Svanberg (1808-1882) who was Professor of astronomy at Uppsala.

In 1799 Svanberg made a preliminary visit to determine whether the mountains were of sufficient size to cause considerable attraction of the vertical. Nothing was found.

The following year Melanderhielm asked Delambre to obtain a Borda repeating circle for him. The money for this he obtained from the Academy on the 18th of July, 1800.

Thus it was that Melanderhielm then put a detailed case for the expedition to the King, who welcomed it and placed the task under the supervision of the Academy. He agreed the financial arrangements on 12 February 1801.

At that time the area of survey was still all in Sweden but in 1809 a large part became Finland. In the peace negotiation of 1809 Russia demanded a frontier in the Kalix river 25 km west of Torneå. Sweden counterdemanded the Kemi river 15 km east of Torneå or at least the old frontier between Västerbotten and Osterbotten at the Kaakamojokki stream some km west of the Kemi. To compromise, the frontier was taken in the Torneå river but the town of Torneå,

situated toward the right bank, on the penninsula of Suensaari, was taken to the grand duchy of Finland which belonged to the Czar of Russia.

Considerable initial investigation was necessary to determine the most likely positions of the stations that Maupertuis had used.

Svanberg and Ofverbom, the chief engineer at the Bureau of Surveys, spent April to October 1801 in this research and the selection of station points. They also added several extra points to both north and south to increase the arc by some 50%. Those in the south used some of the islands that Maupertuis had first thought of incorporating but which he found unsuitable.

Astronomical observations were to be taken at Mallörn, in the south, an island in the Gulf of Bothnia, and at Pahtavaara, in the north.

The Borda repeating circle reached Stockholm in December 1801 and the expedition proper set off in January 1802 with the extra assistance of Holmquist of Uppsala University and Palander of Abo University.

It must be remembered that 1790 marked a new era in history with the introduction of thoughts of the metre and this was one of the first instances of a metre standard going on a survey expedition. Both a metre bar and a toise bar were sent and each of these had been initially standardized by Delambre and Mechain. In fact, under the influence of the new Astronomical School it did not stop there. All angles, for both the horizontal scheme and the astronomy were with a Borda repeating circle graduated in gons. Actually while Svanberg in [240] says a Borda circle [6] and [20] says Lenoir, Leinberg in [142] agrees with Svanberg. This and other instruments arrived in Stockholm in December 1801 in time for the arrival in 1802 of the Academicians.

In [240] Svanberg lists all the measures of each angle and gives the results to 3 decimals of a centesimal second! Yet even the first triangle, of some 100 square kilometres closed to nearly 9 centesimal seconds. Some 80 angles were observed and recorded in this way. Some angles were observed as many as 50 times. Many of the refinements that were incorporated so as not to miss any error source of note were considered by Airy [6] to be useless and even erroneous.

For his baseline Svanberg found and used a mark near the south end of that of Maupertuis but not the north end. In terms of the toise of Peru they measured 7414.4919t or some 8t longer than that of Maupertuis.

The method of base measurement was similar in that it was done in winter-starting on 22nd February 1802, and that for much of its length it was over ice. It was completed on 11th April 1802. They used 4 iron rods of 24.3mm by 31.1mm cross section and a little more than 6m long. They were fashioned such that they could slide together to bring their defining lines adjacent. Each bar was graduated against a 2m standard brought from Paris for the purpose.

Now this new unit of measure was DEFINED as 443.296 lines of the Toise du Perou at 13°R. On the other hand it was also represented as the length of a particular platinum bar at 0°R and 0°C. These different temperatures caused some confusion at the time. To add to the confusion Svanberg was now using a different metal. Did he make his comparison between the iron and platinum at 0°R or 13°R? If both had

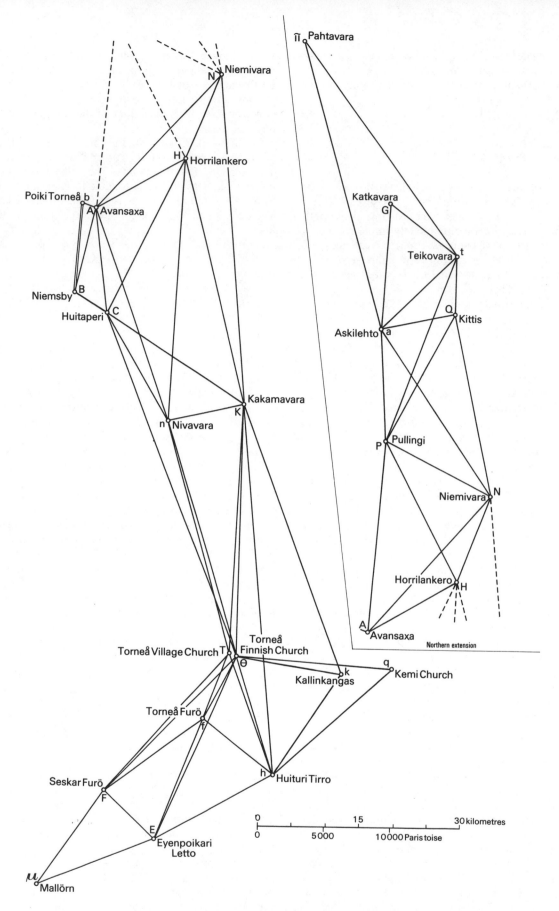

Figure 43. Diagram of the scheme of Svanberg.

been of the same metal this would not have mattered but with different metals, even of the same length at 0 °R they would differ at 13 °R. Thus for correcting the base length there would have been two expansions to be allowed for if the initial comparison had been at 13 °R.

For the measuring the bars were supported on special trestles at three points in such a way that some longitudinal movement was possible to obtain coincidence. Thus each 'span' of bar was some 3m. Could this have contributed any errors?

Holmquist and Ofverbom aligned the marks and placed the bars; Palander and Svanberg obtained coincidence, recorded the temperature and used a small sector to determine any inclination of the bars to the horizontal.

The new north terminal was marked at a dividing fence by burying a piece of charred timber on the top of which was a brass plate and etched, intersecting lines. The whole was covered up. Has it ever been re-located?

Their results were:

Measured length	14 451.912m
Temperature correction —4.3 °C	— 0.713
Inclination correction	—1.409
Reduction to centre of terminals	+1.436
Reduction to sea level	—0.110
	14 451.116m
=	7 414.492t
Or at 16.25 °C	7 413.1124t

Svanberg computed the length TQ as 54,919.2537t against the value of Maupertuis of 54,945.95t. This value was according to Svanberg but it is longer than any of the ten values Maupertuis gave, who favoured 54,944.76t and used a mean of 54,942.57t. Whichever one selected there would still be some 25t difference.

The signals used were formed around tree trunks 20 feet long and so arranged that the sky light could be seen through the bracing trunks. In the top of the central trunk was a rectangular pivoted vane with an open centre. Around the base it was piled with brushwood and rocks. June, July and August were spent in observing the angles of the triangles. Astronomical observations began at Mallörn on the 9th of September and at Torneå on the 5th of October.

Corrections were applied for eccentricity, reduction to the horizontal (since a repeating circle measured in the same inclined manner as a quadrant), and for spherical excess. The triangles and arc were computed by the method of Delambre to give an arc value of
180,827.68m = 92,777.981t
Or as Airy [6] states, if the bars had been compared at 13 °R
180,794.06m = 92,760.731t
Note the large effect (33.3m) of small temperature uncertainties. For terrestrial refraction Svanberg used a value of 0.08 from the work of Bradley. The mean triangular closure was 4.3 ".

While Maupertuis had determined the amplitude of his arc by differences of zenith angles, Svanberg determined latitudes by circum-meridian observations of Polaris. From some 260 observations they found:

Mallörn	72.805 6372^g	=	65° 31' 30.265 "
Pahtavaara	74.607 9723	=	67 08 49.830
Difference			1 37 19.565
Whence		1° =	111 477.408m
		=	57 196.159t
Or using Airy's figures		1° =	111 456.680m
		=	57 185.524t

The latitude values are as quoted by Svanberg although the arithmetic does not exactly tie up. He went further and quoted another set of results based on the refraction according to Prony.
Whence:

Mallörn		65° 31' 31.060 "
Pahtavaara		67 08 51.414
Difference		1 37 20.354
Amplitude		92 777.981t
	1° =	57 188.429t
Or at 16.25 °C		57 177.797t

Airy [6] would reduce the amplitude by 0.28 " to allow for the true position of the moon's node, thus:

1° 37' 19.29 "   and 1° =   111 482.671m
= 57 198.859t
or
1° = 111,461.944m
1° = 57,188.225t

[38] will be found to differ slightly in the decimal places of the above values. Maupertuis had a value of 1° = 57,438t.

Where lines of the new scheme were directly comparable with those of 1736 they were all shorter than before by amounts ranging up to 11.2t. It was suggested that some of this could have been due to non-correction in 1736 for the inclination of the base line which in some parts was said to be 6°.

In solving fully the discrepancies of the 1736 work it was to be very much regretted that Svanberg omitted to observe the latitude at Kittis; the position where the trouble could have occurred.

In comparing his result with that of Bouguer, Svanberg found the flattening as 1/329.246; while when compared with Mechain and Delambre it was 1/307.405. When compared with Peru and France the value was 1/326.674.

Svanberg finally settled on combining with Bouguer and La Condamine to get
1^g = 51,476.543t   f = 1/323.065
and radius at the equator = 327,1452t
or 1° = 57,196.159t

The method of Delambre mentioned in the approach adopted for calculating meridian arcs was first used for the French arc of 1792. In summary it was along the following lines.

The triangles were calculated as spherical ones so that the spherical excess was only computed to determine the observational error. Thus given one side 'a' and the two adjacent angles B and C then

$$\sin b = \frac{\sin a . \sin B}{\sin A} \quad \text{and} \quad \sin c = \frac{\sin a . \sin C}{\sin A}$$

where sin a was the sine in a circle of radius equal to the radius of the earth.

Delambre then wrote sin a = a.sin a/a where the log of 'a' could be easily found from tables and log (sin a/a) varied slowly and could be tabulated.

Then given   - a reasonable value of the earth radius,
                - the length AB
                - angle PAB where P is the pole.
To find   - difference of latitude AB
                - distances between parallels of A and B
                - difference of longitude
                - angle PBA

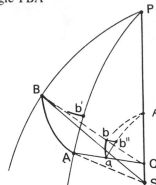

Figure 44.
Delambre's method of reduction.

Draw normals from B and A respectively to S and Q. Assuming unit radius imagine a sphere of centre Q and triangle abρ upon it.

Distance Ab' = AQ × ab" where log AQ can be easily tabulated.

ab" by solution of the spherical triangle.

Azimuth PBA = ρba

Difference of longitude BPA = bρa

Difference of latitudes = distance/radius of curvature at mid point.

$$\text{or} \quad \frac{a^2\,b^2}{\left(b^2.\sin^2\left(\dfrac{\lambda + \lambda'}{2}\right) + a^2.\cos^2\left(\dfrac{\lambda + \lambda'}{2}\right)\right)}$$

$$= \frac{a^2\,b^2}{\left(b^2.\sin^2 \lambda_m + a^2.\cos^2 \lambda_m\right)}$$

The repeating circle.

This instrument was invented about 1756 by Tobias Mayer (1723-1762), a contemporary of Bradley and La Caille, but for a long while it was not adopted. It was first used on part of the arc connecting Paris to Greenwich. Later it was used on the Dunkirk to Barcelona arc by Delambre and Mechain. It could be used for both horizontal angles and astronomical angles.

It was extremely popular in France once it caught on and it was thought by many to provide the ultimate. That all the errors of division and observation were swept away and that no new ones would arise since all possibility of error was removed by repetition. The English were far more reserved and practical in their approach and made little use of it.

Confidence in it was however somewhat shaken around 1792 when Mechain, while repeating a latitude observation, found a difference of some three seconds which he was unable to explain. Later it was agreed that all such instruments were liable to an error that could not be removed by repetition and for which there was no explanation.

In essence the instrument consisted of two circles mounted together each of which had a telescope attached, but only one of which was graduated. The telescope axes were in planes parallel to that of the circles. The mounting was such that the circles could be used in any inclined plane as well as in the horizontal and vertical. There were usually four verniers attached and one telescope had a level vial attached for use with vertical angle measure.

Assume that the angle is required between two distant points P and Q. Point T_1 towards Q and T_2 towards P. Read verniers on T_1.

Turn whole circle until T_2, which is clamped, points towards Q. Release T_1 and point towards P. T_1 has now passed through twice the angle POQ.

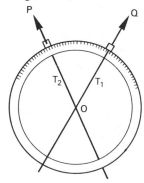

Figure 45.   The repeating circle.

This continued to be repeated as many times as necessary. For vertical angles the bubble is used to set T_2 horizontal and T_1 sighted on the object. It was then possible to do a change of setting such that again a double value was obtained.

This was obviously a technique that was a forerunner of the repetition method used in 20th century theodolites in, for example, subtense bar measurement. Such repeating circles were usually only about 24 inches diameter.

Observations.

The angular observations by Svanberg were recorded in gons to 3 decimals of a centesimal second! The lengths were measured and computed in metres but were also listed in toise. For comparison purposes only these angular values have here been converted to sexagesimal degrees. Where relevant the same lettering has been used as for Maupertuis.

FμE	32° 59′ 29″	θnK	87° 42′ 43″
		CnK	105° 55′ 50″
μEF	64° 39′ 50″	KnH	73° 58′ 40″
FEf	62° 37′ 12″	AnK	95° 30′ 43″
FEθ	65° 50′ 49″	TnK	87° 43′ 55″
fEh	36° 50′ 51″		
		nCK	28° 14′ 13″
μFE	82° 20′ 43″	θCK	37° 21′ 36″

fFE	81° 51' 36"	BCA	54° 39' 03"
ΘFE	91° 56' 44"	ACK	131° 06' 40"
hFE	53° 27' 16"	ACH	30° 56' 37"
TFE	96° 04' 28"		
		bBC	111° 56' 40"
FfE	35° 30' 54"	ABC	102° 45' 39"
hfF	110° 05' 55"	AbB	76° 57' 37"
Θfh	96° 22' 43"		
Tfh	103° 40' 09"	BAn	32° 49' 33"
		CAb	116° 26' 39"
Khf	53° 55' 53"	nAb	126° 41' 00"
fhE	68° 34' 08"	PAH	53° 45' 45"
Θhk	58° 22' 01"	PAb	77° 26' 15"
fhk	96° 25' 07"		
Thf	38° 38' 06"	AHC	36° 42' 07"
fhq	100° 23' 54"	AHn	56° 20' 48"
		AHP	94° 54' 25"
KkΘ	63° 16' 02"	CHK	36° 04' 41"
Θkh	56° 04' 32"	PHN	49° 10' 25"
qΘh	63° 34' 19"	APH	31° 19' 50"
KΘk	91° 09' 15"	NPQ	87° 50' 51"
hΘf	45° 34' 06"	NPt	95° 38' 14"
kΘh	65° 33' 37"	NPa	120° 03' 56"
FΘh	61° 58' 02"	NPH	37° 23' 39"
EΘh	39° 45' 35"		
CΘK	22° 51' 56"	PaN	31° 13' 54"
nΘK	18° 20' 56"	PaQ	99° 09' 40"
		Qaπ	92° 44' 22"
nKΘ	73° 56' 28"	πat	61° 28' 48"
kKn	99° 31' 09"	Gat	42° 37' 52"
nKC	45° 50' 07"		
NKC	53° 26' 45"	aπt	21° 27' 24"
CKH	43° 45' 23"		
nKT	72° 38' 03"	Pta	25° 09' 05"
		Qta	45° 13' 32"
HNa	122° 08' 13"	Ptπ	122° 12' 48"
HNP	93° 26' 05"	GtP	107° 22' 52"
HNK	27° 10' 09"		
PNQ	51° 55' 05"		
PQN	40° 14' 03"		
PQa	48° 37' 21"		
tQa	103° 30' 54"		

HA	7 447.876	θf	5 204.991	TF	13 940.798
HC	13 396.134	θh	8 392.245	θq	9 178.960
HK	19 066.538	fhθ	6 030.182	hq	9 279.726
Hn	19 836.965	θE	14 523.590		

Conclusions.

[55] acknowledged the appearance of considerable accuracy in Maupertuis's arc but indicated the discordance between the terrestrial and astronomical results. It suggested either the arc was too long by 200 t or the amplitude too small by about 12 seconds.

The attempt to unscramble this was by Svanberg in 1801 when he reobserved and extended the arc.

[6] quoted Rosenberger who had made a detailed study of the results and concluded that such large errors as 12 seconds were impossible and this was agreed by Airy. The result suggested by Rosenberger was

$$\text{Arc} \quad 55\ 020.16\ t$$
$$\text{Amplitude} \quad 57' 30.44''$$
$$1° \quad 57\ 405.02\ t$$

This included a correction of $+ 0.7''$ for refraction.

Because of the discordance between the results of Maupertuis and Svanberg further work was done in the same territory in the mid-19th century by Struve and Tenner [38]. In general they occupied the same stations as Maupertuis and Svanberg but the base line was on land rather than on ice. The site was between Torneå Elf and Mount Aavasaksa. Measured in 1851 it was about 1519 t long.

This work formed part of an arc from North Cape to the Dnestre (Danube) river, an amplitude of some 25°. The Scandinavian section was from North Cape via Kautokeino to Torneå, which was observed between 1846 and 1850 with a mixture of Swedish and Norwegian surveyors. The northernmost station was Fuglenaes on an island near Hammerfest. The section from Torneå, north to Kautokeino was by Selander and that further north by Hansteen.

They used a toise bar by Fortin of Paris that was said to be the same length as the toise of Peru as checked by Arago in 1821. Their field standards were

$$P = 1727.99440 \pm 0.000\ 19 \text{ lines at } 13°R$$
and
$$Q = 1728.01991 \pm 0.000\ 77 \text{ lines at } 13°R$$

The Over Torneå base was 1 519. 85006t ± 0.729 ppm or when reduced to the level of the Gulf of Bothnia.

$$1\ 519.83864t \pm 0.00113t.$$

It is interesting to note that the toise was still used here some 50 years after the introduction of the metre.

Other pertinent results from this arc by Struve were:

Sections in Scandinavia:

Finland	60° 05 'N to 65° 50' N		by Struve in 1830-51
Lapland	65 50	68° 54	by Selander 1845-52
Finmark	68 54	70 40	by Hansteen 1845-50

Fuglenaes	70° 40' 11.23"			
Stuor-oivi	68 40 58.40		727 403.2ft	113 753.906t
Torneå	65 49 44.57		1 043 728.0	163 221.904
Kilpi-maki	62 38 05.25		1 168 884.2	182 794.304

The Finnish geodesist, Y.Leinberg [142] in investigations

Computed sides.

πt	18 960.371t	Ab	1 186.015t	kh	9 207.859t
πa	21 415.170	AB	7 239.690	ΘF	13 259.643
tQ	4 213.097	AC	8 656.857	fF	8 396.231
ta	7 893.612	An	16 888.856	fE	9 360.335
tP	14 531.037	bB	7 414.496	hF	11 901.980
Qa	5 763.228	BC	3 409.412	hE	9 693.790
QP	10 672.298	CK	11 406.755	FE	5 492.642
QN	13 549.116	Cn	8 509.641	Fμ	9 117.008
aP	8 111.765	CΘ	25 480.537	Eμ	9 997.394
aN	14 617.827	Kn	5 611.987	Ga	9 531.732
PN	8 757.597	KΘ	17 813.549	Gt	6 515.423
PH	11 552.860	Kk	19 941.373	nT	15 940.240
PA	14 270.970	Kh	25 736.940	KT	16 688.469
NH	7 028.417	nΘ	17 131.778	Tf	6 156.805
NK	25 047.243t	Θk	8 611.403	Th	9 805.440

of the results of Maupertuis suggests that

Error in length of meridian	1.43 "
Error of refraction	0.96
Error of astronomical observations	7.83
Error of plumb line	2.45
Total	12.67 "

A conclusion which closely agrees with the earlier suggestion of an amplitude 12 " short.

Voltaire said of Maupertuis "il avait aplati la Terre et les Cassinis".

Maupertuis, in a letter to Bradley said, "Thus, Sir, You See the Earth is Oblate, according to the Actual Measurements, as it has been already found by the Law of Staticks: and this flatness appears even more considerable than Sir Isaac Newton thought it." [256]

References.
See in particular the following: 1, 2, 3, 6, 9, 17, 28, 31, 32, 35, 38, 40, 55, 78, 94, 98, 104, 113, 114, 142, 143, 148, 153, 163, 198, 211, 232, 240, 252, 256, 260, 266 and 267.

Brief Chronology

20 April	1736	Party left Paris
27 April	1736	Arrived Dunkirk
29 April	1736	Celsius arrived in Dunkirk from London
2 May	1736	All embarked in Prudent
13 May	1736	Anchored off Helsingör
		Celsius and Le Monnier then went overland to Stockholm
21 May	1736	Prudent entered Stockholm
23 May	1736	The party was presented to His Majesty the King
3 June	1736	Sailed from Stockholm
5 June	1736	Maupertuis, with Celsius and Clairaut left Stockholm by road
19 June	1736	Maupertuis arrived in Torneå by sea—having re-embarked at Onska
21 June	1736	Overland group arrived in Torneå
25 June	1736	Reconnaisance begun around islands of Gulf of Bothnia
2 July	1736	Returned to Torneå
6 July	1736	Party left Torneå and proceeded up-river
12, 13 and 17 July	1736	Angular observations at Nivavaara
14-16 July	1736	Angular observations at Kaakamavaara
18-20 July	1736	Angular observations at Huitaperi
22-30 July	1736	Angular observations at Aavasaksa
1-5 August	1736	Angular observations at Pullinki
6-7 August	1736	Angular observations at Kittisvaara
11 August	1736	Angular observations at Niemisvaara
14-17 August	1736	Angular observations at Horrilankero
19-21 August	1736	Angular observations at Aavasaksa
22-24 August	1736	Angular observations at baseline terminals
26 August	1736	Angular observations at Huitaperi
2 September	1736	Angular observations at Torneå
5 September	1736	Angular observations at Kaakamavaara
29 Sept.-10 Oct.	1736	Sector observations at Pello
28 October	1736	Arrived back in Torneå
30 Oct.-17 Nov.	1736	Astronomical observations at Torneå
14-20 December	1736	Preparation of baseline site
21-30 December	1736	Measurement of baseline
17-19 March	1737	New astronomical observations at Torneå
25 March	1737	Left Torneå for Pello
4-11 April	1737	Observations at Pello
9 June	1737	Party left for Stockholm by sea — Maupertuis, Sommereux, Herbelot and Clairaut
10 June	1737	Others left by road — Camus, Celsius, Outhier and Le Monnier
10 June	1737	Ship wreck and salvage near Piteå
14 June	1737	Maupertuis continued by road with Clairaut, Celsius and Camus
21 June	1737	Le Monnier, Outhier and Sommereux continued by road
11 July	1737	Arrived in Stockholm
18 July	1737	Maupertuis and Sommereux sailed for Amsterdam. Others went by coach and sea
13 August	1737	Arrived in Amsterdam
20 August	1737	Arrived in Paris
21 August	1737	Visited the King and others at Versailles to conclude the work of the expedition

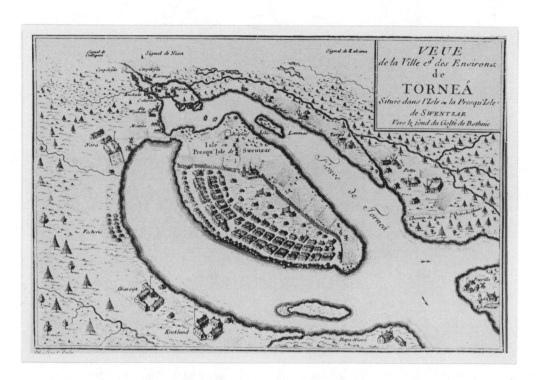

31. *View of the town of Torneå*

32. *The neighborhood of the observatory at Kittis*

193

BIOGRAPHIES

To keep the main text uncluttered with numerous dates and biographical material it has all been brought together here in the form of potted biographies. As throughout the whole text, in general it is only those aspects relevant to the theme of the work that are itemized. Dates up to about 1200 A.D. are distinguished by either B.C. or A.D. Later ones have the A.D. omitted.

Al-Asturlâbî
fl.c 830 A.D.
Muslim astronomer and famous instrument maker - particularly astrolabes. Wrote treatises on astrolabe. Assisted on Arabian arc measure.

Al-Badi'al Asturlâbî
fl.c 1116 A.D.
d. 1139-40 A.D. in Baghdad.
Muslim astronomer and maker of astrolabes.

Al-Battânî
b. before 858 A.D. in Harran, Mesopotamia
d. 929 A.D. Samarra
Muslim astronomer. Worked on spherical trigonometry and tables of cots and sines. Known in Europe as Albategnius. Credited with introduction of shadow square on astrolabe and to setting of sights in a tube. Introduced term 'sinus' as half chord.

Al-Bîrûnî
b. 973 A.D. Khwarizm (Khiva)
d. 1048 A.D. Ghazna, Afghanistan.
Wrote on gnomon lengths, astrolabe and shadow square. One of the greatest Persian astronomers although also able in mathematics, physics, geography and history. Used a simplified stereographic projection.

Al-Bitrûjî
fl. late 12th century. From Pedroche, near Cordova. Hispano-Muslim astronomer and possible inventor of the torquetum — but see Al-Tûsî.

Al Buhtari
Assisted on Arabian arc measure, 9th century A.D.

Al Falaki.
Assisted on Arabian arc measure, 9th century A.D.

Al-Farghânî.
b. Farghans, Transoxiana.
d. after 861 A.D. Egypt
Assisted on Arabian arc measure. Wrote on astronomy and the astrolabe. His 'Elements' when printed in Ferrara in 1493 aided scientific reawakening in Europe. Quoted earth radius as 3,250 miles i.e. Al-Mamun's value.

Al-Fazari
d. 777 A.D.
First Muslim astronomer to construct an astrolabe. Wrote on astrology and mathematical instruments. Calculated sines to every ½ degree. Translated Indian 'Siddhanta' into Arabic.

Al-Hasib
b.c 770 A.D. fl in Baghdad
d.c 870 A.D.
Astronomer. Introduced idea of tangent and compiled tables of values.

Al-Jauhari
fl. 9th century A.D.
Muslim astronomer and mathematician. Made notable observations near Baghdad and Damascus.

Al-Khowârizmi
b. before 800 A.D. Qutrubbull near Baghdad.
d. after 847 A.D.
Mathematician, astronomer, geographer. Member of House of Wisdom in Baghdad. Numerous writings and tables of sine and tan. Introduced term 'algebra'. Assisted on Arabian arc measure.

Al-Kindî
b. 786 A.D. fl. Baghdad
d. c 873 A.D.
Known as philosopher of the Arabs. Wrote on many scientific topics. Constructed a quadrant.

Al-Kûsgi
A Persian astronomer of Ulugh Beg's observatory. fl. 15th century Gave earth circumference as 8,000 parasangs.

Al-Mamun
b. 786 A.D. Baghdad
d. 833 A.D. near Tarsus
Known as Caliph Al-Mamun (The Trustworthy). Son of Al-Rashid (Aaron the Just). Reigned from 809 to 833. Founded an observatory in Baghdad to supercede that of Damascus and formed the House of Wisdom or Scientific Academy. Supervised the Arabian arc measures.

Al-Mansûr
b. c 712 A.D.
d. 775 A.D. Bir Maimun near Mecca
Reigned 754 to 775 A.D. Founded Baghdad in 762 A.D. Restored Arabian rule over all Spain.

Al-Marrakushi
fl. c 1260 A.D. in Morocco
Moroccan astronomer and mathematician. Compiled tables of sines by ½ degree increments. Described forms of level, quadrant and shadow square.

Al-Marwarrudhi
fl.c 800 A.D.
Muslim astronomer. Assisted on Arabian arc measure. Constructed astrolabes.

Al-Muzaffar
d. c 1213 A.D. From Tus in Khurasan.
Muslim astronomer and mathematician. Wrote on the astrolabe. Invented the Tusi staff.

Al-Nairizi
fl. 892-902 A.D. Nairiz near Shiraz
d. c 922 A.D.
Wrote on astronomy and geometry. Utilised tangents.

Al-Rahman
b. 903 A.D. at Ray
d. 986 A.D.

Al Rashîd
b. 763 A.D. at Ray
d. 809 A.D. at Tus
Developed Baghdad to assume former scientific role of Alex andria. Known as Aaron The Just, he reigned from 786 to 809 A.D.

Al-Sûfî
d. c 777 A.D.
Wrote on the astrolabe.

Al-Tûsî
b. 18.2.1201 Tus, Persia
d. 26.6.1274 Kadhimain near Baghdad.
The last great Arab astronomer. He founded the observatory at Maragha in 1259 A.D. and wrote on all aspects of astronomy and mathematics. He was one of those said to have invented the torquetum. Built a 12 ft. radius mural quadrant. He was kidnapped by Al-Rahman and sent to Alamut until its capture by the Mongols in 1256. He then returned to Maragha.

Al-Urdi
fl. 13th century.
Syrian astronomer and engineer. Built a mural quadrant. Developed improved method of observing the meridian.

Abu-l-Wafa.
b. 940 A.D. Buzjan, Quhistan
d. 998 A.D. Baghdad.
Great Muslim astronomer and mathematician. Produced sin tables to 8 decimal places. Studied tangents. Derived various trigonometric relationships.

Alexander the Great
b. 356 B.C. Macedon
d. 13.6.323 B.C. Babylon; buried in Alexandria.
Son of Philip II, he became King of Macedon in succession to his father in 336 B.C. He was instrumental in the founding of Alexandria in 332 B.C. His education was under Aristotle. In 331 he defeated Darius near the Tigris and destroyed Persepolis. Several years were spent in the Hindu Kush and the Punjab. When he died of fever in 323 the Empire disintegrated and the territory was divided by his successors.

Anaxagoras of Clazomenae
b. c. 499 B.C. Clazomenae, Ionia
d. c 428 B.C. Lampsacos
The last philosopher of the Ionian school. He considered the earth as flat and lying on air in an hemisphere. His end came when he was condemned for supporting the Persians.

Anaximandros of Miletus
b. c 610 B.C. Miletus
d. c 545 B.C.
The son of Praxiades, he was a friend of Thales. As a pioneer of exact sciences he is said to have erected the first gnomon in Greece. As a member of the Ionian school he led it after the death of Thales. He drew a map of the earth with Europe and Asia of equal size and with a surround of water.

Anaximenes of Miletus
b. c 585 B.C. Miletus
d. c 528 B.C.
A contemporary of Pythagoras, he was probably a pupil of Anaximandros and philosopher of the Ionian school. He was said by some to have invented the gnomon but this is doubted by others. Had ideas of the stars fixed to a crystalline vault.

Archimedes of Syracuse
b. c. 287 B.C. Syracuse - modern Siracusa, Sicily
d. 212 B.C. in the sacking of Syracuse by the Romans under
Marcellus.
Son of astronomer Pheidias he was a kinsman and friend of
Hieron II, King of Syracuse. He was the greatest of the early mathematicians. By use of a 96 sided polygon in a circle
he found $3.142 > \pi > 3.141$ (although the Greeks never used the symbol π). Used a staff to measure the diameter of
the sun. Anticipated by nearly 2000 years some of the ideas
of Newton. Estimated size of earth as 300,000 stade circumference. Derived relation equivalent to expanding sin
$(x - y)$. In working with large numbers he mentioned what
turned out to be the basis of logs - the addition of orders
of numbers. Constructed an orrery for planetary motions and
used a Jacob staff in its early form. During the second Punic
war and the besieging of Syracuse (214-212 B.C.) he invented various war machines for use against the Romans.
He was killed by a Roman soldier.

Archytas of Tarentum
b. 428 B.C.
d. 347 B.C.
A statesman and general. Was the only great geometer among
the Greeks when Plato founded his school. He was the first
to apply geometry to mechanics.

Aristarchos of Samos
b. c 310 B.C.
d. c 230 B.C.
He was a pupil of Strato, probably in Athens, and later he
taught in Alexandria. Has been claimed as 'the ancient
Copernicus'. He made many astronomical observations and
put forward an heliocentric hypothesis. He devised the scaphe
and, via the triangle of Pythagoras, found the relative
distances of the sun and moon from the earth. These he defined in terms of parts of a quadrant since the 360° circle had
not then been introduced.

Aristotle
b. 384 B.C. Stagira in Thrace
d. 322 B.C. Chalcis in Euboea
Son of a court physician in Macedon, he was a Greek
philosopher and scientist who studied in Athens under Plato.
He was influenced by Ionian, Macedon and medical ideals.
In astronomy some of his conclusions tended to be his ideas
of what he thought ought to be the situation. He married
Pythias of Assos and became tutor to Alexander, son of Philip
II at Pella 343-340. He had one daughter by Pythias and with
a second wife Herpyllis he had a son Nicomachos. In 335-4
he founded a new philosophic school in the Lyceum of
Athens. He gave proofs of the earth as a sphere with a circumference of 400,000 stade. Said to have been the oldest
attempt to estimate the size although it could have been due
to Eudoxus who was the first scientific astronomer. It could
only have been by using the altitude of the sun or a star and
the distance between the observing points—possibly by some
form of pacing.

Aryabhata the Elder
b. 476 A.D. Kusumapura near Patna on the river Jumma
d. c 550 A.D.
Hindu mathematician and astronomer who taught that the
earth rotated daily on its axis. First to compile sine tables

for each degree. Author of one of the oldest Indian
mathematical texts—'Aryabhatiya' which summarised all
developments in mathematics and astronomy up to that date.
He used a vertical staff to measure heights by the shadow
method.

Autolycus of Pitane
b. 2nd half of 3rd century B.C. Pitane, Asia Minor

Astronomer and mathematician who assumed all stars to be
on a sphere. Wrote two treatises on astronomy which are
the oldest extant mathematical texts. One of the most widely learned scholars of his time. His death is usually taken
to mark the end of the Hellenic age.

Adrien Auzout
b. 28.1.1622 Rouen
d. 23.5.1691 Rome
With Picard he introduced the use of telescopes in conjunction with graduated circles. Son of a clerk of the Court in
Rouen he was a great mathematician and philosopher. He
drew up the plans for the founding of the Academy of
Sciences and the Paris Observatory. 1664-66 worked with
Picard on the micrometer. July 1668 he had a quarrel in the
Academy which resulted in his going to Italy although he
did later return to France for a while and also spent some
time in England. His claim to have actually invented the
micrometer was denounced by Townley in favour of
Gascoigne. He introduced the use of aerial telescopes to
dispense with very long tubes.

Venerable Bede
b. c 673 A.D. Wearmouth, England
d. 26.5.735 A.D. Jarrow
Most learned man of his time. Wrote on mathematics and
the representation of numbers. Spent much of his life in two
monasteries. In his 'De natura rerum' he mentioned the
spherical form of the earth and in 'De temporum ratione'
discussed the motion of the sun and other bodies.

Bhâskara (Bhaskaracarya)
b. 1114 A.D. Biddur in the Deccan
d. c 1185 A.D.
Hindu mathematician and astronomer who wrote on many
subjects. Produced sine tables based on 1/24 right angle.
Gave diameter of earth as 1600 yojans (of 7.6 miles each).
Used a horizontal staff and mirror to measure heights of objects. Gave π as $3927/1250 = 3.1416$

Willem Janszoon Blaeu
b. 1571 Alkmaar
d. 21.10.1638 Amsterdam
Said to have measured an early meridian arc in Holland in
17th century but virtually no details exist, other than that
it differed from that of Picard by 60 Rhenish feet—or 60 rods
but these are very different units. Renowned cartographer,
and maker of instruments and globes.

Pierre Bouguer
See page 168 of main text.

Tycho Brahe
b. 14.12.1546 Knudstrup, 30 km east of Hälsingborg, Sweden.
d. 13.10.1601 Prague. Buried in Teyn church.
At the time of his birth Knudstrup was in Denmark but that is now in Sweden. Son of a Danish nobleman. 1559 at Copenhagen University and 1562 at Leipsic University. In 1565 he returned to Denmark which was then at war with Sweden. Thus travelled to Rostock and Basle. His 'wife' died in 1604 but they were not married. They had several children. Founded observatory at Uraniborg (Tower of Heaven) on Hven, which had fine gardens, and was extremely well furnished. Besides an annual pension from the King he also had incomes from several farms and estates and from a prebend of Roskilde Cathedral. He was there from 1576 to 1597. Developed the use of transversals and tried out the nonius. Adopted the prosthaphaeresis method of multiplication using trigonometric tables that were good to 12 or 15 places. When his mass of observations finally came to the hands of Kepler it proved to be the rounding off of the work of Copernicus. From Uraniborg he went to Copenhagen then Hamburg. He was invited by the German Emperor Rudolph II to go to Prague where he was granted a pension and installed in a castle near the city to use as an observatory. Early in 1600 he was joined by Kepler. Brahe produced his biography 'Astronomiae instauratae Mechanica' in 1598.

Henry Briggs
b. Feb. 1561 Warley Wood, Yorkshire
d. 26.1.1630 Oxford
First Professor of geometry at Gresham College London 1596-1619. Later Professor of astronomy at Oxford. First Savilian Professor of geometry at Oxford in 1619. Proposed to Napier the use of powers of ten. They later collaborated. On death of Napier he was able to make the first table of common or Briggsian logs, to 14 places. Derived, or introduced the terms 'mantissa' and 'characteristic'. 1624 published 'Arithmetica logarithmica' with values to 14 places for 1 to 20,000 and 90,000 to 100,000. The gap was later filled by Vlacq. Briggs divided the degree into 100 parts but Vlacq adopted the sexagesimal system.

Mathurin Jacques Brisson
b. 30.4.1723 Fontenay-le-Comte
d. 23.6.1806 Brouessy
Member of French Academy of Sciences. 14.4.1790 he proposed to the Academy the idea of a new measurement system based on a natural standard. Scheme originally included decimal sub-division of a quadrant of a circle but although this appeared in the log tables by Callet in 1795 it did not prevail.

Jobst Bürgi
b. 28.2.1552 Lichtensteig, Switzerland
d. 31.1.1632 Cassel
Court watchmaker to Landgraf Wilhelm IV of Hesse in Cassel and later to Kaiser Rudolph II. He wrote on proportional compasses and on astronomy but is better known for his independent invention of logs. Published 'Arithmetische und Geometrische Progresstabulen' Prague., 1620. Logs were printed in red and antilogs in black. He is said to have been the first to introduce full lines rather than lines of dots on transversals.

Celio Calcagnini
b. 1479 Ferrara
d. 1541
Professor in University of Ferrara. He travelled widely in Germany, Poland and Hungary. Taught of daily rotation of the earth.

Callipos of Cyzicos
b. c 370 B.C. Cyzicos (Sea of Marmara) fl Athens

Astronomer and friend of Aristotle. Criticized theory of homocentric spheres. Estimated the lengths of the seasons.

Charles-Etienne-Louis Camus
See page 171 of main text.

Carpos of Antioch
Dates unknown.
Invented a type of level called the chorobates, which was said to be similar to the alpharion or the diabetes invented by Theodoros of Samos in 6th century B.C.

Gian Domenico Cassini I
b. 8.6.1625 Perinaldo, Imperia, Italy
d. 14.9.1712 Paris
Educated with the Jesuits in Genoa, in 1663 he became Inspector General of Fortifications at Urbino Castle and superintendent of rivers. 1650 became Professor of astronomy in Bologna and 1671 First Astronomer Royal in France. Became a naturalised Frenchman. From 1712 he was involved, with his son, in the measurement of the arc from Dunkirk to Perpignan. Introduced a telescope where the object glass and eyepiece were in separate sections some way apart. Towards the end of his life he was very much involved in the controversy on the shape of the earth.

Jacques Cassini II
b. 18.2.1677 Paris
d. 15.4.1756 Thury near Clermont, Oise.
Succeeded his father in 1712 as Astronomer Royal and wrote numerous items on astronomy. During 1733-4 he measured an arc of a parallel from St. Malo to Strasbourg and the results of this tended to confirm his belief that the earth was prolate. During this time he was joined by his son on the survey.

Cesar-François Cassini de Thruy III
b. 17.6.1714 Thury
d. 4.9.1784 Paris
Succeeded his father in 1756 as Astronomer Royal. Was requested by Louis XV to extend the mapping of France to large scales and this was combined with extensive levelling. This 'Carte de Cassini' was completed 5 years after his death. A year or so before his death he suggested the idea of cooperation between England and France to verify the separation of Greenwich and Paris.

Jean-Dominique Cassini IV
b. 30.6.1748 Paris
d. 18.10.1845 Thury
Succeeded his father as Astronomer Royal. In 1792 he made experiments with the Borda pendulum. Member of the

Academy of Sciences. Among those who carried out the cross-channel connection but the problem it was designed to solve was not immediately unravelled. In the 1790s he was involved with others in the problem of defining a standard unit of measure. In 1794 he was arrested and imprisoned during the French revolution but released after some 6 months to return to the family seat at Thury.

Anders Celsius.
See page 170 of main text.

Chang Heng
b. 78 A.D. Nan-yang
d. 139 A.D.
Chinese astronomer who constructed a celestial hemisphere turning about an equatorial plane. Was chief astrologer and minister under emperor An-Ti. He wrote on astronomy and geometry.

Richard Chanzler (Chancellor)
b. c 1525
d. 1556 Drowned at sea
He was instrumental in the development of transversals. In 1553 was with Hugh Willoughby in seeking the N.E. Passage.

Chhiu Chhang Chhun
b. 1148 A.D. Hsi-hsia, Shantung
d. 1227 A.D. Peking
In 1221 extended the gnomon arc of I. Hsing. Observed summer solstice on banks of Kerulen river in N.Mongolia. He was a Taoist monk of considerable repute and made a long journey from Peking to Persia and throughout central Asia.

Alexis-Claude Clairaut
See page 170 of main text.

Christoph Clavius
b. 1537 Bamberg, Germany
d. 6.2.1612 Rome
Entered the Jesuit order in 1555. At Collegio Romano teaching mathematics in 1565, and was Professor of Mathematics for all except 2 of the next 47 years. One of the two contenders for introducing the decimal point (The other was Magini). He used it in a table of sines in 1593. Described various instruments including a hand quadrant and plumb-bob level. He was an opponent of the ideas of Copernicus although he modified his stance in 1611. Was advisor to Pope Gregory XIII on reform of the calendar and was instrumental in putting October 15th after October 4th in 1582.

Cleomedes
Dates very variable. Possibly born at Lysimachia.
At least thought to have been after Poseidonius (135-50 B.C.) and before Ptolemy (100-178 A.D.). He is the only modern source of knowledge of the works of Eratosthenes and Poseidonius. Appreciated refraction.

Nicholaus Copernicus
b. 19.2.1473 Torun, Poland
d. 24.5.1543 Frauenburg. Buried in the cathedral
Father was a rich citizen of Cracow who became magistrate of Torun. Mother was Barbara Watzelrode of an aristocratic family. Both had died by the time Copernicus was 10 and he was adopted by his maternal uncle Lucas Watzelrode,

Canon of Frauenburg Cathedral. He went to the University of Cracow 1491-5 and in 1497 became Canon of Frauenburg Cathedral. From 1501-3 he travelled in Italy. On his return he lived with his uncle at palace of Heilsberg until 1512 and the remainder of his life was spent at Frauenburg. Was one of the founders of modern science and first to suggest that the earth revolved round a fixed sun. He wrote on both plane and spherical trigonometry and best known for his *De Revolutionibus Orbium Coelestium'* of 1543. Assisted the Pope in the reform of the calendar.

Cosmas Indicopleustes
b. Probably in Alexandria. fl. mid 6th century A.D.
An Alexandrian monk, he refuted the idea of a round earth. Constructed his own cosmology based on the scriptures. Wrote a book on Christian Topography. Travelled widely as a merchant before becoming a monk in Sinai c. 548 A.D.

Couplet
d. 17.9.1736
Newphew of a French Academician of the same name, he died on the plains near Quito after only two days illness. Was to have assisted in the Peru expedition.

Crates of Mallos
fl. early 2nd century B.C. Mallos was a very old Greek settlement. Spent his life in Pergamon as chief of the Philological school and director of the library. He was a Stoic and constructed a terrestrial globe around 168 B.C. while in Rome.

Johannes Curtius

Chancellor to Rudolph II he devised an alternative to the nonius.

Nicolaus de Cusa
b. 1401 Cues. A village on the river Moselle.
d. 11.8..1464 Todi, Umbria
A great logician although his fallacies were exposed by Regiomontanus. He had various strange notions about the earth. His father was Johannes Chrypffs, a ship owner and wine grower. Cusa fled from home and entered the service of a nobleman who had him educated at Deventer. He rose to rank of cardinal and had the Bishopric of Brixen in the Tyrol. His mathematical ability was widely criticized.

Jean André De Luc
b. 8.2.1727 Geneva
d. 7.11.1817 Windsor
Did a considerable amount of work on the barometer and thermometer details of which will be found in [267]. One of his particular interests was the determination of height by use of a barometer.

Democritus of Abdera
b. c 460 B.C.
d. 370 B.C.
A pupil of Anaxagoras he followed the Pythagorean philosophy. Boasted of his skill in geometry and the construction of plane figures.

René Descartes

b. 31.3.1596 La Haye, Touraine
d. 11.2.1650 Stockholm

His father, Joachim was a counsellor in the Parlement of Bretagne from 1586. René went to a Jesuit school in La Fléche from 1604 to 1612. He enlisted in the army but it was not the life for him and he left in 1621. He spent some years travelling before returning to Holland in 1628 and thence on to Stockholm in 1649. He took nothing on trust and built up his own Cartesian philosophy. He extended the early ideas of latitude and longitude to the definition of any point by two coordinates. In 1637 he published 'La Géométrie' in Leyden as an appendix to 'Discours de la Méthode'. This contained the first mention of analytical geometry.

Dicaiarchos of Messina

fl. c 320 B.C. in Athens and Peloponnesos
d. c 285 B.C.

A pupil of Aristotle whose main interests were history and geography. He taught of a spherical earth and some think he was the source of Archimedes value of the earth circumference as 300,000 stade.

Leonard Digges

b. 1510 Barham, Kent
d. c 1570

He was from a long established Kentish family and studied at University College Oxford. He applied his interest in mathematics and science to surveying and in 1556 published 'Tectonicon'. He described the instrument topographicall, theodolitus, and plane table. A further work 'Pantometria' was published posthumously by his son in 1571.

Thomas Digges

b. 1543 Kent
d. August 1595 London

Son of Leonard Digges, he became Member of Parliament in 1572 and 1584. Matriculated at Cambridge. He wrote extensively on the telescope and some suggest that it was invented by his father.

Empedocles of Acragas

b. c 490 B.C. Sicily
d. c 435 B.C. Pelopennese

Maintained many strange ideas on the universe. Considered the universe to be spherical and the earth as egg-shaped.

Eratosthenes

b. c 276 B.C. Cyrene, now Shahhat, Libya
d. c 195 B.C. Alexandria

Son of Aglaos, he was educated in Athens. The greatest mathematical geographer of antiquity, he became head librarian of the museum in Alexandria from c 235 until his death. This was as the result of an invitation by Ptolemaios III of Egypt to become tutor to his son as well as run the library. This post gave him an involvement in all the scientific projects of the time. He made corrections to the Ionian map and drew a baseline of a parallel through Gibraltar. His work has only come down to the present age through the writings of Cleomedes. In old age he became blind and starved himself to death.

Eudoxus of Cnidos

b. c 408 B.C. Cnidos
d. c 355 B.C. Cnidos

The son of Aischines. He was extremely poor and walked considerable distances to get education. He was the founder of scientific astronomy and built an observatory at Cnidos. c 381 he went to Egypt for two years and then to Cyzicus where he collected a group of followers who founded a school in Athens c 368. Most of Euclid book V was due to Eudoxus. Thought by some that the earth circumference given by Aristotle was really the work of Eudoxus. He was credited with a new form of sundial. Was the most capable mathematician of the Hellenic age.

Daniel Gabriel Fahrenheit

b. 24.5.1686 Danzig
b. 16.9.1736 The Hague

In 1714 he introduced the thermometer that takes his name. He was the son of a wealthy Danzig merchant but spent most of his life in Amsterdam.

Jean François Fernel

b. 1485 Montdidier near Paris
d. 25.4.1558 Fontainbleau

Son of a well-to-do innkeeper. At 12 he moved to Clermont. He became Astronomer and doctor to Henry II. In 1524 he had quartan fever and was forced to convalesce. In 1527 he published his first book and another the next year called 'Cosmotheoria'. He obtained a degree in medicine in Paris, 1530 and the following year took the chair in medicine. Made a notable measure of the length of a degree of arc.

John Flamsteed

b. 19.8.1646 near Denby
d. 31.12.1719 Greenwich

Ill health interrupted his schooling so much that his study was by his own initiative. He went to Cambridge University where he obtained an MA and took Orders for a church career. He built a small observatory in Derby where he made numerous observations. In 1675 he was summoned to London to take control of the Greenwich observatory as the First Astronomer Royal. Salary £100. His equipment was mostly based on that used in Paris by the Cassinis. He spent a great amount of time on graduating and calibrating micrometer screws and their scales. He spent some 45 years at Greenwich.

Reiner Gemma Frisius

b. 8.12.1508 Dokkum, Friesland
d. 25.5.1555 Louvain, Belgium

Had to use crutches until the age of 6 because of deformed feet and as a result was educated by friends at Groningen. He went to University at Louvain and became Professor there 1537-9. In 1533 he published the first account of the possibilities of using triangulation, but by use of astrolabe and compass. 1540 he published 'Arithmeticae Practicae Methodus Facilis' in Antwerp. He suggested a method of obtaining longitude from difference in time. Wrote at length on the Jacob staff. He had a son Cornelius (1535-1577) who continued his work. Gemma had married Barbara in 1534. He died of kidney stones which plagued him for 7 years. He is buried in the church of the Dominicans.

Galileo Galilei

b. 18.2.1564 Pisa

d. 8.1.1642 Florence

Son of a Florentine nobleman of reducing estate. Galileo was sent to a convent at Vallombrosa but his father later relented and let him study medicine. Went to Pisa university in 1581 and in 1589 became Professor of mathematics there. Resigned in 1591 and became Professor at Padua the following year. Whilst he did not invent the telescope he was the first to put it to scientific use. He developed proportional compasses; established first law of motion; pursued many aspects of mathematics. He became convinced of doubtfulness of much of Aristotle's teachings in physics. He was a devout follower of the ideas of Copernicus although the works of Copernicus were prohibited at that time. In 1609 he was warned against his heretical utterings on the earth motion. He had no success in trying to get the decree of Pope Urban VIII revoked and the heliocentric theory adopted. In 1638 he became blind.

William Gascoigne

b. 1612 Middleton, Yorkshire

d. 2.7.1644 Marston Moor. In the Royalist disaster.

He was the eldest son of Henry Gascoigne and his first wife Margaret Cartwright. He had little formal education. In 1640 he invented the micrometer and also introduced stadia hairs into it. His untimely death in the battle of Marston Moor precluded his writing of his invention and it was some years later before it became widely known.

Levi Ben Gershon

b. 1288 Bagnols, Gard, France

d. 20.4.1344

An outstanding Jewish mathematician. He was also known as Rabbi Levi ben Gerson, Gerschom, or Gersonides and sometimes Ralbag for short from his initials. He introduced the baculum or Jacob staff. Described use of transversals and diagonal scale. Used chords and sines but not tangents. In 1321 he published a work on arithmetic in Hebrew. Wrote also 'De sinibus, chordis et arcubus'.

Louis Godin

See page 95 of main text.

Jean Godin des Odonais

Cousin of Louis Godin. Was a member of the Peru expedition and was away from France for 38 years. When he returned on 26th June 1773 La Condamine was at La Rochelle to meet him and his wife. He gave general help to the expedition but is best known for the terrible journey his wife made down the Amazon to meet him.

George Graham

b. 1675 Horsgill, Cumberland

d. 16.11.1751 London

A Quaker who began as a watchmaker's apprentice he came to be regarded as the finest instrument builder of his time.

He was responsible for Halley's quadrant and transit instrument and Bradley's zenith sector. He also constructed pendulums. He made many of the instruments used on the expeditions.

Edmund Gunter

b. 1581 Hertfordshire

d. 10.12.1626 London

Initially trained for the ministry but he left that to later become Professor of astronomy in 1619 at Gresham College. He was a colleague of Briggs and this led to the publication of the first table of log sines and tangents for every 1' to 7 places, in 1620. He invented the words cosine and cotangent. In 1630 he produced a log slide rule based on log $(A+B) = \log A + \log B$. It had no sliding parts so was only a forerunner of the 20th century version. He developed the Gunter chain.

Edmund Halley

b. 8.11.1656 Shoreditch, London

d. 14.1.1742 Greenwich

He attended St. Pauls School London and excelled in mathematics and the classics. In 1673 he went to Queen's College Oxford. Before he reached 20 he communicated a paper to the Royal Society. In November 1676 he went to St. Helena to make astronomical observations. 1678 was elected to the Royal Society. 1703 became Savilian Professor of geometry and in 1721 was Astronomer Royal. He was very interested in the construction of log tables and was the one who persuaded Newton to write his 'Principia'.

Hecataeus of Miletus

b. c 550 B.C.

d. c 475 B.C.

The son of Hegesandros, he became known as the father of geography. He produced a map divided into the three parts of Europe, Asia and Libya. His book 'Circuit of the Earth' contained all the Ionian knowledge of the earth at that time. He produced an improved version of the map of Anaximandros.

Anders Hellant

See page 171 of main text.

Heracleides of Pontos

b. c 388 B.C. Heracleia Pontica on the Black Sea

d. 315-310 B.C.

He taught of the rotation of the earth on its axis and founded the geoheliocentric system and theory of epicycles.

Herodotus

b. c 484 B.C. Halicarnassos

d. c 425 B.C. Thurii in Lucania, S.Italy

Was the first great historian and introduced the idea of a meridian. He appreciated the requirement for surveyors in the re-establishment of boundaries destroyed by the floods of the Nile.

Heron of Alexandria

His dates are very variable from 150 B.C. to 250 A.D. As a prolific writer he laid the scientific foundation for engineering and surveying. A formula for area retains his name. He worked on tables of chords and invented a forerunner of the thermometer. In particular he developed the dioptra or forerunner of the theodolite. In addition he developed a level rather akin to the modern water level. Wrote 'A treatise on the Dioptra' and a work on geodesy that could be termed the first surveying textbook. Among the problems he solved was the determination of the distance between two points where one was inaccessible.

Johann Hevelius
b. 28.1.1611 Danzig
d. 28.1.1687 Danzig
Developed a telescope 150 ft. long but nevertheless said he could observe as accurately with open sights as others could with a telescope. He built an observatory in Danzig. Invented a means of grinding a lens mechanically. On 26 September 1679 a fire destroyed his house and all his books and instruments. Luckily he had by then completed his book that represented 50 years of observations. In 1668 he had written 'Cometographia'.

Hicetas of Syracuse

A philosopher and astronomer of the Pythagorean school he taught that the earth rotated on its own axis every 24 hours.

Hipparchus
b. c 180 B.C. Bithynia, N.W. Turkey
d. c 125 B.C. Rhodes
As the father of trigonometry and an unrivalled scientist he was credited with the invention of latitude and longitude. He developed tables of chords, invented a plane astrolabe and constructed celestial globes. He founded both plane and spherical trigonometry and introduced the stereographic projection. He discovered precession of the equinoxes by comparing his results with those of 150 years previously. Spent most of his life in Rhodes. The introduction of a 360° circle is said to have been mostly due to him.

Homer
b. 8th or 9th century B.C.
Many cities claim to have been his birthplace—mostly in Ionia. There is reason to think that he considered the earth as a sphere although some say he considered it as a convex disk.

Robert Hooke
b. 18.7.1635, Freshwater
d. 3.3.1702 London
Professor of geometry at Gresham College. Became secretary of the Royal Society, and from 1662 was curator of experiments. Used telescopic sights in 1667. Had the opportunity to investigate the means of measuring the figure of the earth but did not do so.

I-Hsing
b. 682 A.D.
d. 727 A.D.
A Tantric Buddhist monk of the Thang Dynasty (618-906 A.D.), he was an able mathematician and astronomer and became famous in China. He invented the first escapement for a mechanical clock and constructed a model of the celestial bodies where movement was in relation to the eclip-

tic. Organised a line of gnomon observations through China. 721-7 he prepared a calendar.

Hugot
Little is known about this man other than that he was the instrument mechanic on the Peru expedition. With the rebuilding of instruments that were carried out there he was obviously in an important position. He married while in Quito. Said to have died 1744.

Christiaan Huygens
b. 14.4.1629 Hague
d. 8.6.1695 Hague
One of the most notable scientists of the 17th century. He studied at Leyden. He went to Paris and London in 1660 and 1663 and in 1666 was appointed a member of the French Academy of Sciences. He stayed in Paris until 1681. He invented the pendulum clock and investigated the catenary. He published his work on involutes and evolutes in 1673 as 'Horologium oscillatorium'.

Hyginus Gromaticus
fl. c 120 A.D.
Was a surveyor and user of the groma. Described the use of Indian circles as method of determining a north-south direction. This he did as result of his work on the layout of boundaries.

Hypsicles of Alexandria
b. c 2nd century B.C. fl c 180 B.C.
He wrote the oldest known Greek work on the ecliptic and its division into 360 parts. At about the same time the use of sexagesimal fractions was becoming popular.

Zacharias Jansen
b. 1588 The Hague
d. 1628? Amsterdam
A rival to Lippershey as inventor of the telescope. He was a spectacle maker from Middleberg.

Jorge Juan
See page 97 of main text.

Ibn Junis
fl 1000 A.D.
d. 1009 A.D.
An astronomer of the Fatimite Kingdom of Egypt. He was at Cairo observatory for some while. Wrote an account of the Arabian arc measure.

Joseph de Jussieu
See page 101 of main text.

Charles Marie de la Condamine
See page 97 of main text.

Lactantius
fl. early 4th century A.D.
Wrote a book 'Divine Institutions' which ridiculed the idea of a spherical earth. This was thought to have been written around 315 A.D. How could men possibly hang upside down on the earth?

Philippe de La Hire
b. 18.3.1640 Paris
d. 21.4.1718 Paris
A Pensionnaire of the French Academy of Science he was

Professor of mathematics in the College Royal. From 1679 he was involved in the geodetic survey of France with the Cassinis. Wrote several treatises on mathematics.

Claude Langlois
b. 1700
d. 1756
First French instrument maker of repute. He made several of the instruments used on the expeditions.

Gottfried Wilhelm Leibniz
b. 21.6.1646 in Leipzig
d. 14.11.1716 in Hannover
At 15 he entered the University of Leipzig principally to study law but he was interested in many subjects. After graduating in 1663 he took a doctorate at Nuremberg. He entered the diplomatic service as an elector at Mainz, then with a family in Brunswick and finally he became librarian to the duke in Hannover where he stayed for some 40 years. In 1676 he had developed a calculating machine and was demonstrating it in London. In 1700 he founded the Berlin Academy. He was the first to use a barometer without fluid—i.e., an aneroid type. He introduced use of a dot to represent multiplication. He was a member of the French Academy of Sciences.

Pierre-Charles Le Monnier
b. 20.11.1715 in Paris
d. 3.4.1799 in Herils, Calvedos, France
One time astronomer to Louis XV and father-in-law of the well known mathematician Lagrange. He was a member of the Lapland expedition and did some similar work in France after the return from northern latitudes. He attempted unsuccessfully to combine the advantages of mural and portable quadrants in 1753. He wrote an 'Histoire Celeste' published in Paris 1741. He was a member of the French Academy of Sciences from 1736.

William Leybourn
b. 1626
d. 1716
He was both a teacher of mathematics and a surveyor. He wrote on many topics including astronomy, surveying arithmetic, logarithms and computing rods. Best known for his 'Compleat Surveyor' of 1653 where he described use of the water level, theodolite, circumferentor and plain table.

Hans Lippershey
fl Middleburg early 17th century
He was a Dutch spectacle maker and one of the contenders in 1608 to the invention of the telescope.

Mabillon
A member of the Peru expedition about whom little is known. It is said that at one stage he lost his memory but that it did partially return later.

Giovanni Antonio Magini
b. 13.6.1555 in Padua
d. 11.2.1615 (1617) in Bologna
A well known astronomer and mapmaker he became Professor of astrology, astronomy and mathematics at the University of Bologna from 1588 until his death. He introduced the use of a comma as a decimal separator in his 'De planis triangulis' Venice 1592 but its use did not really become popular until Napier twenty years later.

Pedro Vincente Maldonado y Sotomayor
b. 1704 Riobamba
d. 17.11.1748 London
A mathematician and cartographer, he attended the Jesuit college in Quito. He built a road from Ibarra down the side of the Andes to the Esmeraldas river. At 27 he became Governor of Esmeraldas. He was fluent in French, Spanish and Quechua. When Bouguer and La Condamine were at Guayaquil in 1736 Maldonado joined them and requested to be allowed to become a member of the party. This he did and partnered La Condamine on his mapping work prior to the journey to Quito. He stayed with the party for the duration of the survey work and in 1743 left Para for Lisbon where he arrived in February 1744.

Marinus of Tyre
fl c 150 A.D. possibly at Ptolemais
d. aged 78
A Greek scientist and founder of ancient mathematical geography. He located places by a two coordinate system of latitude and longitude where his base was around the 36°N parallel and the meridian through Fortunatae Insulae or the Islands of the Blessed. Later he moved the meridian to go through Ferro, one of the Canary Islands.

Pierre Louis Moreau de Maupertuis
See page 168 of main text.

Hrabanus Maurus
b. 784 A.D.
d. 856 A.D.
Abbot of Fulda and later Archbishop of Mainz. He was particularly interested in mathematics and astronomy and followed the teaching of Bede. He considered the earth to be at the middle of the world and that it must be square because of the scriptures where the four corners were mentioned.

Menelaus of Alexandria
fl. 100 A.D. in Alexandria
Did much work on the use of chords and wrote 6 books on their calculation. He established a basis for spherical triangles similar to that which Euclid had done for plane triangles. The 'Theorem of Menelaus' for plane triangles was probably known to Euclid but Menelaus produced the equivalent for a spherical triangle.

Pere Marin Mersenne
b. 8.9.1588 Oizé, Maine
d. 1.9.1648 Paris
Educated by the Jesuits at La Fléche, he became a Minorite father. He maintained an extensive correspondence with many eminent European scientists and founded an intellectual centre in Paris around the mid 17th century. His efforts were continued after his death by Abbé Picot and others.

Jacob Metius (Adriaenszoon)
fl. early 17th century, Alkmaar
d. Jan. 1628
Said by some to have been the inventor of the telescope he was only one of several contenders. He used an astrolabe with a decimally divided shadow square and described a quadrant for use at sea. He developed a version of the Jacob

202

staff with a variable cross piece and wrote about the backstaff.

de Morainville
d. 1744
A member of the Peru expedition about whom little is known. He remained in the Province of Quito after the main party had left and met his death by falling from scantling around a church at Cicalpa near Riobamba for which he was architect. On the expedition he acted as draftsman and natural historian.

Jean Baptiste Morin
b. 23.2.1583 Villefranche
d. 6.11.1656 Paris
Professor of mathematics and astronomy at the College Royal in Paris. In 1634 he introduced the use of the telescope to replace open sights.

Gabriel Mouton
b. 1618 in Lyons
d. 28.9.1694 in Lyons
Around 1670 he suggested the adoption of a standard of length that would be universally accepted. His idea was to use a meridian arc of 1 minute with decimal subdivisions.

Petrus van Musschenbroek
b. 14.3.1692 Leiden
d. 19.9.1761 Leiden
Initially a Professor at Utrecht he later became Professor of mathematics and philosophy at Leiden. He checked the work of Snellius and made various re-observations and re-calculations. Unfortunately there is rather strong evidence that he falsified some of the results so that his work has no reliable worth.

John Napier
b. 1550 Merchiston Castle
d. 4.4.1617 Merchiston Castle
He was of notable descent and at 13 went to St. Andrews University. Was a Scottish Laird and Baron of Merchiston who had mathematics as a hobby. For some 20 years he worked on his idea of logarithms before publishing two works in 1614. His 'Mirifici logarithmorum canonis descriptio' explained the nature of logs and gave tables of log sines for every 1 minute. A further posthumous work of 1619 'Mirifici logarithmorum canonis constructio' gave the method of calculating the logs.

Sir Isaac Newton
b. 25.12.1642 Woolsthorpe, Grantham, Lincolnshire
d. 20.3.1727 London.
Buried in Westminster Abbey 28.3.1727
A small, feeble child of little promise who went to school in Grantham from the age of 12. By 1660 he was at Trinity College Cambridge with a BA in 1665 and MA in 1668. At this time he was one of the most able mathematicians in the country. In 1669 he became Lucasian Professor of mathematics at Cambridge and in 1672 was elected to the Royal Society. 1689 he entered Parliament and in 1696 was Warden of the Mint. He was knighted by Queen Anne in 1705. He made numerous contributions to mathematics and science but in particular here he should be remembered for his work on the shape of the earth and his production of instruments such as an octant based on the quadrant and a first form of reflecting instrument similar to a sextant.

Richard Norwood
b. 1590 Stevenage
d. Oct. 1675 Bermuda
A teacher of mathematics and navigation he published several books. In 1635 he measured the arc from London to York but in doing so rather turned the clock.back by not using triangulation which had been in being for some while and about which he must have been aware.

Pedro Nuñez Salaciense
b. 1502 Alcacer do Sol, Portugal
d. 11.8.1578 Coimbra, Portugal
He studied at the University of Lisbon and in 1530 became Professor there. From 1544 to 1562 he was Professor of mathematics at Coimbra. In 1572 he became court advisor on the reform of weights and measures, the results of which were promulgated in 1575. He is best known for his work in navigation and astronomy and in particular for the introduction of the nonius. This was a means of reading subdivisions of angles and was effectively the forerunner of the vernier. Among his navigation observations was the realisation that a ship sailing in a manner to make equal angles with each meridian did not sail in a straight line but on what is now termed a loxodrome.

Oenipedes of Chios
fl. c 465 B.C. Chios
d. c 430 B.C.
A notable astronomer and follower of Pythagoras who discovered—or maybe rediscovered, the obliquity of the ecliptic.

L. Valentin Otto (Otho)
b. c 1550
d. c 1605
He finished the computations of the Rheticus trigonometric tables that were to 15 places. He published them in 1596 as 'Opus Palatinum de Triangulis'.

William Oughtred
b. 5.3.1574 Eton
d. 30.6.1660 Albury
Invented the slide rule and published his results in 1632. Produced the circular as well as straight versions although there was some controversy over the former as one of his students was the first to describe it in print.

Réginald Outhier
See page 170 of main text.

Parmenides of Elea
b. c 515 B.C. at Elea, South Italy
d. after 450 B.C.
Founded the Eleatic school. He was the only philosopher of that time other than Pythagoras, who recognised the earth as a spherical shape.

Blaise Pascal
b. 19.6.1623 Clermont-Ferrand, France
d. 19.8.1662 Paris
When Blaise was only 3 his father retired and devoted all his time to educating his son. Even by the age of 12 he was proving something of a genius at mathematics as well as being able at Latin and Greek. At 14 he went with his father

to the informal meeting of the Mersenne Academy and by 16 was publishing an essay on conics. At 18 he designed and built a calculating machine and had soon sold 50 of them. Soon his interest waned and he turned first to hydrostatics and then theology.

Petrus Peregrinus (Pierre de Maricourt)
b. c 1265

Little is known about him except that he invented and described the box compass as a pivoted needle revolving against a graduated disc akin to an astrolabe.

Georg Peurbach
b. 30.5.1423 Peurbach, Austria—Bavarian border
d. 8.4.1461 Vienna
A lecturer in mathematics at Ferrara, Bologna and Padua and Professor at Vienna University. Court astronomer to the King of Bohemia. Wrote on astronomy, mathematics and the Ptolemaic system. Referred to as the founder of observational and mathematical astronomy in the West. Appreciated the use of sines instead of chords and computed tables at $10'$ increments. Made an angle measuring instrument by dividing a Geometric Square into 1200 parts.

Philolaos
b. mid 5th century B.C. Crotona or Tarentum, S.Italy
d. c 390 B.C.
Mathematician and astronomer. Follower of Pythagoras. Said that at the centre of the universe there was a fire around which the earth and planets revolved daily.

L'Abbé Jean Picard
b. 21.7.1620 La Fléche, Anjou
d. 12.10.1682 Paris
Gardener for the Duke of Crequi then Prior at Rille in Anjou. Later Professor of astronomy at College de France in Paris, 1655. Made the first use of a telescope in surveying as part of a levelling instrument in 1660. Measured an important arc in France and published results as 'Mesure de la Terre' in Paris, 1669. He developed a quadrant of 10 ft. radius with telescopic sights and cross hairs. He used stadia measuring techniques towards the end of the century. He was among the first astronomers at the Paris observatory. Appreciated the change in the rate of a pendulum due to temperature changes. As a result he found the length of a seconds pendulum by adjusting the length of a suspension thread.

Plato
b. c 427 B.C. Aegina or Athens
d. 348-7 B.C. Athens
Founder of a philosophic school that lasted until 529 A.D. Followed the teachings of Pythagoras. Held that the earth was immovable and that the heavens rotated daily. Wrote on the power of the lodestone.

Poseidonius
b. c 135 B.C. Apamea on the river Orantes in Syria
d. c 50 B.C. Rome
A stoic philosopher and most learned man of his time. He became head of the Stoic school. Interested in geography and astronomy, he was the first to appreciate atmospheric refraction. He explained the effects of the tides in terms of the actions of the sun and moon. His measurement of the earth's circumference is only known through the writings of Cleomedes.

Claude Ptolemy
b. 100 A.D. Egypt
d. 178 A.D.
A celebrated astronomer best known for his 'Syntaxis Mathematica' or Almagest which covered all known aspects of astronomy up to that time. He produced a famous atlas that was current until the 16th century. Described the problem of representing a sphere on a flat piece of paper. He developed various survey instruments that were still being copied in the time of Copernicus. He used sexagesimal fractions and advanced the chord tables of Hipparchus. He explained both plane and spherical trigonometry and used a rectangular grid on his maps.

Pythagoras of Samos
b. c 580 B.C. at Samos
d. c 500 B.C. Metapontum
Spent much of his early life travelling and then he settled in Crotona, S. Italy. He gathered a group of young men around him and formed the Pythagorean School. This was a sort of brotherhood and the members were forbidden to divulge their discoveries and teachings individually so all references were to them as a body of Pythagoreans. When there was a revolt in Italy the School was destroyed and Pythagoras fled to Metapontum where he was later murdered.

René Antoine Ferchault de Réaumur
b. 28.2.1683 La Rochelle
d. 18.10.1757 St. Julien du Terroux
Inventor of the Réaumur thermometer that was used on the Peru expedition. Member of the French Academy of Sciences.

Johannes Müller von Königsberg (Regiomontanus)
b. 6.6.1436 Konigsberg, Franconia, Germany
d. 8.7.1476 Rome
He was an infant prodigy who published a book on astronomy at the age of 12. He matriculated from Vienna in 1450. At 16 he became a pupil of Peurbach for astronomy and trigonometry. He later had a position in the Court of King Mathias Corvinus in Hungary. In 1471 he built the first European Observatory in Nuremberg. In 1475 he went to Rome to aid the Pope—Sixtus IV—in reform of the calendar but he died before this was completed. In 1464 he wrote 'De triangulis omnimodis libri V' on trigonometry. He used such survey instruments as the astrolabe, quadrant, geometric square and Jacob staff. He developed the torquetum. He produced a table of sines with a radius of 600,000 and tangents with 100,000.

Georg Joachim Rheticus
b. 16.2.1516 Feldkirch in Austrian Tyrol
d. 4.12.1576 Cassovia, Hungary
Studied at Zurich and then in 1532 went to Wittenberg where he became Professor of mathematics. In 1541 he went to Frauenburg and 1542 took a chair at the University of Leipsig, until 1551. His time at Frauenburg was spent assisting Copernicus. When in his 50s he became physician to a Polish

Prince and then migrated to Cassovia. Particularly interested in trigonometry and tables. Started tables to 15 decimal places at 10″ increments which were completed after his death by Valentin Otto.

Giovanni Battista Riccioli
b. 1598 Ferrara
d. 21.1.1671 Bologna
A Jesuit astronomer. Wrote 'Almagestum Novum' in two volumes in 1651. Gave arguments both for and against the motion of the earth. He developed a level akin to the modern water level and in 1651 determined the length of a seconds pendulum. In 1645 he determined a value for a meridian degree but his method was somewhat suspect.

Jean Richer
b. 1630
d. 1696 Paris
Known particularly for his pendulum observations in Cayenne, 1672, where his results agreed with the theories of Newton rather than with the measurements of the Cassinis. Member of the French Academy of Sciences.

Robert of Chester
fl. 1140 A.D.
In 1145 he translated Al-Khowârizmi's *'Algebra'* into Latin. He was first to use the word 'sine' when translating from the Arabic. Prepared several astronomical tables.

Robert the Englishman
fl. c 1270 A.D. in Paris and Montpelier
Wrote on the quadrant and the astrolabe.

Gilles Personne de Roberval
b. 8.8.1602 Roberval near Beauvais
d. 27.10.1675 Paris
Professor of philosophy in Paris and later Professor of mathematics. He wrote on astronomy and developed the systematic use of telescopic sights. A member of the Mersenne group and later of the French Academy of Sciences.

Sanad ibn 'Ali
fl. c 860 A.D.
Muslim astronomer and mathematician. Helped on the arc measurement near Palmyra. He constructed the Kanisa observatory at Baghdad.

Dr. Jean Seniergues

d. 28.8.1739 Cuença, Peru
Surgeon to the Peru expedition. Murdered by a mob in Cuença.

Willibrord Snellius
b. 1580 Pieterskerkhof, Leiden
d. 30.10.1626 Leiden
Son of Rudolph Snel van Royen and Machteld Cornelisdochter of Oudewater. The property they owned in Doezastraat near Leiden and which passed to Willibrord, was destroyed by an explosion on 12th January 1807. Although not the first to think of the idea of triangulation he was the first to put it into practice. Additionally he developed the resection method of position fixing. Introduced the term loxodrome.

John Speidell
fl 1600-1634
At the time that Briggs was compiling a table of common logarithms so Speidell was concentrating on the natural logs of trigonometric functions. He published 'New Logarithmes' in 1619 in London. He improved Napier's logs by making all the values positive. This he did by subtracting them all from 10^8 and discarding the last two digits.

Strabo
b. c 63 B.C. Amasya Pontus on the Black Sea
d. 19 A.D.
A Greek historian and geographer, he was a member of the stoic school. Was the author of 'Geography'.

Jöns Svanberg
b. 1771 Kalix, west of Torneå.
Studied in Uppsala from 1787 and became Professor of mathematics in 1811. Led the team on the reobservation of the work of Maupertuis 1799-1801. He had a love for decimals and complicated calculations.

Thales of Miletus
b. 640 B.C. Miletus
d. 547 B.C.
Made early observations of the heights of the pyramids—probably by means of the lengths of their shadows. Was the founder of Greek science and philosophy. Particularly adept at mathematics. He derived various theorems in geometry. Founded the Ionic School.

Theodoros of Samos
fl. 550 B.C. Ephesus
One of the earliest known engravers, he was a Greek architect and engineer. Invented an instrument called the 'Diabetes' or 'Libella'.

Theon of Alexandria
fl. early 4th century A.D.
Explained a method for finding square roots by means of sexagesimal fractions. Wrote 11 books of commentary on the work of Ptolemy.

Melchisédech Thévenot
b. 1620 Paris
d. 29.10.1692 Issy
Inventor of the spirit bubble in 1661. Famous traveller and author. Member of the French Academy of Sciences.

Antoine Thomas (An To)
b. 1644 Namur, Belgium
d. 1709
Joined the Jesuits in 1660 and took up residence in China in 1685. He was instrumental in the measurement of a degree of the meridian.

Antonio de Ulloa de la Torre Giral
See page[100]of main text.

Ulugh Beg
b. 22.3.1394 Sultaniyya, Central Asia
d. 27.10.1449 near Samarkand, now Uzbek
His original name was Muhammad Taraqay. A Persian

astronomer, from 1409 he was ruler of Maverannakhr. He became Royal astronomer and founded the observatory in Samarkand in 1423. He was the last great oriental astronomer and his assassination led to the end of the Islamic involvement in astronomy. He prepared planetary tables, and compiled tables of sines and tangents.

Varahamihira

fl c. 505 A.D. near Ujjain

d. c 587 A.D.

Hindu astronomer and poet as well as mathematician and historian. He derived various trigonometric expressions.

Capt. Verguin

Naval engineer from Toulon who was a member of the Peru expedition. Little is reported about his work.

Pierre Vernier

b. 19.8.1584 Ornans, France

d. 14.9.1638 Ornans

Invented the vernier system and described it in a publication of 1631.

Françoise Viète

b. 1540 Fontenay-le-Comte

d. 13.12.1603 Paris

Introduced the practice of using vowels for unknown quantities and consonants for known ones. He was among the greatest French mathematicians of the time and was member of the Bretagne Parlement and member of the King's Council under Henry III and Henry IV. Mathematics was only a hobby with him but he added much to the subject. He produced trig. tables of all six functions at $1'$ increment. Derived many trigonometric identities including the expansion of such relations as $\sin \alpha = \sin (60 + \alpha) - \sin(60 - \alpha)$. In 1579 he published 'Canon mathematicus seu ad triangula cum appendicibus', a treatise on trigonometry.

Adriaen Vlacq

b. c 1600 Gouda

d. 1667 Hague

Completed the gap left in Briggs logarithmic tables. He published the results in 1628.

Edward Wright

b. Oct. 1561 Garveston, Norfolk

d. Nov. 1615 London

Said to have measured a baseline near Plymouth Sound in 1589 and used it together with some measured angles to determine the height of a nearby hill. From this and other data he computed a value for the earth's diameter.

REFERENCES

1. Biographical Encyclopedia of Science and Technology. Asimov,I 1964
2. Dictionary of Scientific Biography. 1978
3. Encyclopedia Britannica. 1971
4. Encyclopedia Britannica Knowledge in Depth. 1977
5. Aaboe,A.,1964. Episodes from the early history of mathematics. Random House.
6. Airy,G.B.,1845. Figure of the earth. Encyclopedia Metropolitana London. Vol.V. 165-240
7. Anon., 1907. Notes on the work of the French geodetic commission to measure the Quito arc. Geog.J. 29:211-212
8. Anon., 1928. Blaeu's measurement of an arc of meridian. Geog.J. 72:528
9. Anon., 1942. The story of the repeating circle. Empire Survey Reviews 6 (43 Jan):318-319; 6 (44 Apr):364-367
10. Anon., 1967. Absolute gravity measurements. Bull.D'Information. (17 Dec)
11. Ariotti,E., 1971-2. Aspects of the conception & development of the pendulum in the 17th century. Arch. Hist.of Exact Sc. 8:126-142
12. Armitage,A., 1954. Jean Picard and his circle. Endeavor. (Jan):17-21
13. Armitage,A., 1951. The world of Copernicus. Mentor Books.
14. Beer,A., Ho Ping-Yu; Lu Gwei-Djen., Needham, J., Pulleybank, E. G., and Thompson, G. I., 1961. An 8th century meridian line: I-Hsing's chain of gnomons and the pre-history of the metric system. Vistas in Astronomy. 4: 3-28. Pergamon Press, Oxford.
15. Bell,E.T., 1945. The development of mathematics.
16. Berriman,A.E., 1953. Historical metrology. Dent.
17. Besterman,T., 1958. Les Lettres de la Marquise du Châtelet. Institut et Musee Voltaire. Geneva. 2 Vols.
18. Bevis, ., 1753. On Mr Gascoigne's invention of a micrometer. Phil.Trans.Roy.Soc. XLVIII : 190
19. Bialas,V., 1982. Erdegestaltung, Kosmologie und Weltanschauwung. Stuttgart.
20. Bigourdan.G., 1912. Grandeur et figure de la terre, Ouvrage augmente....Paris.
21. Bion,N., 1752. Traite de la construction et des principaux usages des instrumens de mathematique. Paris.
22. Blondel,M.,and Geoffroy,M., 1702. An account of what passed in the last Public Assembly of the Royal Academy of Science in Paris, 1701. Phil.Trans.Roy.Soc. XXIII (278): 1097
23. Bond,J.D., 1921-2. The development of trigonometric methods down to the close of the XVth century. ISIS 4: 295-323.
24. Bosmans,H., 1924,1926. L'Oeuvre scientifique d'Antoine Thomas de Namur,S.J. Annales Soc.Sc.Bruxelles. 44: 169-208. 46: 154-181.
25. Bouguer,P., 1744. Relation abrégée du voyage fait au Pérou. Mem.Acad.Sc. Paris.
26. Bouguer,P., 1749. La figure de la Terre. Tombert. Paris.
27. Bouguer,P., et al, 1754. Operations faites par l'ordre de L'Academie pour mesure l'intervalle.... Mem.Acad.Sc. Paris. 172-186
28. Boyer,C.B., 1968. A history of mathematics. Wiley.
29. Bronowski,J., 1973. The ascent of man. BBC Publications.
30. Brown,H., 1934. Scientific organisations in seventeenth century France. Baltimore Press.
31. Brown,H., 1975. From London to Lapland. In Literature & History in the age of ideas. Ed.C.G.S.Williams. Ohio State University Press. 69-94.
32. Brown,H., 1976. Science and the human comedy. Univ.Toronto.
33. Brown,L.A., 1951. The story of maps. Cresset Press. London.
34. Brunet,P., 1929. Maupertuis étude biographique. 33-64.Paris.
35. Brunet,P., 1952. La vie et l'oeuvre de Clairaut. Paris.
36. Bunbury,E.H., 1879. A history of ancient geography. London. 2 Vols.
37. Burden,W.W., 1960. 66 centuries of measurement. The Sheffield Corpn. Dayton, Ohio.
38. Butterfield,A.D., 1906. History of the determination of the figure of the earth from arc measurements. Worcester, USA.
39. Bychawski,T., 1959. The measurements of a degree executed by the Arabs in the 9th century. IXth Int.Congress Hist. of Sc. Barcelona. 635-638
40. Cajori,F., 1922. A history of mathematics. Macmillan. N.York.
41. Cajori,F., 1929. History of determination of the heights of mountains. ISIS 12: 482-514.
42. Cajori,F., 1966. Sir Isaac Newton's Mathematical Principia. Univ. California Press, Berkeley. 2 Volumes.
43. Cassini,J., 1665. Biographical Note. Phil.Trans.Roy.Soc. Abridged. Vol. 1: 8
44. Cassini,J., 1720. Traité de la grandeur et de la figure de la terre. Mem. Acad. Sc. pour 1718. Paris
45. Cassini de Thury,C.F., 1744. La meridienne d'observatoire royal de Paris verifiee dans toute l'entendue du royaume par de nouvelles observations, pour en deduire le vraye grandeur. Paris.

46. Cassini de Thury,C.F., 1783. Description geometrique de la France. Paris.

47. Celsius,A., 1738. De observationibus pro figura telluris determinanda. Uppsala.

48. Clairaut,A-C., 1735. Détermination géométrique de la perpendiculaire à la méridienne tracée par M.Cassini. Mem. Acad. Sc. pour 1733. Paris. 406-416

49. Clairaut,A-C., 1737. Some investigations...of the figure of the earth. Phil.Trans.Roy.Soc.XL (445):19

50. Clairaut,A-C., 1738. Sur la nouvelle méthode de M.Cassini pour connoître la figure de la terre. Mem.Acad.Sc. pour 1735. :117-122

51. Clairaut,A-C., 1739. Sur la mesure de la terre par plusieurs arcs de méridien pris à différentes latitudes. Mem.Acad.Sc. pour 1736. :111-120

52. Clairaut,A-C., 1741. Suite d'un mémoire donné en 1733 qui a pour titre: Détermination géométrique de la perpendiculaire à la méridienne etc. Mem.Acad.Sc. pour 1739. :83-96

53. Clairaut,A-C., 1743. Théorie de la figure de la terre. Paris.

54. Clairaut,A-C., 1737. Biographical note. Phil.Trans.Roy.Soc. Abridged Vol 8: 118.

55. Clarke,A,R., 1880. Geodesy. Clarendon Press.

56. Collinder,P., 1956. On the oldest earth measurements. Congrès Int.d'Hist.des Sciences. 8 Actas 1: 456-462

57. Collinder,P., 1962. Dicaearchus and the Lysimachian measurement of the earth. Congrès Int.d'Hist.des Sciences. 10. Actas 1: 475-477

58. Colcord,J.E., 1976. Surveying before 1776. Proc.Amer.Congress Surveying & Mapping.(Feb.) : 318-332.

59. Cook,A.H., 1961. The shape of the earth. New Scientist. Aug. 3rd: 286-288

60. Datta,B. & Singh,A.N., 1962. History of Hindu mathematics. Bombay Asia Publishing House.

61. Daumas,M., 1953. Les instruments scientifiques aux XVII et XVIII siècles. Paris.

62. David,J.P., 1769. Dissertation sur la figure de la terre.

63. David.J.P., 1769. Replique à la lettre de la Condamine.

64. Delambre,J.B.J., 1799. Determination d'un arc du meridien.

65. Delambre,J.B.J., 1806-1810. Base du systeme metrique decimal. Vol.I 1806. Vol.II 1807. Vol.III 1810. Paris

66. Delambre,J.B.J., 1817. Histoire de l'astronomie ancienne. 2 Vols. Paris. Reprinted New York 1965.

67. Delambre,J.B.J., 1819. Histoire de l'astronomie du moyen age. Paris.

68. Delambre,J.B.J., 1827. Histoire de l'astronomie au dix-huitieme siècle. Paris.

69. Delambre,J.B.J., 1821. Histoire de l'astronomie moderne. Paris.

70. Derham,Rev.W., 1704. An instrument for meridian observations. Phil.Trans.Roy. Soc. XXIV (291):1578

71. Derham,Rev.W., 1717. Extracts from letters by Gascoigne and Crabtree proving Gascoigne invented telescopic sights. Phil.Trans.Roy.Soc. XXX (352):603

72. Desaguliers,Rev.J.T., 1724. A dissertation concerning the figure of the earth. Phil.Trans.Roy.Soc.XXXIII (386):201.

73. Digges,L., 1556. A book named Tectonicon. London.

74. Dilke,O.A.W., 1971. The Roman land surveyors. David & Charles.

75. Diller,A., 1949. Ancient measures of the earth. ISIS (40): 6-9.

76. Dreyer,J.L.E., 1906. History of the planetary systems from Thales to Kepler. Cambridge. 2nd Ed. 1953.

77. Dreyer,J.L.E., 1914. Criticism of paper by Payn on The Well of Erathosthenes. Observatory. (37):352-353.

78. Eames,J., 1740. An account of Celsius's figure of the earth. Phil.Trans.Roy.Soc. XLI (457):371.

79. Engelsberger,M., 1969. Beitrag zur Entwicklungsgeschichte des Theodolits. Deutsche Geod.Comm. Heft 134. Munich.

80. Eves,H., 1953. An introduction to the history of mathematics. Holt,Rinehart & Winston. 3rd Ed. 1969.

81. Feather,N., 1959. Mass, length and time. Edinburgh Univ.Press

82. Fernel,J.F., 1528. Cosmotheoria

83. Fischer,I., 1981. At the dawn of geodesy. Bulletin Géodésique (2): 132-142

84. Fleet,J.F., 1912. Some Hindu values of the dimensions of the earth. J.R.Asiatic S.:463-470

85. Fletcher,E.N.R., 1968. Ancient metrology. Survey Review. XIX (148):270-277

86. Fritz,S.S.J., 1922. Journal of the travels and labours of Father Samuel Fritz in the R. of the Amazons between 1686 and 1723. Hakeuyt Soc.London.

87. Gade.J.A., 1947. The life and time of Tycho Brahe. Princeton University.

88. Gandz,S., 1929. The origin of angle geometry. ISIS.12 :452-481

89. Garrett,A. & Rowlinson,J.S., 1969. The metric system. Trans. of original by M.Danloux-Dumesnils. Athlone Press.

90. Garstin,W.A., 1952. A note on Islamic astronomy. Empire Survey Review. XI (85):306-309

91. Garstin,W.A., 1952. A note on Hispano-Muslim astronomy. Empire Survey Review. XI (86):356-359

92. Gibbs, S.L., 1976. Greek and Roman sundials. New Haven. Yale University Press.

93. Glaisher,J.W.L., 1920. On early tables of logarithms and early history of logarithms. Q.Jnl.Pure & Applied Maths 48:151-192

94. Glass,B., 1955. Maupertuis, a forgotten genius. Scientific American. 193:100-110

95. Godin,L., 1733. Méthode pratique de tracer sur terre un parallèle par un degré de latitude donnée;..... Mem.Acad. Sc. Paris. :223-232

96. Goguel,J., 1971. Bouguer au Chimborazo. Bulletin Géodésique (101):329-334

97. Goodman,E.J., 1972. The explorers of South America. Macmillan.

98. Gore,J.H., 1891. Geodesy. Heinemann Scientific Handbook.

99. Greaves,J. 1647. A discourse of the Roman foot and denarius. London.

100. Groueff,S., 1974. L'Homme et la terre. Larousse Paris-Match. :14-39

101. Gunther,R.T., 1923. Early science in Oxford. Vol.1. Oxford.

102. Haasbroek,N.D., 1968. Gemma Frisius, Tycho Brahe and Snellius and their triangulations. Netherland Geod.Comm.

103. Hagen, V.W. von., 1949. South America called them. Series called "Explorations of the great naturalists" Hale.

104. Hahn, R., 1971. The anatomy of a scientific Institution. The Paris Academy of Sciences 1666-1803. Univ. of California Press.

105. Hall, A.R. & Hall, M.B., 1971. The correspondence of Henry Oldenberg. University of Wisconsin Press.

106. Hanke, L., 1936. Dos Palabras on Antonio de Ulloa's Noticias Secretas. Hispanic American Hist. Rev. 16:479-514

107. Heath, T.L., 1921. History of Greek mathematics. O.U.P. 2 Vol.

108. Hendrikz, D.R., 1944. South African units of length and area. Trig. Survey of S.Africa. Publ. 2

109. Hinks, A.R., 1927. Figure of the earth. A graphical discussion. Geog.Jnl. 69:557-575

110. Hook, 1716. Description of Mr. Gascoigne's micrometer. Phil.Trans.Roy.Soc.(29):541

111. Hutton, C., 1821. On the mean density of the earth. Phil.Trans.Roy.Soc.:276

112. Jackson, J.E., 1952. The quantity of a degree in our English measure. Empire Survey Review. XI (86):359-362

113. Jones, T.B., 1967. The figure of the earth. Coronado Press

114. Jones, T.P., 1977. Oblate or prolate? The Ontario Land Surveyor Fall: 34-36

115. Juan, J. & Ulloa, A., 1747. Relación histórica del viaje a América meridional...Madrid. 4 Vols.

116. Juan, J. & Ulloa, A., 1748. Observaciones astronomicas y Physicas...Madrid. 2 Vols.

117. Juan, J. & Ulloa, A., 1752 Voyage historique de l'Amerique...Paris.

118. Juan, J., & Ulloa, A., 1758. Voyage to S.America. London. 2 Vols. Reprint 1972 Milford House, Boston, USA.

119. Juan, J., & Ulloa, A., 1767. A voyage to S.America. In "New collection of voyages" Vol. 1. by Knox.

120. Juan, J. & Ulloa, A., 1810. A voyage to S.America. In "Voyages and travels" Vol. 2. by Philips, R.

121. Juan, J. & Ulloa, A. 1813. A voyage to S.America. In "Voyages and travels" Vol. XIV :313-696. by Pinkerton, J. London.

122. Juan, J. & Ulloa, A., 1978. Discourse and political reflections of the Kingdom of Peru. Univ. Oklahoma Press. Ed. J.J. Telaske.

123. Kiely, E.R., 1953. Surveying instruments, their history and classroom use. Bur. of Pub. Teachers Coll. Columbia Univ. New York. Reprint by Carben.

124. King, H.C., 1955. The history of the telescope. Griffin, London

125 King-Hele, D.G., 1964. The shape of the earth. Jnl. Inst. Navigation. 17 (1)

126. Kirkpatrick, F.A., 1935. Noticias Secretas. Hispanic American Hist. Rev. 15: 492-3

127. Koestler, A., 1968. The sleepwalkers. Hutchinson.

128. Koyré, A., 1973. The astronomical revolution. Trans. R.E.W. Maddison. Methuen.

129. Kuhn, T.S., 1974. The Copernican revolution. Harvard Univ. Press Cambridge.

130. La Condamine, C.M.de., 1747. A succinct abridgement of a voyage made within the inland parts of S.America. London. This was translated from the original French.

131. La Condamine, C.M.de., 1746. Lettre sur l''emeute populaire de Cuença. Paris.

132. La Condamine, C.M.de., 1751. Journal du voyage fait par ordre du Roi ā 'l'equateur. Paris.

133. La Condamine, C.M.de., 1752. Supplement au Journal historique du voyage ā 'l'equateur. Paris.

134. La Condamine, C.M.de., 1751. Histoire des pyramides de Quito. Paris.

135. La Condamine, C.M.de., 1751. Mesures des trois premiers degrés du méridien. Paris.

136. La Condamine, C.M.de., 1755. On the curvature of the earth. Phil.Trans.Roy.Soc. XLIX :622

137. La Condamine, C.M.de., 1813. Interior of S.America. In 'Voyages and travels' Vol. XIV :211-269. by Pinkerton, J. London.

138. La Condamine, C.M.de., 1749. Biographical note. Abr.Phil.Trans.Roy.Soc. Vol.9:664

139. Lalande, M.de., 1761. Remarks on Norwood's measure. Phil.Trans.Roy.Soc.LVIII: 369

140. Lalande, M.de., 1789. Sur la mesure de la terre, que Fernel publia en 1528. Mem.Acad.Sc. Paris for 1787. :216-222

141. Leake, W.M., 1839. On the stade as a linear measure. Jnl.Roy.Geog.Soc. IX:1-25

142. Leinberg, Y. 1928. Maupertuis' in astemittanksen virheistä Maanmittans.

143. Levallois, J-J., 1969. Geodesie Generale. Edition Eyrolles. Vol. 1:78-99

144. Levallois, J-J., 1983. La determination du Rayon terrestre par Picard 1669-1671. Bulletin Géodésique (57):312-331

145. Lewis, N. & Reinhold, M., 1955. Roman civilisation. Source book II, The Empire. Harper.

146. Leybourn, W., 1653. The compleat surveyor. London.

147. Liesganig, J., 1768. Measure of the meridian. Phil.Trans.Roy.Soc. LVIII :15

148. Lindroth, S., 1952. Swedish men of science 1650-1950. The Swedish Institute, Stockholm.

149. Littlehayes, G.W., 1907. The recent scientific missions for the measurement of arcs of the meridian in Spitzbergen and Ecuador. B.Amer.G.S. 39:641-653

150. Lloyd, G.E.R., 1970. Early Greek Science. Thales to Aristotle. Chatto and Windus.

151. Lloyd, G.E.R., 1973. Greek Science after Aristotle. Chatto and Windus.

152. Lyons, H.G., 1927. Ancient surveying instruments. Geog.Jnl. 69:132-143

153. Maillet, P.L., 1960. P. Maupertuis pour le bicentenaire de sa mort. Paris Univ.

154. Mathieson, J., 1926. Geodesy. A brief historical sketch. Scot.Geog.Mag. 42:328-347

155. Maupertius, P.L., 1731. De Figuris quas Fluida... Phil.Trans.Roy.Soc. XXXVIII (422):

156. Maupertuis, P.L., 1732. Discours sur les différentes figures des astres.

157. Maupertuis, P.L., 1735. Sur la figure de la terre et sur les moyens que l'astronomie et la geographie fournissent pour la determiner. Mem.Acad.Sc. Paris for 1733:153-164

158. Maupertuis, P.L., Degré du meridien entre Paris et Amiens

159. Maupertuis, P.L., 1736. Sur les figures des corps celestes. Mem. Acad. Sc.Paris. for 1734,:55-109

160. Maupertuis, P.L., 1738. Examen desintéressé des différens ouvrages qui ont été faits pour determiner la figure de la terre.

161. Maupertuis, P.L., 1738. Examen des trois dissertations que M.Desaguliers a publiees sur la figure de la terre, dans les Trans.Phil.Nos.386, 387 and 388 Amsterdam.

162. Maupertuis, P.L., 1738. Examen figure de la terre. Mem.Acad.Sc.Paris for 1735.

163. Maupertuis, P.L., 1738. The Figure of the Earth. London. Trans. from the French.

164. Maupertuis, P.L., 1740. La figure de la terre determinee... Mem.Acad.Sc.Paris for 1738:90-96 and 389-466

165. Maupertuis, P.L., 1767. Travels of Maupertuis and associates. In 'New Collection of voyages, discoveries and travels'. Vol. 4

166. Maupertuis, P.L., 1767. Travels made in 1736. In 'The World displayed.' Vol. 20

167. Maupertuis, P.L., 1810. Journey to the arctic. In 'General collection of voyages.' Vol. 16. by Philips, R.

168. Maupertuis, P.L., 1732. Biographical note. Abr.Phil.Trans.Roy.Soc. Vol.7:519

169. McBride, B.St.C., 1965. La Condamine. Measurement at the Equator. History Today. 15 (8):567-575

170. McCaw, G.T., 1907. The progress of geodesy. Inst.Civ.Eng.Dublin

171. McCaw, G.T., 1932. Reform of the calendar. Empire Survey Review 1.(3):115-121

172. McCaw, G.T., 1932. Standards of length in question. Empire Survey Review. I(6):277-284

173. McCaw, G.T., 1933. Note on foreign measures. Empire Survey Review II(8):122-123

174. McCaw, G.T., 1935. Notes on survey apparatus and invention. Empire Survey Review. III(17):169-175

175. McCaw, G.T., 1939. The two metres-the story of an African foot. Empire Survey Review. V.(32):96-105

176. McCaw, G.T., 1939-42. Linear units old and new. Empire Survey Review. 5.(34):236-259. 6.(40):85-93 6.(44):339-340

177. McCaw, G.T., 1942. Latitude and the solstitial day. Empire Survey Review. 6(43):286-297

178. Melvill, E.H.V., 1977. Derivation of the word theodolite. S.African Survey Jnl. :8-10

179. Menut and Denomy. 1968. Du ciel et du monde. Oresme. Univ. of Wisconsin Press.

180. Menze, E.A., 1974. The Penguin Atlas of World History. Vol.1.

181. Merriman, M., 1881. The figure of the earth. New York.

182. Merriman, M., 1907. Elements of precise surveying and geodesy.

183. Mieli, A., 1939. La Science Arabe, et son rôle dans l'évolution Scientifique mondiale. E.J.Brill. Leiden.

184. Mouton, F.R., 1903. The shape of the earth. Geog.Jnl. 2:481-486 and 521-527

185. Musschenbrock, P.van., 1729. Physicae experimentalis et geometricae de magnete. Leiden.

186. Needham, J., 1954-1971. Science and civilisation in China. C.U.P. 4 Vols.

187. Neugebauer, O., 1951. The exact sciences in antiquity. Copenhagen. New edition Dover. 1969

188. Neugebauer, O., 1975. A history of ancient mathematical astronomy. Springer. 3 Vols.

189. Newton, I., 1687. Philosophiae Naturalis Principia Mathematica.

190. Norwood, R., 1637. The seaman's Practice, contayning a fundamentall Probleme in Navigation...

191. Norwood, R., 1676. Concerning quantity of a degree. Phil.Trans.Roy.Soc. XI (126):636

192. Ornstein, M., 1913. Role of the scientific societies in the seventeenth century. New York.

193. Outhier, R., 1744. Journal d'un voyage au nord en 1736-1737. Paris.

194. Outhier, R., 1808. Voyage to the North. Trans. of (193) in 'A general collection of the best and most interesting voyages and travels' London. Vol. 1. :259-336 by Pinkerton, J.

195. Papworth, K.M., 1936. Early measurements of arc in Iraq. Empire Survey Review. 3(20):334-336

196. Payn, H., 1914. The well of Eratosthenes. Observatory. 37:287-288

197. Pederson, O., 1974. A survey of the Almagest. Odense Univ.Press

198. Perrier, G., 1908. La figure de la terre. L'ancienne et la nouvelle mesure de l'arc meridien de Quito. Revue de Geographie. Tome 2:201-508

199. Perrier, G. 1911 Les Academiciens au Peru (1735-1744) Extract from Bull.de la Soc.Astro. de France. Mar/Apr.

200. Perrier, G., 1925. Arc de Meridien Equatorial en Amerique du Sud, 1899-1906. Gauthier-Villars, Paris. 14 Vols.

201. Perrier, G. 1929 La Mission Géodésique Française de l'Equateur, 1899-1906. Conf.Dec.1925.Prague.

202. Perrier, G., 1929. Les Academiciens au Peru 1735-1744. Conf. Dec.1925. Prague.

203. Perrier, G., 1939. Petit Histoire de la Géodésie. Alcan, Paris.

204. Peterson, I., 1984. Private communication.

205. Petrie, F., 1940. Linear units old and new. Empire Survey Review 5(36):373

206. Picard, J., 1671. La mesure de la terre. Paris.

207. Picard, J., 1675. Account of the measure of the earth's meridian Phil.Trans.Roy.Soc. IX (111):261

208. Pinkerton, J., 1813. Voyages and Travels. Vol. XIV contains abridged trans. of Voyage to Peru by Bouguer, P. see [25]

209. Pinkerton, J., 1813. Voyages and Travels. Vol. XIV contains trans. of A voyage to S.America, by Juan J. and Ulloa,A. See [121]

210. Pinkertòn, J., 1813. Voyages and Travels. Vol. XIV contains trans. of Interior of S.America by La Condamine, C.M.de., see [137]

211. Pinkerton, J., 1808. Voyages and Travels. Vol. I contains trans. of Voyage to the North, by Outhier, R., see [194]

212. Potter, A.J., 1932. The earliest geodetic triangulation. Empire Survey Review. I (3):100-109

213. Preston, E.D., 1889. The need of a remeasurement of the Peruvian arc. US Coast & Geodetic Survey Report App. 7 :199-208

214. Radice, B., 1971. Who's who in the ancient World. Penguin.

215. Raeder, H. et al., 1946. Trans. of 'Description of his instruments and scientific work' by Tycho Brahe, 1598.

216. Raisz, E., 1956. Mapping the World. Abelard-Schuman, London.

217. Raper, M., 1761. Remarks on Norwood's measure. Phil.Trans.Roy.Soc. LII :366

218. Rennell, J., 1791. On the rate of travel of camels. Phil.Trans.Roy.Soc. LXXXI :129

219. Roberts, P.H., 1964. The figures of the earth and other rotating bodies. Newcastle-upon-Tyne Univ.

220. Robson, I., 1821. A treatise on geodetic operations. Durham.

221. Rodriguez, D.J., 1812. Figure of the earth. Phil.Trans.Roy.Soc.

222. Sarton, G., 1914. Le monument des missions géodésiques françaises á Quito. ISIS 2:163-164

223. Sarton, G., 1927. Introduction to the history of science. Williams & Wilkins, Baltimore. 3 Vols.

224. Sarton, G., 1935. The first explanation of decimal fractions and measures. ISIS 23:153-244

225. Sarton, G., 1954. Ancient science and modern civilisation. Univ. of Nebraska Press

226. Sarton, G., 1959. A history of science. Ancient science through the Golden Age of Greece. Harvard Univ. Press

227. Sarton, G., 1959. A history of science. Hellenistic science and culture in the last three centuries B.C.. Harvard Univ. Press.

228. Schott, C.A., 1898. Inquiry into the relative value and need to check the Peruvian arc of 1736-1743. US Coast and Geodetic Survey Report. App. 4 :231-232

229. Schott, C.A., 1900. The transcontinental triangulation and the American arc of the parallel. US Coast and Geodetic Survey. Sp.Pub.4 :869-871

230. Seidenberg, A., 1962-6. The sixty system of Sumer. Arc. for Hist. of Exact Sc. 2:436-440

231. Singer, C., 1941. A short history of science to the nineteenth century. Oxford.

232. Smith, D.E., 1923-5. History of mathematics. Ginn & Co. Boston.

233. Smith, J.R., 1969. The development of two standards. Survey Review. XX (153):133-146

234. Snellio, W., 1617. Eratosthenes Batavus de Terrae ambitus vera quantitate...Leyden.

235. Stirling, J. 1735. Of the figure of the earth and the variation of gravity on the surface. Phil.Trans.Roy.Soc. XXXIX (438):98-105

236. Stones, E., 1766. Some reflexions on the uncertainty of many astronomical and geographical positions.

237. Strasser, G., 1957. Ellipsoidische parameter der Erdfigur 1800-1950. Munchen. Deutsches Geodetische Kommission.

238. Strasser, G., 1974. The toise, the yard and the metre. 17th Australian Survey Congress. Melbourne.

239. Struve, F.G.W., 1860. Arc du Méridien de 25°20' entre le Danube et la Mer Glaciale mesuré depuis 1816 jusqu'en 1855. Band 1. St. Petersbourg. LXXIII

240. Svanberg, J., 1805. Exposition des Opérations faites en Lapponie...Stockholm.

241. Svanberg, J., 1805. Determination d'un arc du meridien.

242. Taylor, E.G.R., 1930. Tudor geography. 1485-1583. Methuen.

243. Taylor, E.G.R., 1954. Mathematical practitioners of Tudor and Stuart England. C.U.P.

244. Taylor, E.G.R., 1966. Mathematical practitioners of Hanoverian England, C.U.P.

245. Taylor, E.G.R., 1971. The haven finding art. Hollis and Carter.

246. Taylor, E.W., 1942. Standards of measure. Empire Survey Review. 7(48):50-56

247. Taylor, E.W., 1943. The art of original circular dividing. 1650-1850. Empire Survey Review. 7(49):99-107

248. Taylor, E.W., 1944. The evolution of the dividing engine. Empire Survey Review. 7(52):226-234

249. Thomson, J.O., 1948. History of ancient geography. Cambridge.

250. Thureau-Dangin, F., 1939. Sketch of a history of the sexagesimal system. Osiris. 7:95-141.

251. Tobé, E., 1984-5. Private communications.

252. Todhunter, I., 1873. A history of the mathematical theories of attraction and the figure of the earth. Reprint 1962 Dover.

253. Todhunter, I., On the arc of the meridian measured in Lapland. Trans.Camb.Phil.Soc. XII

254. Toomer, G.J., 1984. Ptolemy's Almagest. Duckworth.

255. Townley, R., On Mr. Gascoigne's micrometer. Phil.Trans.Roy.Soc. (25):457

256. Tweedie, C., 1922. James Stirling. Clarendon Press, Oxford.

257. Wallis, J. & Hevelius, M., Extract of a letter ...concerning divisions by diagonals. Phil.Trans.Roy.Soc. (111):243

258. Walters, R.C.S., 1922. Greek and Roman engineering instruments. Trans.Newcomen Soc. 2:45

259. Waugh, A.E., 1973. Sundials. Their theory and construction. Dover.

260. Werner, A.P.H., 1966-1968. A calendar of the development of surveying. Various issues of Australian Surveyor.

261. Werner, A.P.H., 1984. The philosophy of progress. Any lessons from the history of surveying? Australian Surveyor. 32(3):167-185

262. Whitaker, A.P., 1935. Antonio de Ulloa. Hispanic Amer.Hist.Rev. 15:155-194

263. Whitaker, A.P., 1938. Jorge Juan and Antonio Ulloa's prologue to their secret report of 1749 on Peru. Hispanic Amer.Hist.Rev. 18:507-513

264. Wilford, J.N., 1981. The mapmakers. New York.

265. Wingate, E., 1626. The construction and use of the logarithmeticall tables.

266. Wolf., A., 1959. History of science...in the 16th and 17th centuries. Harper Torchbook. New York. 2 Vols.

267. Wolf, A., 1961. History of science...in the 18th century. Harper Torchbook, New York. 2 Vols.

268. Wolf, C., 1902. Histoire de l'Observatoire de Paris. Paris.

269. Zach, F.X.von., 1811. Memoire sur le degre du meridien.

INDEX

Aavasaksa 172, 174, 175, 176, 181, 191
Abacus 14, 52
Acragas 2
Academy of Sciences, Paris 41, 69, 70, 71, 79, 80, 88, 93, 94, 95,
    97, 100, 101, 152, 168, 170, 171, 177
Accademia dei Lincei 70
Accademia del Cimento 70
Agrimensores 4, 42, 43
Ailparoupachca 113
Akkad 1
Al-Asturlâbî 19, 21, 22, 194
Al-Battâni 21, 28, 194
Al-Bîrûni 194
Al-Bitrûji 22, 194
Al-Buhtari 19, 194
Al Daula 21
Alexandria 10, 11, 12, 13, 21
Al Falaki 19, 20, 194
Al-Farghâni 19, 21, 194, 195
Al-Fazari 21, 194, 195
Algebra 21
Al-Hasib 21, 28, 194, 195
Al Jauhari 195
Alkmaar 39, 40, 63, 64
Al-Khowârizmi 19, 21, 22, 28, 194, 195
Al Kindi 19, 20, 194, 195
Al Kufah 19
Al Kûsgi 194, 195
Almagest 8, 14
Al-Mamun 19, 21, 195
Al-Mansûr 21, 195
Al-Marrakushi 22, 23, 28, 195
Al-Marwarrudhi 19, 21, 22, 195
Al-Muzaffar 22, 195
Al-Nairîzi 21, 22, 195
Al-Rahman 195
Al-Rashîd 19, 21, 195
Al-Sûfi 21, 22, 195
Al-Tûsi 22, 23, 28, 195
Al-Urdi 22, 23, 195
Al-Wafa 22, 28, 195
Amazon, River 161
Amenemhat I 4
Amiens 71, 79
Amoula 112, 113
Amphipolis 13
Anaxagoras 1, 2, 195, 196
Anaximandros 1, 4, 5, 8, 195, 196
Anaximenes 1, 4, 5, 195, 196
Angles, Brahe 57, 63
  Peru scheme Table 122-135
  Picard 74
  Reduction 120
  Rcsolution 17
  Snellius 64, 65, 66
An To 84
Apamea 2, 3
Apollonia 13
Archimedes 5, 6, 195, 196

Archytas 5, 196
Arc Measure, A-Mamun 19, 210
  Archimedes 5, 6, 210
  Aristotle 5, 6, 210
  Blaeu 40, 63, 210
  Brahe 63, 210
  Eratosthenes 10, 210
  Fernel 38, 39, 67, 210
  I-Hsing 16, 210
  Maupertuis 184, 210
  Musschenbroek 66, 210
  Norwood 40, 68, 80, 82, 210
  Peru 143, 210
  Peru recomputation 143
  Peru.Godin & Juan 145, 210
  Picard 40, 41, 63, 64, 71, 210
  Poseidonius 12, 210
  Ptolemy 12, 210
  Riccioli 40, 68, 210
  Svanberg 187, 189, 210
  Thomas 84, 210
Aristarchos 2, 3, 5, 6, 8, 40, 196
Aristotle 2, 3, 5, 196
Armillary sphere 22, 23, 38, 39
Armilles solsticiale 8
Aryabhata 3, 14, 15, 27, 28, 196
Assuan 10, 11, 13
Astrolabe 5, 14, 21, 22, 23, 38, 39, 42, 43, 45
Athens, Lyceum. 5
Athens, School. 5
Atlas 3
Autolycus 2, 3, 5, 6, 196
Auzout, Adrien. 34, 35, 40, 41, 196

Babylon 1
Baculum 22, 23
Baculus Jacob 43
Baghdad 19, 21, 22
  Observatory 19, 21
Baikal, Lake 16, 17
Bannos 110
Barometer 40, 42, 43, 52, 53, 105, 106, 152, 153, 183
  Aneroid 87, 88
Base line 38, 39, 57, 64, 65, 66
  Cuença 145
  La Caille 80
  Lapland 175, 177
  Picard 71, 80
  Tarqui 113, 121, 136, 147, 151, 152
  Yarouqui 106, 108, 118, 147, 151, 152
Bede, The venerable. 3, 21, 196
Bematistes 11, 12
Benoit 55, 56
Bergen op Zoom 63, 64
Berlin Academy 70, 88, 97, 100, 101, 168, 170, 171
Bessel 143
Bhâskara 22, 28, 196
Bignon, Jean Paul 70
Blaeu, Wilhelm 40, 63, 64, 65, 66, 196, 197

Bologna, Academy   100, 101
Borda   91, 187
Borja   161
Borma   113
Boscovitch   55, 56
Bothnia, Gulf of   171, 172, 174
Boueran   110, 113
Bouguer, Pierre   35, 44, 45, 53, 83, 88, 90, 91, 94, 95, 189, 190
Bradley   51, 52, 93, 94, 184, 186, 187, 189, 190, 191, 192
Brahe, Tycho   30, 34, 38, 39, 40, 43, 44, 45, 57, 172, 174, 196, 197
Brahestad   172, 174
Brahmagupta   14, 15
Brasse of Florence   76, 77
Briggs, Henry   36, 37, 39, 40, 196, 197
Brisson, Mathurin   89, 197
Brouncker, Lord   69
Brunnius   175
Bürgi, Jobst   35, 36, 37, 39, 40, 197
Byzantium   10, 11, 13
Cadiz   102, 103
   Observatory   100, 101
Caesar, Julius   28
Cahouapata   113
Calculating machines   40, 41, 52, 87, 88
Calcagnini, Celio   197
Calculus   5, 21
Calendar   28
   Julian   38, 39
   Reform of   38, 39, 70
Callao   163
Callimachus   10
Callipos   2, 197
Camel journey time   11, 12
Campanario   111, 112
Camus, Charles   76, 77, 88, 170, 171, 197
Canones gnomonis   38
Canopus   12
Carabourou   53, 54, 105, 106, 108, 111, 118, 152, 153
Carpos   14, 15
Carriage wheel   40, 67
Carthage   10, 11
Carthagena   101, 102, 103
Cassinis, The   3
Cassini I, G.D.   70, 79, 80, 82, 87, 92, 93, 197
Cassini II J.   79, 88, 90, 92, 93, 94, 153, 197
Cassini III C-F,   77, 79, 89, 94, 197
Cassini IV J.D.   79, 197
Cataclasis   13, 14
Cayamburo   110, 111
Cayenne   54, 55, 70, 80, 82, 83, 87, 92, 93, 162, 164, 165
Celsius, Anders   53, 88, 89, 94, 170, 198
Centre, Reduction to   57, 63
Chain,Gunter   40
   Surveyors   64, 65, 66, 68
Chalapu   112, 113
Chaldeans   28
Changalli   111, 112
Chang Heng   14, 15, 198
Chanzler, Richard   29, 30, 38, 39, 198
Characteristic   36, 37, 39, 40
Châtelet, Marquise de   94. 168, 170
Chazelles   79, 80
Chhiu Chhang Chhun   22, 23, 198

Chiao Chou   16, 17
Chimborazo, Mount   88, 95, 97, 105, 153
Chinan   113
Chinchulagua   110, 111, 112
Chitchitchoco   112, 113
Chord   6, 14, 22, 23, 28, 29
   Table of   13, 14, 27
Chorobates   14, 15
Choujai   113
Chronhelm, Comte de   171, 172
Circle, Chinese   14, 15
   Division of   8, 11, 13, 21,
Circle. Indian   14, 15
   Sexagesimal   14, 15
Circumferentor   45, 68
Clairaut, Alexis-Claude   76, 77, 88, 89, 168, 170, 198
Clavius, Christoph   29, 30, 32, 33, 34, 38, 39, 198
Clazomenae   1, 2
Cleomedes   11, 12, 13, 14, 198
Cnidos   2
   Observatory   5
Coimbra   30
Coimbre   84
Colbert, Jean   69, 70, 79
Collegium Naturae Curiosorum   70
Collioure   79, 80
Columbus, Christopher   92, 103
Conception   163
Coordinates. Brahe   63
   Polar   87, 88
   Rectangular   87, 88
   Snellius   66
Copernicus, Nicholaus   3, 14, 28, 29, 34, 38, 43, 198
Corazon   112
Cord, Knotted   42, 43
Cosin   111, 112
Cosine   22, 23
Cosmas Indicopleustes   3, 198
Cotangents   21, 22
   tables   22, 23, 28, 29
Cotapaxi   110, 112
Cotchesqui   51, 52, 111
   Astronomical observations   142, 143
Counting systems   4
Couplet Jnr.   54, 55, 101, 102, 105, 106, 165,198
Couplet Snr.   79, 80
Crates   13, 14, 198
Cross hairs   34, 40, 41
Cross staff   42, 43
Crotona   4, 5
Cubit   19, 20
Cuença   110, 113, 114, 145, 146, 152, 159, 161
Cuicocha   112
Cuitaperi   176, 177
Curtius, Johannes   30, 198
Cusa, Nicolaus de   3, 198
Cyrene   10, 13
Cyzicos   2

Damascus   19
   Observatory   19
Danzig Observatory   40
Dargatu   28
Davis, John   43, 44

Day, Division of 4
Decimal notation 14, 15, 28, 29, 38, 39, 40
  system 4
Degree, length of 18, 19, 20
  of longitude 79, 80
  value. Al-Mamun 19, 20
    Bouguer 83
    Cassini 102, 103
    Eratosthenes 11, 12
    Fernel 67
    I-Hsing 18, 19
    Lapland 184
    Norwood 68
    Peru 143, 148
    Picard 74
    Poseidonius 12
    Ptolemy 12
    Riccioli 68
    Snellius 66
    Svanberg 189
    Thomas 85
Delamain, Richard 40, 52
Delambre 8, 63, 64, 80, 90, 91, 94, 143, 162, 163, 187, 189, 190
De Luc, Jean André 53, 198
De Mairan 55, 186, 187
Democritus 5, 198
Desaguliers, J.T. 94
Descartes, René 40, 69, 70, 199
Deserts 109, 110
Des Hayes 54, 55, 80, 82
Diabetes 4, 5, 14, 15
Diagonal scale 22, 23
Dicaiarchos 2, 199
Digges, Leonard 39, 45, 199
  Thomas 39, 199
Dioptra 5, 6, 13, 14, 21, 22, 42, 43, 45
Dividing engine 34
Divini, Eustachio 35, 87
Dolomboc 113
Dunkirk 79, 80, 171, 172

Earth, Centre of 1
    Circular Island 1
    Circumference 11, 12, 76, 77
    Convex disc 1
    Diameter 22, 39, 40, 76, 77
    Egg shaped 1, 2, 3, 93
    Flat 1, 2, 3
    Flat disc 1
    Habitable 2, 3, 10
    Non-symmetrical 3
    Oblate 55, 94
    Prolate 3, 79, 80, 87
    Radius 4, 5, 10, 83, 143
    Rectangle 1
    Rotation 1, 2
    Short cylinder 1
    Size of 5, 6, 10, 11, 38, 39, 143
    Sphere 1, 2, 3, 93
    Spheroidal 86
    Square 3
    Tabernacle 3
Eccentricity 91, 143
Eclipse 14

Ecliptic, Obliquity of 1, 4, 5, 8, 12, 14, 15, 17, 18, 28, 45, 143, 185, 186
Ecphantos 2
Eisenschmid 93
Elam 1
El Coraçon 112, 152, 153
Elephantine 11, 12
Ell 28, 145, 172
Empedocles 1, 2, 199
Equinox 8
Eratosthenes 2, 3, 8, 10, 12, 13, 45, 64, 199
Esmeraldas, River 105
Euclid 14
Eudoxus 2, 5, 199
Expedition, Lapland 42, 43, 52, 53, 80, 94
    Peru 51, 52, 53, 55, 80, 94, 95
Fahrenheit, Daniel 53, 199
Fatamid observatory 21, 22
Faustus, Lucius 42, 43
Ferdinand II, Grand Duke 52, 53, 70
Fermat 69, 70
Fernel, Jean 38, 39, 67, 199
Ferro 14
Finger reckoning 21
Flamsteed, John 34, 87, 88, 199
Fontenelle, Bernard 101
Foot 14, 15
    Bologna 76, 77
    Chinese 15, 16, 18, 85
    French 67, 68
    Geodetic 90
    Greek 10, 11
    London 76, 77, 102, 103
    New Chinese 85, 86
    Paris 76, 77, 102, 103
    Rhynland 63, 64, 76, 77
    Royal 102, 103
    Universal 90
Fortunatae Insulae 14
Fractions. Decimal 28, 29, 39, 40
    Sexagesimal 13, 14, 15, 28, 29
    Unit 13, 14
Frisius, Gemma 28, 29, 38, 39, 40, 43, 57, 199
Gabala, Bishop of 3
Galilei, Galileo 3, 30, 39, 40, 52, 53, 54, 200
Gascoigne, William 34, 35, 40, 200
Gavle 172, 175, 176
Geminus 5, 6, 13, 14
Geocentric system. 2, 3
Geodesy 5, 6, 21, 22, 38, 39
Geoheliocentric system 5
Geometric square 38
Geometry, Analytical 40
    Cartesian 40
    Science 4, 5
Gerbillon J.F 85
Gershon, Levi Ben 22, 23, 29, 30, 43, 200
Gesh 4, 28
Girard, Albert 40
Globe. Celestial 13, 170
    Terrestrial 13, 14, 38, 39
Gnarp 172, 175, 176
Gnomon 4, 5, 8, 15, 16, 17, 18, 21, 85, 86

213

Goamani 111, 112
Goapoulo 111, 112
Godin. des Odonnais 101, 102, 106, 108, 164, 200
   Louis 55, 56, 88, 89, 94, 95
    Madame 164
Gotenburg 171, 172
Graham, George 34, 44, 45, 51, 52, 88, 101, 102, 170, 183, 184,
     186, 187, 200
Grand Châtelet 55
Granvik 175, 176
Gravitation 88
Gravity, Change of 95, 97
Greenwich Observatory 44, 45, 87, 88
Grid, Rectangular 14
Grimaldi, Francesco 40, 68
Groma 42, 43
Guanacauri 113, 114, 145
Guayama 113
Guayaquil 103, 104, 105
Gunter, Edmund 36, 37, 40, 52, 200

Halley, Edmund 54, 55, 87, 88, 200
Hansteen 191, 192
Härnösand 172
Harpendonaptae 5
Hayford 143
Hecataeus 1, 200
Heliocentric system 2, 3, 5
Heliometer 35, 95
Hellant, Anders 170, 171, 174
Helsingör 171, 172, 176
Hemisphaerium 8
Hemisphere, Celestial 14
Henry II of France 38, 39
Heracleides 2, 5, 200
Herbelot, d' 171
Hero 21, 22
Herodotus 5, 200
Heron 13, 14, 21, 22, 43, 45, 200
Hevelius, Johann 34, 40, 201
Hicetas 112
Hipparchus 11, 12, 13, 14, 28, 43, 201
Hivicatsou 112, 113
Holmquist 187, 189
Homer 1, 201
Hommel, Johann 30, 38, 39
Homocentric system 2
Hooke, Robert 35, 40, 45, 51, 69, 70, 87, 201
Horace 5
Horrilankero 175, 176, 177
Hsiang-Chou 16, 17
Hsü-Chou 16, 17
Hsü Yueh 14, 15
Hua-Chou 16, 17
Huai-nan-tzu 13, 14
Hudiksvall 172
Hué 16, 17
Hugot 88, 89, 101, 102, 151, 152, 165, 201
Huitaperi 174, 176, 177, 180, 181
Huygens, Christiaan 35, 53, 54, 55, 87, 90, 92, 93, 186, 187, 201
Hyginus Gromaticus 14, 15, 201
Hypsicles 13, 28, 201
I-Hsing 16, 22, 23, 201
   Arc measure 16

Ilimissa 110, 111, 139, 141
Ilmal 113
Inca, Ile de l' 139, 141
Inch, Chinese 16, 18
   Paris 90
Indal 172
Instrument Topographical 39, 45
Ionian map 10
Ionic school 1, 2, 4, 5

Jacob staff 5, 6, 22, 23, 38, 39
Jandi-Shapur observatory 21, 22
Jansen, Zacharias 39, 40
Jesuit College 88, 89, 108, 109, 157, 159
   Order 38, 39, 68, 84, 85, 86, 87, 88, 90, 110, 164
Juan, Jorg 88, 89, 97, 100
Junis, Ibn 201, 202
Jupiter satellites 54, 55, 79, 80, 92, 93, 162
Jussieu, Joseph de 88, 89, 101, 102, 165

Kaakamavaara 174, 175, 176, 181
Kardaja 21
Karungi 174
Ketima 176, 177
Kittis, Mount 174, 183, 189, 190
Kittisvaara 176, 177, 181, 183, 184
Konigsberg, Johannes. See Regiomontanus
Korpikylä 174
Kota-pacsi 112
Kukas 174
Kukkola 174
Kuo shou-ching 18

La Caille 77, 80, 94, 153, 190
La Condamine, Charles de 53, 55, 77, 88, 89, 90, 91, 93, 95, 97,
    189, 190
Lactantius 3, 201, 202
La Hire, Philippe de 33, 34, 79, 87, 88, 201, 202
Lalande 67, 68
Lalangouco 113
Lang-chou 16, 17
Langlois, Claude 45, 55, 88, 101, 102, 174, 177, 179, 201, 202
Lapland scheme. Angular observations 177, 179
   arc length, angular 183, 184
     linear 181
      Svanberg 187, 189
   astronomy, new 187
     Pello 183
     Tornea 183
   base line measure 180
     site 175
   chronology 191, 192
   conclusions 191
   degree length 184
   general details 172, 174
   heights 185
   instrument verification 185
   journey, outward 171
     return 175, 176
   orientation 181
   pendulum experiments 186
   personalities 168
   preparation 171
   reconnaissance 172, 174

refraction   185, 186
Remeasure, Svanberg   187
   length   191
   angles   190, 191
Repeating circle   190
survey stations   176
La Rochelle   54, 55, 101, 102
Latitude   8, 54, 55
League, Marine   76, 77
  Mean French   76, 77, 94, 172
  Paris   76, 77
Leibniz, Gottfried   41, 52, 53, 70, 87, 88, 202
Leiden   63, 64
Leinberg.Y.   191, 192
Le Monnier, Pierre   76, 77, 89, 170
Lenses, combined   39, 40
Leopold, Prince   70
Le Roi, Julien   183
Levelling, barometric   53
Level of Heron   13, 14
  Plumb bob   22, 23
  spirit   87, 88
  water   40
Leybourn, William   30, 32, 36, 37, 45, 87, 202
Libella   4, 5
Limal   112, 113
Limpie-Pongo   112
Lin-I   16, 17
Linné   53, 170
Lippershey, Hans   39, 40, 202
Liu Chhuo   16
Liu Hung   14, 15
Logarithms   5, 6, 35, 36, 39, 40, 105, 106, 152, 153
  Naperian   36
  Natural   36
  Sine tables   36, 37
  Tangent tables   36, 37
  Trigonometric functions   40
Longitude   102, 103
Los Bannos   113, 114, 145
Louisburg   100, 101, 163
Louisiana   164
Louis XIV   69, 70, 79
  XV   70, 94, 170, 171
  XVI   90, 91
Louvain   38, 39
Louville de. Tables   185, 186
Louvois, Marquis de   79
Loxodrome   40
Loyola, Inigo López de   38, 39
Lu Chi   14, 15
Luleá   172
Lythenius Dr   171, 172

Mabillon   101, 102, 165, 202
Magdalena, River   103
Magellan   92
Magini, Giovanni   29, 38, 39, 40, 202
Magnetic needle   22, 23, 45
Mairan   186, 187
Maldonado, Pedro   88, 89, 105, 161, 162, 163, 165, 202
Mallet, A.M.   87, 88
Mallörn   187, 189

Malvasia   35
Malvoisin   71, 79
Mantissa   36, 37, 39, 40
Map, Celestial   14, 15
Maragha Observatory   22
Maraldi   79, 80, 153
Marañon river   161
Marinus   14, 202
Marseilles   13, 43
Martinique   101, 102, 103
Massilia   13
Mathematics, Greek   13, 14
Maurepas, Comte de   101, 102, 171, 176
Maupertuis, Pierre   55, 56, 76, 77, 80, 83, 88, 92, 93, 94, 152, 168
Maurus, Hrabamus   3, 202, 203
Mechain   90, 91, 187, 189, 190
Melanderhielm   187
Meldecreutz   172
Menelaus   14, 202, 203
Mercator, Nicolaus   37
Merkhet   43
Meroé   10, 11
Mersenne, Pere Marin   69, 70, 202, 203
Messina   2
Metius, Jacob   39, 40, 202, 203
Metre   40, 41, 89, 90, 91, 187
  des Archives   91
  French Legal   64
  Legal   91
Meyer, Tobias   190
Micrometer   34, 35, 40, 41, 51, 52, 80, 88, 177, 179, 183, 184
  Eyepiece   34, 35
Middleberg   39, 40
Milan   112
Mile, Arabian   19, 20
  Black   19
  English   10, 11, 63, 76, 77, 94
  Florentine   76, 77
  Geographical   102, 103
  German   63
  Italian   67, 68
  Swedish   172
Miletus   1, 8
Milliar   40, 41, 90
Minerals   110, 111
Mira   111, 112, 145
Montecristi   103, 104, 105
Morainville de   88, 89, 101, 102, 106, 108, 165, 202, 203
Montfort, Hubert de   69, 70
Morin, Jean   87, 204
Morland, Samuel   52, 53
Mosul   19, 20
Moulmoul   112, 113
Mouton, Gabriel   40, 41, 54, 90, 203
Musschenbroek, Petrus van   64, 66, 88, 203

Nabouco   112, 113
Napier, bones   35, 36, 39, 40
  John   29, 35, 36, 37, 39, 40, 203
Naples museum   2, 3
Namurelte   113, 114
Nankung Yüeh   16, 17
Newton, Isaac   3, 5, 6, 40, 41, 55, 79, 80, 82, 87, 88, 93, 94, 168, 186, 187, 203

Nicaea 13
Niemisby 177
Niemisvaara 174, 176, 177, 181, 183
Nishapur observatory 21, 22
Nivavaara 174, 175, 176, 181
Nolet, Abbé 186
Nonius 30, 38, 39
Norwood, Richard 40, 68, 203
Nunez, Pedro 30, 38, 39, 203
Nurnberg observatory 38

Obliquity, see ecliptic
Octant 79, 87, 88
Odonnais J.G. 88, 89
Oenipedes 1, 4, 5, 8, 203
Ofverbom 187, 189
Oikoumene 10
Olbia 10, 11
Oldenberg, Henry 69
Onska 172
Orientation, Peru scheme 136, 137
Orrery 5, 6
Otho, Valentin 28, 29, 39, 203
Ouangotassin 112, 113
Ouaoua Tarqui 113
Oughtred, William 40, 52, 203, 204
Outhier, Réginald 80, 88, 170
Över-Torneå 174, 175, 191
Oyambaro 111, 118

Pace, Bologna 68
Pahtavaara 187, 189
Palander 187, 189
Palmar, Monument at 104, 105, 159
Palmyra 19
Pambamarca 108, 109, 111, 152, 153, 159
Panama, Isthmus 103, 104
Panecillo mount 153
Pan Ku 14
Papa-Ourcou 112, 139
Para 153, 162
Parallel, degree of 79, 80
Paramos 109, 110
Paris observatory 79, 80, 87, 88, 102, 103, 186, 187
Parmenides 1, 2, 203, 204
Pascal, Blaise 40, 41, 52, 53, 69, 70, 203, 204
Pas de Bologna 76, 77
    Geométrique 85
    Old Roman 85, 86
    Universalis 71, 74
Paterno, Nicholosi di 22
Paulisa Siddhantra 14, 15
Pelazzi 38, 39
Pello 174, 175, 176, 177, 183, 186, 187
Pellos, Francesco 38, 39
Pendulum 40, 41, 42, 43, 54, 87
    bimetallic 55
    compensating 88
    seconds 71, 74, 83, 87, 88, 90, 91, 92, 95, 97, 152, 153, 162, 174, 175, 176, 183, 186, 187
Peregrinus, Petrus 22, 23, 203, 204
Perpignan 80

Perrault, Charles 69, 70
Peru scheme, angle reduction 120
    angles. Table after 120
    arc length, Bouguer 138, 143
        La Condamine 137, 138, 143
    arc of Godin & Juan 145
    astronomical observations 142
        Cotchesqui 142, 143
        Tarqui 142
    Base line Cuenca 145
        reduction 145
    Tarqui 121, 136
        length 136
        reduction 136
    Yarouqui 118
        length 118
        height reduction 120
        measure 106, 108
        reconnaissance 106, 108
        reduction 118
        site 105, 106
    commemoration 157, 159
        bronze scale 159
        column 159
        monument 159
    chronology 165, 166
    coordinates, Bouguer 138, 139
        La Condamine 137, 138
    equipment 148, 151
    extra work of Godin & Juan 145
        general details 108
    instrument verification 108
    journey, outward 101, 102
        return 161
    mountains & paramos 110
    orientation 136, 137
    personalities 95
    plumbline deflection 153
    possible errors 153, 157
    preparation 101
    pyramid inscription 159
    recomputation 1984 146, 147
        arc length 148
        results 147
        Bouguer 148
        Juan 148
        La Condamine 147, 148
        Ulloa 148
    results Juan & Godin 146
        Ulloa 145, 146
    sea level link 139
    Seniergues death 159, 161
    station heights 141
    survey stations 111
Peurbach, Georg 28, 29, 30, 38, 204
Philae 10, 11, 13
Phillip V, King 101, 102
Philolaos 1, 2, 5, 204
Philopator 10
Philosophical Transactions 69
Photometry 95, 97
Pi 4, 14, 15, 21, 22

Picard, Jean   34, 35, 40, 41, 54, 55, 63, 64, 68, 70, 71, 74, 79, 80,
    82, 87, 88, 90, 93, 94, 204
Pico Tenerife   102, 103
Pied du Roi   55, 56
Pien-Chou   16, 17
Pillatchiquir   113
Pinnules   39
Pipping   174, 176
Pitane   2, 3
Pitchincha   53, 54, 105, 106, 108, 109, 111, 139, 153, 162,
Piteå   172, 175, 176
Plain table   45, 55, 174
Plato   1, 2, 5, 10, 204
    Philosophical school   5
Plethora   10, 11
Plumb line deviation   88, 101, 146, 147, 153
Poiky-Torneå   177
Pole, English   68
Polimetrum   45
Porchester Castle   163
Porto Bello   103, 104
Poseidonius   2, 3, 12, 13, 14, 204
Pougin   113
Precession of equinoxes   13
Pressure, atmospheric   92, 95
Prince   180, 186
Projection, stereographic   13, 22
Prosthaphaeresis   39, 40
Ptolemy   3, 8, 12, 13, 14, 21, 22, 27, 28, 34, 38, 43, 45, 204
    Almagest   8
    Atlas   3, 14
    III   10
    IV   10
Pu   16, 18, 19, 204
Pueblo Viejo   145, 146, 152
Pugin   159
Pullinki   174, 176, 177, 181, 183
Pyramids, Egypt   10
    Peru   157, 159
Pythagoras   1, 2, 3, 204
    Theorem   4, 5
    School   4, 5

Quadrant   14, 19, 20, 21, 22, 33, 34, 38, 39, 40, 41, 42, 43, 45,
    64, 65, 66, 71, 74, 79, 80, 85, 86, 88, 95, 97, 148,
    151, 171, 172, 174, 176, 183
    Davis   44, 45
    Verification   108
Quadratum Geometricum   38
Quantities, known   39
    unknown   39
Quinoa Loma   113
Quinquina   161
Quito   103, 105, 152, 153, 164

Radius astronomicus   71, 74
Rameses   4
Ramsden, Jesse   34
Raqqah   19
Réaumur, René   53, 183, 186, 204
Refraction, astronomical   39, 40, 87, 104, 105
    atmospheric   12, 13, 14, 77, 88, 120, 146, 147, 184, 185, 186
    tables   57
Regiomontanus   3, 28, 29, 30, 38, 45, 204

Repeating circle   51, 52, 187, 190
Resection   40
Rheticus, Georg   28, 29, 39, 204, 205
Rhodes   10, 11, 12
Riccioli, Giovanni   3, 40, 68, 90, 94, 204, 205
Richer, Jean   54, 70, 80, 82, 83, 87, 92, 162, 163, 204, 205
Riobamba   110, 113, 164
Robert of Chester   22, 205
    the Englishman   22, 23, 205
Robertval, Gilles   40, 205
Roemer, Olaus   34
Rome   10, 11, 13, 14
Rood, Rhynland   64
Rope stretchers   4, 5
Royal Society, The   34, 35, 69, 87, 88, 89, 93, 97, 100, 101,
    164, 168, 170, 171
Rubber   105, 162

Sacha Tian Loma   113
Saint Domingo   101, 102
Saint Joachim   162
Saint Malo   79, 80
Saint Pablo   162
Samos   2, 3
Samarqand   22, 23, 28, 29
    observatory   22, 23
Sanad bin àli   19, 205
Sanquay   110
Schangailli   111, 112
Scaphé   8, 11
Selander   191, 192
Seleucos   2, 3
Senegoalap   113
Seniergues, Jean   88, 89, 101, 102, 159, 161, 165, 205
Series, expansion   37
    infinite   87, 88
Sesgum   113
Sextant   22, 23, 39, 68, 183
    Fakhri   22, 23
    stone   19
Shadow, end of   8, 9
    noon   8, 9
    reckoning   22
    square   21, 22, 23
    variation in length   8, 17
Shamila   22, 23
Sicily   10, 11
Sight, Open   87
    telescopic   40, 41, 87
Sighting tube   13, 14
Sinacaouan   113
Sinasaguan   109, 110
Sine function   13, 14, 15, 21, 22, 23, 27, 28, 29,
    table of   14, 15, 22, 23, 27, 28, 29
    theorem   22, 23
    versed, table of   14, 15, 22, 23
Sinjar, Plain of   19
Sisa-Pongo   113
Sissons, Jonathan   45
Skaphion   8
Skellefteå   172
Skiotheron   8, 11
Slide rule   40, 52
    logarithmic   40

Snellius, Willibrod   40, 57, 71, 88, 94, 205
Snell's law   40
Söderhamn   172
Solstice   5, 6, 8, 11, 13, 17, 18, 22, 23, 45
Sommereux, de   171, 172
Sound, speed of   153, 162, 163
Sourdon   74
Speidell, John   36, 37, 40, 205
Sphere, celestial   14, 15
Spherical excess   40
Square   4, 5, 42, 43
  root   4, 14, 15, 26, 27
Stade   10, 11, 12, 15
  Attic   10, 11, 12, 205
  Chinese   85, 86
  Egyptian   10, 11, 12, 205
  Short   10, 11, 12, 205
Stadia hairs   40, 41
Stars, daily rotation   1, 2, 5
  transit of   14
Stevin, Simon   29, 39, 40
Stick, shadow of   8
Stirling   93
Stockholm   171, 172, 175, 176
  Academy   100, 101, 170, 171, 187
Stoic   2, 3
  School   12
Strabo   2, 3, 12, 205
Strasburg   79, 80
Struve   191, 192
Suensaari   174, 177, 187
Sumer   1
Sumerians   4
Sun, diameter   5, 6
  motion of   5
  movement   4, 5, 6
Sundial   5, 21, 22
Sundsvall   175, 176
Surampalte   113
Svanberg, Jöns   187, 205
Swentzar   177
Syene   10, 11, 12, 13
Syracuse   2, 5, 6

Tadmor   19
Ta-Hsiang   17
Tangent function   21, 22
  table of   28
Tanlagoa   111
Tarabita   110, 111
Taragay, Muhammed   28, 29
Tarqui   51, 52
  astronomical observations   142, 143
Tchoulapou   112, 113
Telescope   34
  introduction of   34, 35, 39, 40, 41
Teneriff   163
Tenner   191, 192
Thai-Yuan   16, 17
Thales   4, 5, 205
Thang, long   16
  short   16
Theodolite   13, 14, 39, 43, 45

Theodolitus   39, 45
Theodoros   4, 5, 14, 15, 205
Theon   14, 15, 205
Thermometer   14, 40, 42, 43, 52, 53, 88, 89, 170, 174, 175, 183, 186
  Réaumur   103, 186
Thévonet, Melchisédech   69, 70, 87, 88, 205, 206
Thieh-Lo   16, 17
Thomas, Antoine   84, 87, 88, 205, 206
Tialoma   113
Tibbon, Jacob   43
Tides   12, 13, 14, 92
Timepiece   14
Timocharis   13
Tivicatsu   112, 113
Toise, Chinese   85
  du Châtelet   76, 77, 90, 152, 180
  by Fortin   191
  Du Mairan   55, 56
  Du Nord   55, 56, 180
  Du Perou   55, 56, 91, 106, 108, 152, 187
  Standard   42, 43, 55, 88, 89, 152, 187
  Universal   76, 77
Toledo   22
Tongouragoa   110
Toricelli   40, 53, 70
Torneå   170, 171, 177, 181, 185, 187
Torneå Elf   191
Torquetum   22, 38, 39, 42, 43, 45
Townley, Richard   34, 35, 40, 41
Transversals   22, 23, 29, 30, 38, 39, 40, 57
Triangle, area of   13, 14, 21, 22
  plane   29
  spherical   29, 120
Triangulation   28, 29, 38, 39, 40
  first   57
Trigonometric functions   28
  identities   39
Trigonometry, plane   13, 14, 15
  spherical   13, 14, 21, 22
Triquetrum   14, 19, 20, 22, 23, 38, 39, 43, 67
Tu   16, 17, 18, 19
Tusi staff   22

Ulloa, Antonio de   88, 89, 100, 101
  capture   163
Ulugh Beg   22, 23, 205, 206
Umeå   172, 175, 176
Universal unit of measure   40, 41, 54, 71, 74, 88, 89, 90, 95, 97
Universe, centre of   3
Uraniborg observatory   39, 57, 70

Varahamihira   14, 15, 27, 28, 205, 206
Varin   54, 55, 80, 82
Velocity of river   161
Vengotasin   109, 110, 112, 113
Verbiest, Peré F.   84
Verguin   88, 89, 101, 102, 106, 108, 162, 163, 165, 205, 206
Vernier, Pierre   33, 34, 205, 206
Vernier.   33, 34, 38, 39, 40, 88
Viète, Francois   29, 39, 205, 206
Virga   40, 41, 90
Vlacq, Adriaen   36, 37, 40, 205, 206

Vojakkala   174
Volcanoes   110
Voltaire   95, 168, 170, 191, 192

Wallis, Dr John   32, 33, 34, 36, 37
Wang Fan   14, 15
Wei-Chou   16, 17
Wren, Christopher   90
Wright, Edward   39, 40, 206

Yang-Chhêng   16, 17
Yassouai   113
Yasuay   109, 110, 113
Ya 'Qub   21
Year, length of   4, 5
  tropical   13
Ygoalata   113
Yuan-Thai   17
Yü-Chou   16, 17

Zach, von   143
Zagroum   113
Zenith sector   42, 43, 45, 51, 71, 74, 79, 80, 145, 151, 152,
        183, 185
Zero symbol   14, 15
Zodiac, signs of   28

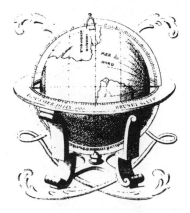